Asymptotic Approaches in Nonlinear Dynamics

W0055371

Springer
Berlin
Heidelberg
New York
Barcelona
Budapest
Hong Kong
London
Milan
Paris
Singapore
Tokyo

Springer Series in Synergetics

Editor: Hermann Haken

An ever increasing number of scientific disciplines deal with complex systems. These are systems that are composed of many parts which interact with one another in a more or less complicated manner. One of the most striking features of many such systems is their ability to spontaneously form spatial or temporal structures. A great variety of these structures are found, in both the inanimate and the living world. In the inanimate world of physics and chemistry, examples include the growth of crystals, coherent oscillations of laser light, and the spiral structures formed in fluids and chemical reactions. In biology we encounter the growth of plants and animals (morphogenesis) and the evolution of species. In medicine we observe, for instance, the electromagnetic activity of the brain with its pronounced spatio-temporal structures. Psychology deals with characteristic features of human behavior ranging from simple pattern recognition tasks to complex patterns of social behavior. Examples from sociology include the formation of public opinion and cooperation or competition between social groups.

In recent decades, it has become increasingly evident that all these seemingly quite different kinds of structure formation have a number of important features in common. The task of studying analogies as well as differences between structure formation in these different fields has proved to be an ambitious but highly rewarding endeavor. The Springer Series in Synergetics provides a forum for interdisciplinary research and discussions on this fascinating new scientific challenge. It deals with both experimental and theoretical aspects. The scientific community and the interested layman are becoming ever more conscious of concepts such as self-organization, instabilities, deterministic chaos, nonlinearity, dynamical systems, stochastic processes, and complexity. All of these concepts are facets of a field that tackles complex systems, namely synergetics. Students, research workers, university teachers, and interested laymen can find the details and latest developments in the Springer Series in Synergetics, which publishes textbooks, monographs and, occasionally, proceedings. As witnessed by the previously published volumes, this series has always been at the forefront of modern research in the above mentioned fields. It includes textbooks on all aspects of this rapidly growing field, books which provide a sound basis for the study of complex systems.

A selection of volumes in the Springer Series in Synergetics:

Synergetics An Introduction
3rd Edition By H. Haken

Handbook of Stochastic Methods
for Physics, Chemistry,
and the Natural Sciences   2nd Edition
By C. W. Gardiner

Noise-Induced Transitions Theory
and Applications in Physics, Chemistry,
and Biology   By W. Horsthemke, R. Lefever

The Fokker-Planck Equation
2nd Edition By H. Risken

Nonequilibrium Phase Transitions
in Semiconductors Self-Organization
Induced by Generation
and Recombination Processes By E. Schöll

Synergetics of Measurement, Prediction
and Control By I. Grabec, W. Sachse

Predictability of Complex Dynamical Systems
By Yu. A. Kravtsov, J. B. Kadtke

Interfacial Wave Theory of Pattern Formation
Selection of Dentritic Growth and Viscous
Fingerings in Hele-Shaw Flow   By Jian-Jun Xu

Cooperative Dynamics in Complex
Physical Systems Editor: H. Takayama

Information and Self-Organization
A Macroscopic Approach
to Complex Systems   By H. Haken

Foundations of Synergetics I
Distributed Active Systems   2nd Edition
By A. S. Mikhailov

Foundations of Synergetics II
Complex Patterns   2nd Edition
By A. S. Mikhailov, A. Yu. Loskutov

Synergetic Economics   By W.-B. Zhang

Quantum Signatures of Chaos   By F. Haake

Nonlinear Nonequilibrium Thermodynamics I
Linear and Nonlinear Fluctuation-Dissipation
Theorems   By R. Stratonovich

Nonlinear Nonequilibrium Thermodynamics II
Advanced Theory By R. Stratonovich

Modelling the Dynamics of Biological Systems
Editors: E. Mosekilde, O. G. Mouritsen

Self-Organization in Optical Systems
and Applications in Information Technology
2nd Edition Editors: M. A. Vorontsov,
W. B. Miller

Jan Awrejcewicz   Igor V. Andrianov
Leonid I. Manevitch

Asymptotic Approaches in Nonlinear Dynamics

New Trends and Applications

With 58 Figures

 Springer

Professor Jan Awrejcewicz

Division of Control and Biomechanics (I-10), Technical University of Lódz
1/15 Stefanowskiego St., PL-90-924 Lódz, Poland

Professor Igor V. Andrianov

Pridneprovye State Academy of Civil Engineering and Architecture
24a Chernyshevskogo St., Dnepropetrovsk 320005, Ukraine

Professor Leonid I. Manevitch

Institute of Chemical Physics, Russian Academy of Sciences
4 Kosygin St., 117977 Moscow, Russia

Series Editor:
Professor Dr. Dr. h.c.mult. Hermann Haken

Institut für Theoretische Physik und Synergetik der Universität Stuttgart
D-70550 Stuttgart, Germany
and
Center for Complex Systems, Florida Atlantic University
Boca Raton, FL 33431, USA

ISSN 0172-7389

Library of Congress Cataloging-in-Publication Data
Awrejcewicz, J. (Jan), 1938–   Asymptotic approaches in nonlinear dynamics : new trends and applications / Jan
Awrejcewicz, Igor V. Andrianov, Leonid I. Manevitch   p.   cm. -- (Springer series in synergetics, ISSN 0172-7389)
Includes bibliographical references and index.
ISBN-13: 978-3-642-72081-9   e-ISBN-13: 978-3-642-72079-6
DOI: 10.1007/ 978-3-642-72079-6
1. Nonlinear oscillations. 2. Asymptotic
expansions. I. Andrianov, I.V. (Igor ´ Vasil´ evich) II. Manevich, L. I. (Leonid Isaakovich) III. Title. IV. Series.
QA867.5.A63   1998   531'.32' 01515355--DC21   98-7666

© Springer-Verlag Berlin Heidelberg 1998
Softcover reprint of the hardcover 1st edition 1998

Typesetting: Data conversion by K. Mattes, Heidelberg
Cover design: *design & production* GmbH, Heidelberg
SPIN 10652100   55/3144 - 5 4 3 2 1 0 - Printed on acid-free paper

Preface

> How well is Nature simulated by the varied
> asymptotic models that imaginative
> scientists have invented?
>
> B. Birkhoff [52]

This book deals with asymptotic methods in nonlinear dynamics. For the first time a detailed and systematic treatment of new asymptotic methods in combination with the Padé approximant method is presented.

Most of the basic results included in this manuscript have not been treated but just mentioned in the literature. Providing a state-of-the-art review of asymptotic applications, this book will prove useful as an introduction to the field for novices as well a reference for specialists.

Asymptotic methods of solving mechanical and physical problems have been developed by many authors. For example, we can refer to the excellent courses by A. Nayfeh [119–122], M. Van Dyke [154], E.J. Hinch [94] and many others [59, 66, 95, 109, 126, 155, 163, 50d, 59d]. The main features of the monograph presented are: 1) it is devoted to the basic principles of asymptotics and its applications, and 2) it deals with both traditional approaches (such as regular and singular perturbations, averaging and homogenization, perturbations of the domain and boundary shape) and less widely used, new approaches such as one- and two-point Padé approximants, the distributional approach, and the method of boundary perturbations.

Many results are reported in English for the first time. The choice of topics reflects the authors' research experience and involvement in industrial applications. The authors hope that this book will introduce the reader to the field of asymptotic simplification of the problems of the theory of oscillations, and will be useful as a handbook of methods of asymptotic integration as well.

The narration is commonly based on examples given by applied mechanics of structures (primarily, plates and shells) and fluid mechanics, but scarcely of quantum physics. Obviously, the methods in question are really versatile in application, covering applied mathematics, physics, mechanics and other basic sciences. The authors have paid special attention to examples and discussion of results rather than to burying the ideas in formalism, notation, and technical details. The aim is to introduce mathematicians – as well as physicists, engineers, and other consumers of asymptotic methods – to the world of ideas and methods in this burgeoning area.

The effect of asymptotic methods (AM) on the theory of oscillations increases multifold. The vitality and prospect of AM becomes obvious from the fact that active interaction between numerical and analytical methods is accomplished via asymptotics. It is a pity that asymptotic mathematics does not occupy the decent place in education programmes of high schools. Certain tutorial aspects, useful for training mechanics, physicists, applied mathematicians and engineers, are presented here.

Let us scan in detail the contents of the chapters.

An introduction the depicting the principal ideas of asymptotic approaches through simple, "transparent" examples is given.

The first part is devoted to discrete systems.

First, an introduction to classical perturbation techniques are presented. The KBM methods and the equivalent linearization are described in some detail. Nonconservative nonautonomous systems are considered and nonresonance oscillations as well as oscillations in the neighbourhood of resonance are analysed. A general approach to the analysis of unstationary nonlinear systems is given. Particular attention is paid to consideration of combined parametric and self-excited oscillations in a three-degree-of-freedom mechanical system. This example includes a derivation of the equation of motion and a determination of instability zones. The so-called modified Poincaré approach is presented and illustrated on the basis of a one-degree-of-freedom system and then this approach is extended to the analysis of general nonlinear systems. Then, the Hopf bifurcation is discussed from the viewpoint of the asymptotic approach. Finally, a method of controlling and improving the stability of periodic orbits of vibro-impact systems is proposed. This method is based on the feedback loop control with a time delay. This subchapter includes two parts of our investigation. In what follows a perturbation technique is applied to estimate delay loop coefficients for the improvement of stability of the vibro-impact motion for one-degree-of-fredom systems. Then, control of the periodic motion of the one-degree-of-freedom vibro-impact oscillator is analysed numerically, showing good agreement with the analytical prediction.

Nonlinear normal vibrations are a generalization of normal (principal) vibrations of linear systems. In the normal mode all position coordinates can be defined from any one of them. Using normal modes of nonlinear systems gives very interesting results, and in Sect. 2.10 we write about some aspects of the asymptotic construction of an object.

Progress in the applications of AM in the theory of oscillations as well as in applied mathematics on the whole is closely linked with the introduction of new small parameters and, respectively, new asymptotic procedures. This is the field of Sect. 2.11.

In Sect. 2.12 we deal with one- and two-point Padé approximants (PA). Usually PAs are used for the extension of the area of applicability of perturbation series. We propose to use PAs in connection with AM in many new cases, in particular:

1. Estimation of the convergence domain for perturbation series. In particular, such estimation may be obtained on the basis of the comparison of the perturbation series and PA. This result is justified by many interesting examples from nonlinear mechanics.
2. Elimination of nonuniformities of asymptotic expansions. The PA eliminates nonuniformities of asymptotic expansions in more important mechanical problems in a simpler way than, for instance, Lighthill's method. Up till now two-point PAs (TPPAs) have not been so widely used in mechanics. We represent new applications of the TPPAs for matching local expansions in nonlinear dynamics.

The second part specifies the most important and useful forms and techniques of asymptotic thinking for the theory of oscillations for continual systems.

Relations between the dynamics of discrete and continual systems are based on the procedures of discretization and continualization. The procedure of discretization is described well in many books, so we have paid some attention only to the continualization (the passage from discrete to continuous systems) in Sect. 3.1.

Section 3.2 is devoted to the homogenization approach. The main problem in this field is in the solution of the so-called cell (or local) problem. This problem has usually been treated by a numerical method. We have used an asymptotic method for solving the cell problem and have constructed an approach in this book. The approach presented fills the substantial gap between numerical methods of the thin shell theory, which lack generality and the possibility of grasping the common features of the behaviour of the structures concerned, and approximate schemes, based on heuristic hypotheses. The methods proposed are wide-ranging in applications and lead to simple and clear design formulae, useful for practical analyses. The averaging approach is one of the most useful tools in the theory of nonlinear oscillations. Usually it is used with respect to time variables, but in Sect. 3.3 we show new perspectives for averaging with respect to spatial variables.

V.V. Bolotin proposed an effective asymptotic method for the investigation of linear continuous elastic system oscillations with complicated boundary conditions. The main idea of this approach is in the separation of the continuous elastic system into two parts. Then the matching procedure permits us to obtain a complete solution of the dynamics problem in a relatively simple form. The idea of Bolotin's asymptotic method was generelized for the nonlinear case in Sect. 3.4.

Regular and singular asymptotics in a wide range of forms are the old, but formidable weapons in the armoury of an asymptotic mathematician. In Sect. 3.5 a lot of interesting problems are solved on this basis, and some interesting aspects of the application of these traditional approaches are no-tified.

A new AM for solving mixed boundary value problems is considered in Sect. 3.6. The parameter ε is introduced into the boundary conditions in such a way that the $\varepsilon = 0$ case corresponds to the simple boundary problems and the case $\varepsilon = 1$ corresponds to the general problem under consideration. Then, the ε-expansion of the solution is obtained. As a rule, the expansion of the solution is divergent just at the point $\varepsilon = 1$. The PAs are used to remove this divergence.

The TPPA in application to nonlinear dynamic problems for a continuous system – a plate on a nonlinear foundation – is displayed in Sect. 3.7.

The discovery of the soliton in 1965 by Kruscal and Zabusky has brought revolutionary changes in nonlinear science, and we describe some uses of the soliton technique in Sect. 3.8.

In Sect. 3.9 a nonlinear analysis of spatial structures is described on the basis of the so-called modified envelope equation.

The third part of the book includes an investigation of discrete-continuous systems. In Sect. 4.1, periodic oscillations of discrete-continuous systems with a time delay are analysed. In Sect. 4.2 a simple perturbation technique is described as it is used in the analysis of discrete-continuous systems with a time delay and with homogeneous boundary conditions. In Sect. 4.3 the nonlinear behaviour of an electromechanical system is investigated on the basis of an averaging technique supported by symbolic computation using the Mathematica package. Then the obtained averaging amplitude differential equations are analysed numerically.

The book is mainly based on the authors' papers [6–24, 28–39, 115, 125, 156, 2d–27d, 55d, 56d, 62d].

Finally, the first author (J.A.) wishes to acknowledge the financial support by the Polish National Scientific Research Committe Grants No 7T07A01710 and No 7T07A00210.

Mr K. Tomczak and Mr G. Wasilewski are thanked for their time and consideration paid to the preparation of this book.

Łódź   *J. Awrejcewicz*
Dnepropetrovsk   *I.V. Andrianov*
Moscow   *L.I. Manevitch*

April 1998

Contents

1. Introduction: Some General Principles of Asymptotology[1]

Almost any physical theory formulated in mathematical terms in a general way is extremely complicated. Therefore, both in creating a theory and in its further development, the simplest limiting cases that admit analytical solutions are of paramount importance. It is quite common that in the limiting case there are fewer equations or the (differential) equation has a lower order or the nonlinear equation is replaced by a linear one or the original system is subjected to a kind of averaging and so on and so forth.

Behind the above-mentioned idealizations, however diverse they may seem, lies a high degree of symmetry inherent in a mathematical model of the phenomenon at issue in its limiting situation. An asymptotic approach to a complex and perhaps "insoluble" problem consists basically in treating an original – insufficiently symmetric – system as an approximation to a given symmetric one. It is basically important that the determination of corrections allows one to study deviations from the limiting case in a way which is much simpler than a direct study of the original system.

At first sight, the potentialities of such an approach are limited by a narrow range of variations in the parameters of the system.

The experience gained in the study of various physical systems has shown, however, that in the case of system parameters varying considerably and the system itself departing from one limiting symmetric pattern, in general another limiting system, often with a less pronounced symmetry, exists and a perturbed solution can now be formed for the latter one.

This enables the system behaviour to be defined over the entire range of the parameter using a finite number of limiting cases. Such an approach makes the most of one's physical intuition and contributes to its further enrichment and also leads to the formation of new physical concepts. Thus the boundary layer – an important concept in fluid mechanics – is of a pronounced asymptotic nature and is related to the localization at the boundaries of a streamlined body in the zone where the viscosity of the fluid cannot be neglected (see Fig. 1.1 and also "Album of boundary layers" [3]). In the mechanics of a deformable rigid body and in the theory of electricity, similar phenomena are known as the *edge effect* and the *skin effect*, respectively.

---

[1] See also [40, 45, 80, 105, 109, 142, 28d, 38d].

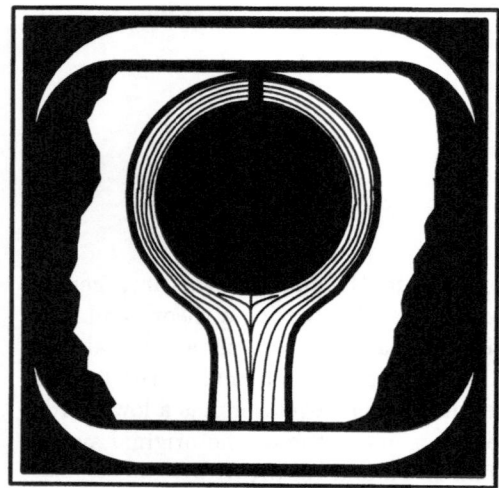

Fig. 1.1. Boundary layer near a streamlined sphere

That the asymptotic method assists in relating different physical theories with one another is of little consequence. Albert Einstein would point out that "the happiest lot of physical theory is to serve as a basis for more general theory while remaining a limiting case thereof".

The above-mentioned problems will be clarified in this chapter.

1.1 An Illustrative Example

As an example of the technical aspect of the method, consider a simple algebraic example. The biquadratic equation

$$x^4 - 2x^2 - 8 = 0 \qquad (1.1.1)$$

is reduced to a quadratic equation and readily solved by setting $z = x^2$. Then we have

$$x_{1,2} = \pm 2, \qquad x_{3,4} = \pm\sqrt{2}\,\mathrm{i}, \qquad \mathrm{i} = \sqrt{-1}.$$

Such a simplification is due to the symmetry of the equation: substituting $(-x)$ for x does not change it. Let us assume that the original equation describes a given physical system with its parameters undergoing small changes and, as a consequence, the equation takes the form

$$y^4 - \varepsilon y^3 - 2y^2 - 8 = 0. \qquad (1.1.2)$$

In this case the system is said to have received a small perturbation; the expression $(-\varepsilon y^3)$ is referred to as the "perturbation term" and ε as the "small parameter". The system becomes asymmetric, and the solution of the new equation can no longer be written in a simple form. The roots of the new equation, y_i $(i = 1, \ldots, 4)$, however, should not differ significantly from

x_i, hence set $y_i = x_i$. The error of such a substitution is determined by the value of the discarded term $(-\varepsilon y^3)$. To make the solution more accurate, let us represent it as a series

$$y_i = x_i + \varepsilon y_i^{(1)} + \varepsilon^2 y_i^{(2)} + \ldots; \qquad i = 1, \ldots, 4. \tag{1.1.3}$$

Substituting this expansion into the perturbation equation and equating the coefficients of the same power of ε we find

$$y_i = \frac{0.25 x_i^2}{x_i^2 - 1}; \qquad i = 1, \ldots, 4. \tag{1.1.4}$$

Evaluation of corrections could be continued without any difficulty, but the deviation from the exact solution will inevitably increase with the increase in the value of ε.

Consider now the opposite case of large perturbations. Then the reciprocal ε^{-1} will be small. Then, the roots of equation (1.1.2) can be divided into two groups. As ε^{-1} tends to zero, three roots tend to zero and the fourth increases indefinitely. The two groups may be found using expansions in the small parameter ε^{-1}.

$$y_1 = \varepsilon + \ldots, \tag{1.1.5}$$
$$y_2 = -2\varepsilon^{-1/3} + \ldots,$$
$$y_{3,4} = a\varepsilon^{-1/3}(1 \pm \sqrt{3}\,i). \tag{1.1.6}$$

There exists, however, a region where the asymptotic approximation produces unsatisfactory results. This is the region where "small" values of ε are already "large" and "large" values of ε are still "small" (see Fig. 1.2). The problem of forming a solution within such a region on the basis of available limiting values is one of the most difficult when employing asymptotic methods as in the problem of deciding as to what is to be considered "small" or "large". This will be considered later.

Besides, it should be noted that perturbation solutions of a problem represented as expansions in series in the power of a small parameter of type (1.1.3) do not necessarily converge to the solution which is being sought. The expansions are often asymptotic. The ratio of each term of the series to the preceding one tends to zero when the expansion parameter approaches its limiting value, say, zero; and the deviation of the sum of the first N terms of such a series from the function represented by the complete series is of the $(N+1)$th order. (In examining a series for convergence, the parameter is regarded as fixed and the limit of the sum of N terms of the series is taken as N tends to infinity.) In particular cases a divergent asymptotic series (with infinite limit) is sometimes more useful than a convergent one as only a few of the initial terms give a fair approximation.

Let us consider some typical situations where the asymptotic approach is effective.

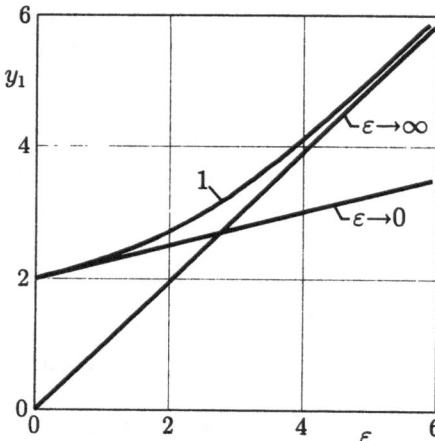

Fig. 1.2. Comparision of exact (numerical) and asymptotic solutions for the algebraic equation of fourth order

1.2 Reducing the Dimensionality of a System

A high order of an algebraic or a differential equation or a large number of such equations are all manifestations of one of the principal difficulties that arise in solving physical problems. This difficulty is sometimes called "the imprecation of dimensionality". In order to get over it, two antithetical approaches have been developed. The first one proves to be effective if individual elements of a system under consideration differ markedly from each other in one or another characteristic. Then, by introducing characteristics of different elements, one is able to carry out an asymptotic reduction of dimensionality, or in other words, a reduction in the degrees of freedom, and then one can try to improve the solution obtained by using the asymptotic approximation. A typical example of such a situation is a three-body problem in elastic mechanics. The masses of celestial bodies (say, those of the Sun, the planet Jupiter and the Earth), as a rule, differ markedly, and a small parameter – the mass ratio – enables an asymptotic reduction of the dimensionality to be achieved. The classical methods of celestial mechanics are based on this, the limiting (high symmetry) case being the exactly solvable two-body problem. Celestial mechanics is the first branch of science where the asymptotic method (the perturbation theory) has played a dominant role, and moreover, this method was originally developed in response to the pressing necessity of solving problems in celestial mechanics.

It should be noted that asymptotic methods are often used without being specifically regarded as such and even without being fully understood. Thus, one-degree-of-freedom models are employed extensively in engineering. Clearly, employing such models always involves an asymptotic reduction in the dimensionality and the possibility, at any rate in principle, of finding the corresponding corrections, but a clear indication that this is the case is rare.

Let us now consider the second way of getting out of the difficulty.

1.3 Continualization

If a system under consideration consists of a set of homogeneous elements, then the asymptotic approach can be used not only for the reduction of dimensionality but also for increasing dimensionality. Thus, we approach a highly important class of physical models where discrete systems are replaced by continuous ones.

As an example, let us consider the longitudinal oscillations of an infinite chain of the similar masses connected by springs of equal length L and rigidity c (Fig. 1.3a). With the smooth oscillation form characterized by the displacement u_k at each point kL ($k = 0, \pm 1, \pm 2, \ldots$), the chain can be replaced by a continuous rod, thus enabling us to change from the infinite system of ordinary differential equations

$$mu_{ktt} = c(u_{k+1} - 2u_k + u_{k-1})$$

to the single partial differential equation

$$mu_{tt} = cL^2 u_{xx}.$$

The degrees of freedom have grown in number (the continuum replacing the countable set), and the relative simplicity of this limiting case of long-wave oscillations is due to the symmetry of the partial differential equation not varying under an arbitrary displacement along the rod.

As the period of oscillations and their wavelengths will decrease, the error of the approximate solution obtained in this way will increase. Another limiting case for the same system is for the minimum possible wavelength oscillations (Fig. 1.3b). Their form can be readily calculated and employed as the first approximation in the study of the short-wave oscillations of the system. In this case the desired solution should have the form of the product of the solutions of the limiting case on the smooth function which is deduced from the partial differential equation (Fig. 1.3c).

The method of transition from discrete models to continuous models has found extensive applications in physics, and the entire mechanics of the continuum is based essentially on this method.

This is not always so, however, as in the case under consideration. Fluids, say, do not lend themselves to the purpose of defining a periodic equilibrium structure in reference to which oscillations are executed. Nevertheless, at a macroscopic level we perceive the fluid flow as a continuum flow which can be simulated by a continuous fluid model. It is true that continuity is provided by the averaging of small-scale (microscopic) movements. The consequences of such an averaging will be discussed below. This will show now that the transition to the differential equations of hydrodynamics becomes possible.

In conclusion let us quote from Erwin Schrödinger who figuratively explains the efficiency of the method: "Let's assume we would tell an ancient Greek that the individual particle path in a fluid could be traced. The ancient Greek would not believe that man's limited intellect could solve such

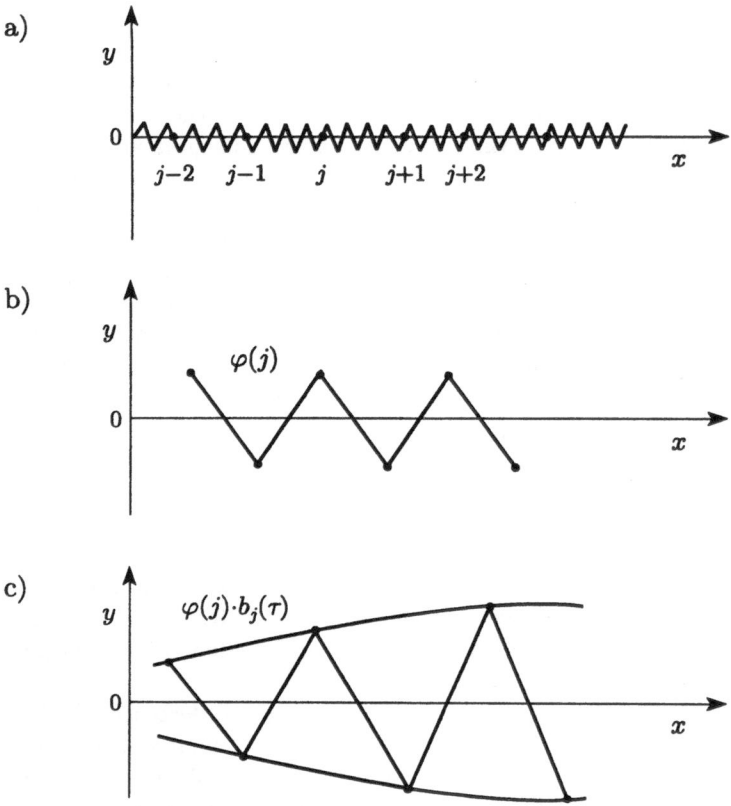

Fig. 1.3. (a) Infinite chain; (b) minimum possible wavelength oscillations; (c) short-wave oscillations

an intricate problem. The point is that we have learned to master the whole of the process using but a single differential equation".

1.4 Averaging

In many physical problems, some variables vary very slowly, others more rapidly. It is natural to bring the question of whether it is appropriate to study first the global structure under consideration, digressing from its local distinctive features, and then to investigate the system locally. It is the averaging method that is aimed at the division of the fast and slow components of the solution. Without going into the details of the method – the more so because it has at present many modifications – it will be noted only that it involves the introduction of the "slow" (macroscopic) and the "rapid"

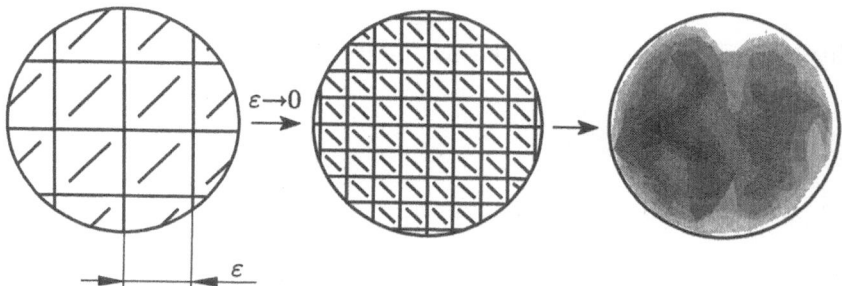

Fig. 1.4. Homogenization procedure: from the periodic inhomogeneity to the homogeneous material with new properties

(microscopic) variables whose equations are separated and can be solved independently, or sequentially (see Fig. 1.4).

This method was developed for and gained wide use in solving problems in celestial mechanics and nonlinear oscillation theory that are defined by common differential equations. At present, the method is used to great advantage for solving variable-coefficient partial differential equations in such disciplines as the theory of composites, or the design of reinforced, corrugated, perforated, etc., shells. An original nonhomogeneous medium or structure is reduced to a homogeneous one (generally speaking, to an isotropic one) with some effective characteristics. The averaging method allows one not only to obtain the effective characteristics but also to investigate the nonhomogeneous distribution of mechanical stresses in different materials and structures, which is of great significance for evaluating their strength.

1.5 Renormalization

Regrettably, the simple averaging of small-scale movements is not always applicable, either. There occur such problems wherein several different-scale movements show up markedly even at the macroscopic level. Among these is, for example, the study of what is known as critical phenomena related to phase transitions, or the study of turbulence. In this case, a number of successive averaging procedures for all scales has to be carried out. This is the very essence of the renormalization procedure which forms the renormalization group method. A rigorous renormalization of the procedure, however, involves considerable technical difficulties. A practical solution of the problem is offered by a quite unexpected asymptotic method.

The fact is that in a four-dimensional imaginary world these problems do not occur, and this makes it possible to carry out an ordinary averaging. This case could be considered as a limiting one, with the quantity $\varepsilon = 4 - d$ (where d is the spatial dimensionality) as a small parameter. In the real three-dimensional world $d = 3$, and $\varepsilon = 1$, which is not small. Nevertheless, an asymptotic expansion in the parameter proved to be quite effective in solving the most complicated problems of critical-phenomena physics.

1.6 Localization

Real system deviations from the limiting (i.e., ideal) system may be of a different nature. Sometimes these deviations are small over the entire range of the system parameter variations: it is not infrequent, however, that the deviations are high, although localized within a small region. This is true for the above instance of a body streamlined by a fluid. Another example is the transition (reducing) from the three-dimensional model of an elastic body to the two-dimensional model (plates, shells), or to the one-dimensional model (rodes, beams). In this case a narrow boundary layer (of the order of the plate or shell wall thickness; or of the cross-sectional characteristic size of the rod or the beam) exists near the body boundaries wherein the three-dimensionality of the original problem manifests itself. Upon reducing the three-dimensional problem to the two-dimensional one, it is still possible to isolate the so-called end effects concentrated at the shell boundaries or its structural inhomogeneities. The concept of the boundary layer is closely related to the so-called St. Venant principle that says that in the analysis of a structure it is possible to digress from the detailed load distribution pattern in fixing its elements. In fact, however, the distribution pattern is essential, but within narrow zones only more extension is defined by the element cross-sectional characteristic sizes or by the load-variation period. Mathematically, defining a boundary layer is due to the fact that a simplified differential equation is of a smaller order than the original one. The asymptotic approach in this case is termed singular.

1.7 Linearization

If the equations of a physical theory are nonlinear, then even a small number of degrees of freedom or a localized solution do not ensure the overcoming of mathematical difficulties. The problem is solved by linearization – an asymptotic method – that relies on the concept of low-intensity processes.

A linear approach (to the problem) allows one to formulate such fundamental concepts as the normal vibration, the eigenfunction and the spectrum. For a linear system with n degrees of freedom with no damping, one can always choose such "normal" coordinates which describe the system by n oscillations for pendulums not linked to one another. In other words, any motion of a linear system is represented by a linear combination of normal oscillations (or waves), that is, by the so-called expansion in a Fourier series.

It is of fundamental importance that the oscillations are singled out not only mathematically but also physically. Thus, it is precisely the normal oscillations that will resonate under the influence of the periodic force.

If we consider a linear system as the first approximation to the nonlinear one (that is the crux of local linearization) then, when taking into account the nonlinear corrections in the equations of the second and following approximations, there appear dummy external loads that bring about the normal oscillation resonances. This can be avoided by "touching up" the parameters of the normal linear oscillation.

However, nonlinear systems, especially high-dimensional ones, quite often do not lend themselves to correct the description in the approximation of the local linearization method. Thus, the combination of high dimensionality with strong nonlinearlity was until recently considered an insurmountable difficulty in carrying out a structural study of a physical system. But a fairly extensive class of multidimensional nonlinear systems that permit such a study has been recently discovered. These systems, known as "integrable systems", have particular solutions as stable solitary waves – solitons – that are in a way analogues of normal oscillations defined in linear systems. Thus, a nonlinear generalization of the Fourier method – the method of the inverse scattering problem – is based on solitons which play a fundamental role taking the place of the usual Fourier components. The method of the inverse problem of scattering can be treated as the nonlocal linearization of the original nonlinear equation. In other words, the latent instability of a nonlinear system makes it possible to find a transformation that reduces the construction of an extensive class of solutions to the analysis of linear equations.

The integrable systems can in their turn act as an approximation in the analysis of the systems that approximate them, but are nonintegrable within the framework of an asymptotic approach.

1.8 Padé Approximants

So far we have assured ourselves that practically any physical problem, whose parameters include the variable parameter ε, can be approximately solved as ε approaches zero or infinity. How is this "limiting" information to be used in the study of a system at intermittent values of ε, say, $\varepsilon = 1$? This problem is one of the most complicated in asymptotic analysis. As yet there is no general answer to the tricky question of how far the parameter ε can be considered small (or large) in the problem involved, though in many instances this problem is alleviated by the so-called two-point Padé approximants.

The two-point Padé approximant is represented by the rational function $F(\varepsilon)$, in which the $m + 1$ coefficients of expansion in the Taylor series when $\varepsilon \to 0$, and the m coefficients in the Lorentz series when $\varepsilon \to \infty$ coincide with the corresponding coefficients of the governing series for $F(\varepsilon)$.

Experience shows that the Padé approximants do indeed quite often allow the limiting expansions to be "sewn together" after defining the regions of "small" and "large" values of ε. This resembles a known interpolation procedure, that is, the reconstruction of the intermediate values of a quantity by its two extreme values. The role of such known values is played in this case by the asymptotics as ε tends to zero and to infinity (see Fig. 1.5).

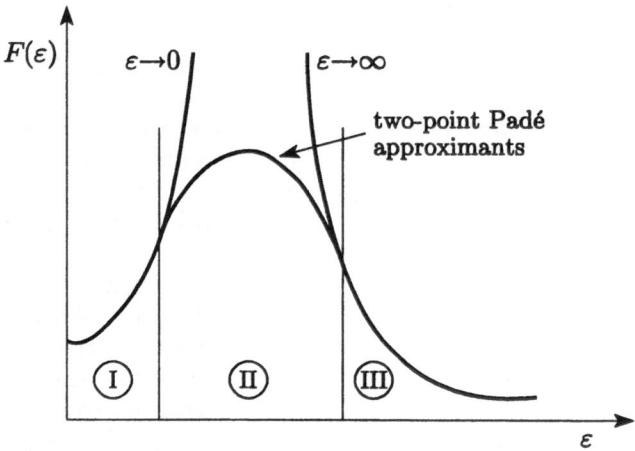

Fig. 1.5. Matching of limiting asymptotics by two-point Padé approximants

For instance, for equation (1.1.2) the Padé approximant for the first root

$$y_1 = (2 + 0.57\varepsilon + 0.12\varepsilon^2)(1 + 0.12\varepsilon)^{-1}$$

derived on the basis of the asymptotics of the form (1.1.3) as ε tends to zero or of the form (1.1.5) as ε tends to infinity, defines satisfactorily the exact solution at any value of ε (Fig. 1.2, curve 1, in this case the exact (numerical) solution and the two-point Padé approximant solution practically coincide).

1.9 Modern Computers and Asymptotic Methods

The reader must have repeatedly asked themself a question: are the asymptotic methods of any practical use at all when there are computers? Is it not simpler to write a program for any original problem to solve it numerically by using standard procedures?

This may be answered as follows: first, asymptotic methods are very useful in the preliminary stage of solving a problem even in cases where the principal aim is to obtain numerical results. The asymptotic analysis makes it possible to choose the best numerical method and gain an understanding of a vast body of numerical material, though not properly arranged. Secondly, asymptotic methods are especially effective in those regions of parameter values where machine computations are faced with serious difficulties. Laplace used to say, not without reason, that asymptotic methods are the "more accurate, the more they are needed". Moreover, there is the possibility of developing such algorithms wherein smooth portions of solutions are obtained numerically, and the asymptotic approaches are applied to those parameter value regions where these solutions change drastically, say, within boundary layers. Thirdly, the asymptotic methods develop our intuition in every possible way and play, as noted above, an important role in shaping the mentality of, say, a contemporary scientist or engineer. Therefore, it would be more proper to consider the asymptotic and numerical methods not as competing, but as mutually complementary.

Again, computers further considerably the development of the asymptotic method. For instance, defining higher approximations is a major difficulty in applying asymptotic methods. In solving complex problems by manual calculations one may succeed in defining two or three approximations at the most. Now the burden of manual calculations can be shouldered by the computer.

1.10 Asymptotic Methods and Teaching Physics

"Few of the equations of physics have exact solutions which are manageable, and one usually has to have recourse either to approximate methods or to numerical solutions. Numerical work becomes cumbersome if the problem has a great number of variables, or if one is interested in a general survey of possible solutions. In those cases the natural approach is by approximation. In teaching physics we probably overemphasize the exceptional problems which have closed solutions in terms of elementary functions, and do not give enough attention to the more common situation in which approximations have to be used. Beginners are usually uncomfortable with approximations, and, even if only an approximate answer is required, often prefer to find the exact answer, if this is possible, and then to approximate. This is understandable because the art of choosing a suitable approximation, of checking its consistency (e.g. ensuring there are no oscillations) and finding at least intuitive reasons for expecting the approximation to be satisfactory, is much more than solving an equation exactly" [130].

1.11 Problems and Perspectives

Naturally, we do not describe all new asymptotic approaches successfully used in the theory of nonlinear oscillations. We only mention algorithms of conjugate operators and equations [115, 57d], methods on the basis of the Lie group [165], renormalization group and intermediate asymptotics [46–48], and the asymptotic-numerical approach [60d]. The last one has, in our opinion, a great future, because it gives the possibility of using the merits of the asymptotical and numerical approaches simultaneously.

The most interesting discoveries in the field under consideration, in our opinion, are closely connected with developments in the theory of summation of the extrapolation and interpolation of asymptotic series, especially in the theory of Padé approximants and its generalization, and in searching for new parameters of asymptotic expansions. We have analysed the merits and demerits of Padé approximants in Sect. 2.12. The following parameters are used as asymptotics in this book: the difference of frequencies $\omega_1 - \omega_2$ (Sect. 2.10.2); the amplitude A, A small and A large (Sect. 2.12.4); the parameter δ ($X^{1+\delta}$, Sect. 2.11.1); the power of nonlinearity N (X^N, Sect. 2.11.3); the ratio of flexible to extension rigidities $(EI)/(EFR^2)$ (Sect. 3.5.1); the ratio of the thickness to the radius of a shell, h/R (Sect. 3.5.2); the ratio of the typical size of nonhomogenity to the cell size (Sect. 3.3); the ratio of the structure size, a/b (Sect. 3.5.3); $1/\lambda$, where λ is a large frequency (Sect. 3.4); the ratio of foundation and plate rigidities, $(cL^5)/D \to 0$ and $(cL^5)/D \to \infty$ (Sect. 3.7); and the formal small parameter ε (Sect. 3.6).

Successive use of the asymptotic approach in nonlinear dynamics is closely linked with the choice of parameters for asymptotic expansions, and this problem is one of the most interesting and attractive in this field.

2. Discrete Systems

2.1 The Classical Perturbation Technique: an Introduction

The classical method of perturbation is based on the assumption that the influence of the nonlinear part of the considered differential equations is small in comparison to the influence of the linear part of the equations, or that the oscillation amplitude is small. The perturbation technique can also be used even if the deviations from the true (sought) solution are not small, but are localized in a small space. This is emphasized by the formal or natural introduction of the "small" perturbation parameter ε to the differential equation. The solution of the equations are sought in the form of power series because of the parameter ε (for $\varepsilon = 0$ the fundamental solution – the first term of the required series – is the solution to the linear differential equation). The next solution components, standing by the successive powers of ε, are obtained from the recurrent sequence of linear differential equations with constant coefficients.

The main idea of the perturbation technique is focused on the asymptotic reduction of the dimension of the dynamical system, or in the language of mechanics, on the reduction of the degrees of freedom of the system. Generally, such an approach can be initiated when the system can be divided into subsystems which are different because of their dynamical characteristics (slow and quick motions, soft and hard types of stiffness characteristics, small and large damping of the system), which allows for the introduction of one or a few perturbation parameters.

Another positive aspect of the perturbation technique is connected with the general properties of nonlinear dynamical systems. Only in very rare cases are we able to find the analytical solutions of the systems governed by a nonlinear differential equation. For instance, up to now it has been impossible to describe analytically a chaotic solution. Even when we sometimes have such a solution, it is approximated by complex functions, and in practice the benefit of such an approach is doubtful. Additionally, supposing that we have some of the particular solutions of the analysed systems, the superposition rule does not work for nonlinear systems, and it is impossible to find a solution for the arbitrarily taken initial conditions. The asymptotic series, however, sometimes allows us to overcome these problems.

Further we introduce so-called local asymptotic linearization, which is based on the linear first approximation to the nonlinear system. Recently other interesting nonlinear behaviour has been detected during an analysis of strong high-dimensional nonlinear systems. In spite of their complicated form, they sometimes possess partial solutions, called solitons, which can serve as a start to the asymptotic (non local) approach.

The main advantages of asymptotic (perturbation) analysis are as follows: a) an analytical form of the solution; b) the solutions can serve as the initial solutions for numerical simulations; c) the perturbation approach can serve as a tool for establishing the physical and engineering meaning of the dynamics.

The main weak point of the asymptotical approach lies in the fact that formally it is a very difficult task to prove a strict connection between an exact solution and that described by the asymptotic series. It can happen that because of the character of nonlinearities a periodic solution of the system for $\varepsilon = 0$ does not occur for $\varepsilon \neq 0$. It can also happen that a few or infinitely many solutions for $\varepsilon \neq 0$ correspond to a periodic solution for $\varepsilon = 0$.

The most difficult task connected with the application of the asymptotic series is to prove the convergence of the series in a wide enough interval, as well as to estimate the validity of the assumption that ε is a small enough parameter. Even if such a proof is successfully done, usually it is based on inequality chains, which is somehow not convenient enough for further analysis. Usually such problems are omitted, using instead numerical computations to test the validity of the results. The introductory results discussed below can be found in many books devoted to perturbation techniques, among others in [45, 48, 55, 59, 122, 23d, 26d, 60d].

An algorithm applying the perturbation technique will be explained by the consideration of an autonomous conservative one-degree-of-freedom system (this method is called the Krylov method).

Consider the system governed by the equation

$$\ddot{y} + \alpha_0^2 y = \varepsilon Q(y). \tag{2.1.1}$$

In order to simplify the calculations, we consider a trivial solution at $y = 0$, which leads to $Q(0) = 0$. We develop a function $Q(y)$ in a Taylor series in the vicinity of the equilibrium point $y = 0$, where $(\mathrm{d}Q/\mathrm{d}y)_{y=0} = 0$. The solution to (2.1.1) is sought in the form

$$y = y_0(t) + \sum_{k=1}^{K} \varepsilon^k y_k(t). \tag{2.1.2}$$

In order to eliminate the secular terms (unrestrictedly growing in time), we introduce the additional series

$$\alpha^2 = \alpha_0^2 + \sum_{k=1}^{K} \varepsilon^k \alpha_k. \tag{2.1.3}$$

Taking into account (2.1.2) and (2.1.3) in (2.1.1), we get

$$\sum_{k=1}^{K} \varepsilon^k \ddot{y}_k + \left(\alpha^2 - \sum_{k=1}^{K} \varepsilon^k \alpha_k \right) \sum_{k=0}^{K} \varepsilon^k y_k = \varepsilon Q \left(\sum_{k=0}^{K} \varepsilon^k y_k \right). \tag{2.1.4}$$

Then we develop the right-hand side of (2.1.4) into a power series because of the small parameter ε in the vicinity of $\varepsilon = 0$ and we get

$$\varepsilon Q(y) = \varepsilon Q \left(\sum_{k=0}^{K} \varepsilon^k y_k \right) |_{\varepsilon=0} + \left\{ Q \left(\sum_{k=0}^{K} \varepsilon^k y_k \right) + \varepsilon Q' \left(\sum_{k=0}^{K} \varepsilon^k y_k \right) \right.$$

$$\cdot \sum_{k=1}^{K} k \varepsilon^{k-1} y_k \bigg\} |_{\varepsilon=0} \varepsilon + \left\{ Q' \left(\sum_{k=0}^{K} \varepsilon^k y_k \right) \sum_{k=1}^{K} k \varepsilon^{k-1} y_k \right.$$

$$+ Q' \left(\sum_{k=0}^{K} \varepsilon^k y_k \right) \sum_{k=1}^{K} k \varepsilon^{k-1} y_k + \varepsilon Q'' \left(\sum_{k=1}^{K} \varepsilon^k y_k \right) \sum_{k=1}^{K} k \varepsilon^{k-1} y_k$$

$$+ \varepsilon Q' \left(\sum_{k=1}^{K} \varepsilon^k y_k \right) \sum_{k=2}^{K} k(k-1) \varepsilon^{k-2} y_k \bigg\} |_{\varepsilon=0} \frac{\varepsilon^2}{2} + \dots$$

$$= Q(y_0)\varepsilon + 2Q'(y_0)y_1 \frac{\varepsilon^2}{2} + \dots, \tag{2.1.5}$$

where: $Q' = dQ/dy$, $Q'' = d^2Q/dy^2$, ...

Taking into account (2.1.5) and having compared the terms standing near the same powers on the left and right hand sides of (2.1.4), we obtain

$$\begin{aligned}
\varepsilon^0 &: \ddot{y}_0 + \alpha^2 y_0 = 0, \\
\varepsilon^1 &: \ddot{y}_1 + \alpha^2 y_1 = \alpha_1 y_0 + Q(y_0), \\
\varepsilon^2 &: \ddot{y}_2 + \alpha^2 y_2 = \alpha_2 y_0 + \alpha_1 y_1 + y_1 Q'(y_0),
\end{aligned} \tag{2.1.6}$$

$$\dots \qquad \dots\dots$$

The solution to the first equation of the recurrent set (2.1.6) is the function

$$y_0 = a_0 \cos \Psi, \tag{2.1.7}$$

where

$$\Psi = \alpha t + \Theta_0. \tag{2.1.8}$$

Taking into account (2.1.7) in the second equation of (2.1.6), we get

$$\ddot{y}_1 + \alpha^2 y_1 = \alpha_0 a_0 \cos \Psi + Q(a_0 \cos \Psi). \tag{2.1.9}$$

The function $Q(a_0 \cos \Psi)$ is a periodic function because of Ψ with the period 2π, therefore we can develop it into the Fourier series

$$Q(a_0 \cos \Psi) = \frac{1}{2} b_0 + \sum_{n=1}^{\infty} b_n \cos n\Psi, \tag{2.1.10}$$

where

$$b_n = \frac{2}{\pi} \int_0^\pi Q(a_0 \cos \Psi) \cos n\Psi \, d\Psi. \tag{2.1.11}$$

Taking into account (2.1.10) in (2.1.9), we get

$$\ddot{y}_1 + \alpha^2 y_1 = \frac{1}{2} b_0 + (\alpha_1 a_0 + b_1(a_0)) \cos \Psi + \sum_{n=2}^\infty b_n \cos n\Psi. \tag{2.1.12}$$

In order to get a periodic solution we should eliminate the secular term from $y_1(t)$, which leads to the condition

$$\alpha_1 a_0 + b_1(a_0) = 0. \tag{2.1.13}$$

From this equation we get

$$\alpha_1 = -\frac{b_1(a_0)}{a_0}, \tag{2.1.14}$$

and the first unknown coefficient of (2.1.13) is estimated. Therefore, a solution to (2.1.12) has the form

$$y_1 = a_1 \cos(\alpha t + \Theta_1) + \frac{b_0}{2\alpha^2} + \sum_{n=2}^\infty \frac{b_n}{\alpha^2 - (n\alpha)^2} \cos n\Psi. \tag{2.1.15}$$

The above constants a_0 and Θ_0 are defined by the initial conditions, and we take $a_1 = \Theta_1 = 0$. This leads to the determination of a_0 and Θ_0 from (2.1.2), which serves to obtain the constants a_0 and Θ_0 on the basis of the assumption that $y_0(t)$ fulfil the initial conditions, while a_k and Θ_k are found from the condition that $y_k(t_0) = 0$.

Further, we take $a_1 = \Theta_1 = 0$ and from (2.1.15) we get

$$y_1(t) = \frac{b_0}{2\alpha^2} + \sum_{n=2}^\infty \frac{b_n}{\alpha^2 - (n\alpha)^2} \cos n\Psi. \tag{2.1.16}$$

Therefore, the right-hand side of the third equation of (2.1.5) is defined. On the basis of the condition of avoiding a secular term in the solution $y_2(t)$, we get the coefficient α_2, and the solution of that equation is the next term of the sequence (2.1.2). After a limitation to the first approximation $O(\varepsilon^2)$, we get

$$y = a_0 \cos(\alpha t + \Theta_0) + \varepsilon \frac{1}{\alpha^2} \left[\frac{1}{2} b_0 + \sum_{n=2}^\infty \frac{b_n}{1 - n^2} \cos n(\alpha t + n\Theta_0) \right], \tag{2.1.17}$$

where

$$\alpha = \sqrt{\alpha_0^2 + \varepsilon \alpha_1(a_0)} = \sqrt{\alpha_0^2 - \varepsilon \frac{b_1(a_0)}{a_0}}. \tag{2.1.18}$$

The period of the sought solution is

$$T = \frac{2\pi}{\sqrt{\alpha_0^2 - \varepsilon \frac{b_1(a_0)}{a_0}}} \qquad (2.1.19)$$

and depends on the oscillation amplitude a_0.

Example 2.1.1. For the mechanical system presented in Figure 2.1 calculate the oscillation period and find its oscillations analytically.

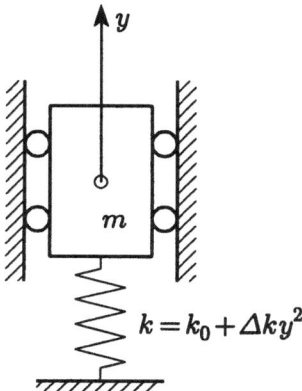

$k = k_0 + \Delta k y^2$

Fig. 2.1. One-degree-of-freedom conservative oscillator

The equation of motion has the form

$$\ddot{y} + \alpha_0^2 y = -\varepsilon y^3, \qquad (2.1.20)$$

where

$$\alpha_0^2 = \frac{k}{m}, \quad \varepsilon = \frac{\Delta k}{m}. \qquad (2.1.21)$$

The recurrent set of equations is given below

$$\ddot{y}_0 + \alpha^2 y_0 = 0, \qquad (2.1.22)$$

$$\ddot{y}_1 + \alpha^2 y_1 = \alpha_1 y_0 - y_0^3, \qquad (2.1.23)$$

$$\ddot{y}_2 + \alpha^2 y_2 = \alpha_2 y_0 - \alpha_1 y_1 - 3y_0^2 y_1. \qquad (2.1.24)$$

The solution to (2.1.22) is

$$y_0 = a_0 \cos \Psi, \quad \Psi = \alpha t + \Theta_0. \qquad (2.1.25)$$

Taking into account (2.1.25) in (2.1.23) and on the basis of the equation

$$\cos^3 \Psi = \frac{3}{4} \cos \Psi + \frac{1}{4} \cos 3\Psi, \qquad (2.1.26)$$

we get

$$\ddot{y}_1 + \alpha^2 y_1 = a_0 \left(\alpha_1 - \frac{3}{4} a_0^2 \right) \cos \Psi - \frac{1}{4} a_0^3 \cos 3\Psi. \qquad (2.1.27)$$

From the condition of avoiding the secular form we find that

$$\alpha_1 = \frac{3}{4}a_0^2, \tag{2.1.28}$$

and taking into account $a_1 = \Theta_1 = 0$, we get

$$y_1 = \frac{a_0^3}{32\alpha^2}\cos 3\Psi. \tag{2.1.29}$$

Taking into account (2.1.28) and (2.1.29) in (2.1.24), we obtain

$$\ddot{y}_2 + \alpha^2 y_2 = \alpha_2 a_0 \cos\Psi + \frac{3a_0^5}{128\alpha^2}\cos 3\Psi - 3\frac{a_0^5}{32\alpha^2}\cos 3\Psi \cos^2\Psi. \tag{2.1.30}$$

Because

$$\cos^2\Psi \cos 3\Psi = \frac{1}{4}(\cos 5\Psi + 2\cos 3\Psi + \cos\Psi), \tag{2.1.31}$$

therefore

$$\ddot{y}_2 + \alpha^2 y_2 = \left(\alpha_2 a_0 + \frac{3a_0^5}{128\alpha^2}\right)\cos 3\Psi - \frac{3a_0^5}{128\alpha^2}\cos 3\Psi$$
$$- \frac{3a_0^2}{128\alpha^2}\cos 5\Psi. \tag{2.1.32}$$

From (2.1.32) we get

$$\alpha_2 = \frac{3a_0^4}{128\alpha^2}, \tag{2.1.33}$$

and for $a_2 = \Theta_2 = 0$ we obtain

$$y_2 = \frac{3}{1024}\frac{a_0^5}{\alpha^2}. \tag{2.1.34}$$

Finally, we get

$$y = a_0 \cos\Psi + \varepsilon\frac{a_0^3}{32\alpha^2}\cos 3\Psi + \varepsilon^2\frac{a_0^5}{1024\alpha^2}(3\cos 3\Psi + \cos 5\Psi), \tag{2.1.35}$$

where

$$\alpha = \sqrt{\alpha_0^2 + \varepsilon\frac{3}{4}a_0^2 + \varepsilon^2\frac{3}{128}\frac{a_0^4}{a_0^2 + \varepsilon\frac{3}{4}a_0^2}}, \quad \Psi = \alpha t + \Theta_0. \tag{2.1.36}$$

The oscillation amplitude, which corresponds here to the maximum hang-out from the equilibrium position, is obtained for the following time moments

$$t_n = \frac{n\pi - \Theta_0}{\alpha} \quad (\Psi = \alpha t + \Theta_0 = n\pi). \tag{2.1.37}$$

The period of oscillation is given by the formula

$$T = \frac{2\pi}{\sqrt{\alpha_0^2 + \varepsilon\frac{3}{4}a_0^2 + \varepsilon^2\frac{3}{128}\frac{a_0^4}{a_0^2 + \varepsilon\frac{3}{4}a_0^2}}}. \tag{2.1.38}$$

In order to estimate the corrections introduced by the successive approximation, we take $\varepsilon = a_0 = \alpha_0 = 1$ and we calculate:

A. For $\varepsilon^{(0)}$, we have $T^{(0)} = 6.28$ for every amplitude $a^{(0)}$;

B. For $\varepsilon^{(1)}$, we have $T^{(1)} = 2\pi/\sqrt{\alpha_0^2 + \frac{3}{4}\varepsilon\alpha_0^4} = 4.7472$ for $A^{(1)} = a_0 + \varepsilon a_0^3/32\alpha_0^2 = 1.01778$;

C. For $\varepsilon^{(2)}$, we have $T^{(2)} = 2\pi/\alpha = 4.7293$ for $A^{(2)} = 1.035156$.

2.2 Krylov–Bogolubov–Mitropolskij Method

This method can be applied to systems of second-order ordinary differential equations. Its fundamental parts will be outlined on the basis of an example of one-degree-of-freedom autonomous systems of the form

$$\ddot{y} + \alpha_0^2 y = \varepsilon Q(y, \dot{y}). \tag{2.2.1}$$

Let us suppose that $y = \dot{y} = 0$ is the equilibrium position and the function $Q(y, \dot{y})$ is analytical because of its variables. The main difference between this and the Krylov method lies in the assumption that the amplitude and the phase are functions of time. We are looking for the solution

$$y(t) = a \cos \Psi + \sum_{k=1}^{K} \varepsilon^k y_k[a(t), \Psi(t)]. \tag{2.2.2}$$

For conservative systems the amplitude is constant, and

$$\frac{da}{dt} = 0. \tag{2.2.3}$$

The derivative of Ψ with respect to time is also constant for conservative systems and can be approximated by the series

$$\dot{\Psi} = \alpha = \alpha_0 + \sum_{k=1}^{K} \varepsilon^k \alpha_k(a). \tag{2.2.4}$$

However, for nonconservative systems we introduce the following series

$$\dot{a} = \sum_{k=1}^{K} \varepsilon^k A_k[a(t)], \tag{2.2.5}$$

$$\dot{\Psi} = \alpha_0 + \sum_{k=1}^{K} \varepsilon^k B_k[a(t)]. \tag{2.2.6}$$

From (2.2.2) one obtains

$$\dot{y} = \dot{a} \cos \Psi - a\dot{\Psi} \sin \Psi + \sum_{k=1}^{K} \varepsilon^k \left(\frac{\partial y_k}{\partial a}\dot{a} + \frac{\partial y_k}{\partial \Psi}\dot{\Psi} \right), \tag{2.2.7}$$

$$\ddot{y} = \ddot{a}\cos\Psi - 2\dot{a}\dot{\Psi}\sin\Psi - a\ddot{\Psi}\sin\Psi - a\dot{\Psi}^2\cos\Psi$$

$$+ \sum_{k=1}^{K} \varepsilon^k \left(\frac{\partial^2 y_k}{\partial a^2}\dot{a}^2 + \frac{\partial^2 y_k}{\partial a\partial\Psi}\dot{a}\dot{\Psi} + \frac{\partial y_k}{\partial a}\ddot{a} + \frac{\partial^2 y_k}{\partial\Psi\partial a}\dot{\Psi}\dot{a} \right.$$

$$\left. + \frac{\partial^2 y_k}{\partial\Psi^2}\dot{\Psi}^2 + \frac{\partial y_k}{\partial\Psi}\ddot{\Psi} \right). \tag{2.2.8}$$

Taking into account (2.2.2) and (2.2.8) for the left-hand of side (2.2.1), we get

$$L = \left[\ddot{a} - (\dot{\Psi}^2 - \alpha_0^2)a\right]\cos\Psi - (2\dot{a}\dot{\Psi} + a\ddot{\Psi})\sin\Psi$$

$$+ \sum_{k=1}^{K} \varepsilon^k \left(\frac{\partial^2 y_k}{\partial a^2}\dot{a}^2 + \frac{\partial^2 y_k}{\partial\Psi^2}\dot{\Psi}^2 + 2\dot{a}\dot{\Psi}\frac{\partial^2 y_k}{\partial a\partial\Psi} + \frac{\partial y_k}{\partial a}\ddot{a} \right.$$

$$\left. + \frac{\partial y_k}{\partial\Psi}\ddot{\Psi} + \alpha_0^2 y_k \right). \tag{2.2.9}$$

According to (2.2.5) and (2.2.6), we get

$$\ddot{\Psi} = \sum_{k=1}^{K} \varepsilon^k \frac{dB_k}{da}\dot{a} = \sum_{k=1}^{K} \varepsilon^k \frac{dB_k}{da} \sum_{k=1}^{K} \varepsilon^k A_k = \varepsilon^2 \frac{dB_1}{da}A_1 + O(\varepsilon^3), \tag{2.2.10}$$

and after some transformations one obtains

$$L = \varepsilon \left[\alpha_0^2 \left(\frac{\partial^2 y_1}{\partial\Psi^2} + y_1 \right) - 2\alpha_0 B_1 a\cos\Psi - 2\alpha_0 A_1 \sin\Psi \right]$$

$$+ \varepsilon^2 \left[\alpha_0^2 \left(\frac{\partial^2 y_2}{\partial\Psi^2} + y_2 \right) + 2\alpha_0 B_1 \frac{\partial^2 y_1}{\partial\Psi^2} + 2\alpha_0 A_1 \frac{\partial^2 y_1}{\partial a\partial\Psi} \right.$$

$$+ \left(A_1 \frac{dA_1}{da} - (2\alpha_0 B_2 + B_1^2)a \right)\cos\Psi$$

$$\left. - \left(2\alpha_0 A_2 + 2A_1 B_1 + \frac{dB_1}{da}A_1 a \right)\sin\Psi \right] + O(\varepsilon^2). \tag{2.2.11}$$

We develop the right-hand side P of (2.2.1) into a power series because of the parameter ε in the vicinity of $\varepsilon = 0$ of the form

$$P = \varepsilon Q(y, \dot{y})$$

$$= \varepsilon \left\{ Q(y, \dot{y})\mid_{\varepsilon=0} + \left(\frac{\partial Q}{\partial y}\frac{dy}{d\varepsilon}\mid_{\varepsilon=0} + \frac{\partial Q}{\partial\dot{y}}\frac{d\dot{y}}{d\varepsilon}\mid_{\varepsilon=0} \right) + O(\varepsilon^2) \right\}$$

$$= \varepsilon Q(a\cos\Psi, -a\alpha_0\sin\Psi) + \varepsilon^2 \left[\frac{\partial Q}{\partial y}(a\cos\Psi, -a\alpha_0\sin\Psi)y_1 \right.$$

$$\left. + \frac{\partial Q}{\partial\dot{y}}(a\cos\Psi - a\alpha_0\sin\Psi)\left(A_1\cos\Psi, -aB_1\sin\Psi + \alpha_0\frac{\partial y_1}{\partial\Psi} \right) \right]$$

$$+ O(\varepsilon^2). \tag{2.2.12}$$

Equating the terms of the same powers of ε, we get the following recurrent set of differential equations

$$\frac{\partial^2 y_1}{\partial \Psi^2} + y_1 = \frac{1}{\alpha_0^2}(f_0 + 2A_1\alpha_0 \sin \Psi + 2B_1\alpha_0 a \cos \Psi), \qquad (2.2.13)$$

$$\frac{\partial^2 y_2}{\partial \Psi^2} + y_2 = \frac{1}{\alpha_0^2}(f_1 + 2A_2\alpha_0 \sin \Psi + 2B_2\alpha_0 a \cos \Psi). \qquad (2.2.14)$$

and so on, where:

$$f_0 = Q(a \cos \Psi, -a\alpha_0 \sin \Psi), \qquad (2.2.15)$$

$$f_1 = -2B_1\alpha_0 \frac{\partial^2 y_1}{\partial \Psi^2} - 2A_1\alpha_0 \frac{\partial^2 y_1}{\partial a \partial \Psi} - \left(A_1 \frac{dA_1}{da} - B_1^2 a\right) \cos \Psi \qquad (2.2.16)$$

$$+ \left(2A_1 B_1 + aA_1 \frac{dB_1}{da}\right) \sin \Psi + \frac{\partial Q}{\partial y}(a \cos \Psi, -a\alpha_0 \sin \Psi)y_1$$

$$+ \frac{\partial Q}{\partial \dot{y}}(a \cos \Psi, -a\alpha_0 \sin \Psi)\left(A_1 \cos \Psi - aB_1 \sin \Psi + \alpha_0 \frac{\partial y_1}{\partial \Psi}\right).$$

The function defined by (2.2.15) is periodic with period 2π and it can be approximated by the Fourier series of the form

$$f_0(a, \Psi) = \frac{1}{2}b_{00}(a) + \sum_{n=1}^{\infty}(b_{0n}(a) \cos n\Psi + c_{0n}(a) \sin n\Psi), \qquad (2.2.17)$$

where the coefficients are defined below

$$b_{0n}(a) = \frac{1}{\pi} \int_0^{2\pi} Q(a \cos \Psi, -a\alpha_0 \sin \Psi) \cos n\Psi \, d\Psi, \qquad (2.2.18)$$

$$c_{0n}(a) = \frac{1}{\pi} \int_0^{2\pi} Q(a \cos \Psi, -a\alpha_0 \sin \Psi) \sin n\Psi \, d\Psi, \quad n = 0, 1, 2, \ldots . \ (2.2.19)$$

Taking into account (2.2.17), (2.2.18) and (2.2.19), we have

$$\frac{\partial^2 y_1}{\partial \Psi^2} + y_1 = \frac{1}{\alpha_0^2}(c_{01} + 2A_1\alpha_0) \sin \Psi + \frac{1}{\alpha_0^2}(b_{01} + 2B_1\alpha_0 a) \cos \Psi$$

$$+ \frac{b_{00}}{2\alpha_0^2} + \frac{1}{\alpha_0^2} \sum_{n=2}^{\infty}(b_{0n} \cos n\Psi + c_{0n} \sin n\Psi). \qquad (2.2.20)$$

The first two terms standing on the right-hand side of (2.2.20) grow unrestrictedly in time and their occurrence contradicts the assumptions of the asymptotic method. Therefore, we equate them to zero, which leads to the determination of the unknown function $A_1(a)$ and $B_1(a)$ according to the formulas

$$A_1(a) = -\frac{c_{01}(a)}{2\alpha_0}, \qquad (2.2.21)$$

$$B_1(a) = -\frac{b_{01}(a)}{2\alpha_0 a}.$$ (2.2.22)

On the basis of these results, the general solution to (2.2.13) is

$$y_1 = a_1 \cos(\Psi + \theta_1) + \frac{b_{00}}{2\alpha_0^2}$$

$$+ \frac{1}{\alpha_0^2} \sum_{n=2}^{\infty} \frac{1}{1-n^2} (b_{0n} \cos n\Psi + c_{0n} \sin n\Psi).$$ (2.2.23)

Taking $a_1 = \theta_1 = 0$, the above equation can be reduced to the form

$$y_1 = \frac{b_{00}}{2\alpha_0^2} + \frac{1}{\alpha_0^2} \sum_{n=2}^{\infty} (b_{0n} \cos n\Psi + c_{0n} \sin n\Psi).$$ (2.2.24)

It is not difficult to see that now the function y_1 defined by (2.2.24) is known thanks to (2.2.18) and (2.2.19) and it has period 2π because of the variable Ψ. Now we develop the function f_1 into the Fourier series

$$f_1(a, \Psi) = \frac{1}{2}b_{10}(a) + \sum_{n=1}^{\infty}(b_{1n}(a) \cos n\Psi + c_{1n}(a) \sin n\Psi),$$ (2.2.25)

where

$$b_{1n}(a) = \frac{1}{\pi} \int_0^{2\pi} f_1(a, \Psi) \cos n\Psi \, d\Psi,$$ (2.2.26)

$$c_{1n}(a) = \frac{1}{\pi} \int_0^{2\pi} f_1(a, \Psi) \sin n\Psi \, d\Psi, \quad n = 0, 1, 2, \ldots.$$ (2.2.27)

Making calculations in a similar way, we define

$$A_2(a) = -\frac{c_{11}(a)}{2\alpha_0},$$ (2.2.28)

$$B_2(a) = -\frac{b_{11}(a)}{2\alpha_0 a}.$$ (2.2.29)

and the function

$$y_2 = \frac{b_{10}}{2\alpha_0^2} + \frac{1}{\alpha_0^2} \sum_{n=2}^{\infty} \frac{1}{1-n^2} (b_{1n} \cos n\Psi + c_{1n} \sin n\Psi).$$ (2.2.30)

We can define the unknown function $a(t)$ and $\Psi(t)$ occurring in the series (2.2.2) by solving the differential equations (2.2.5) and (2.2.6). Separating the variables in (2.2.5), we get

$$dt = \frac{da}{\sum_{m=1}^{M} \varepsilon^k A_k(a)},$$ (2.2.31)

and then, after integration, we obtain

$$t = \int \frac{da}{\sum_{m=1}^{M} \varepsilon^k A_k(a)} + a_0, \tag{2.2.32}$$

where a_0 is a constant dependent on the initial conditions.

However, even in the case when the integral given by (2.2.32) can be defined using elementary functions it will be difficult to find explicit formulas necessary to solve (2.2.5). If it is possible to get an explicit expression for $a(t)$, then from (2.2.6) one obtains

$$\Psi = \int \left(\alpha_0 + \sum_{k=1}^{K} \varepsilon^k B_k[a(t)] \right) dt + \Theta_0. \tag{2.2.33}$$

Taking into account the above results we get the general solution defined by (2.2.2) with the constants a_0 and Θ_0 dependent on the initial conditions.

Example 2.2.1. Determine an analytical form of the oscillations governed by the Rayleigh equation of the form

$$\ddot{y} + \alpha_0^2 y = (2h - g\dot{y}^2)\dot{y}. \tag{2.2.34}$$

We formally introduce the parameter ε, and the above equation has the form (2.2.1) already discussed, where

$$\varepsilon Q(y, \dot{y}) = \varepsilon(2h - g\dot{y}^2)\dot{y}. \tag{2.2.35}$$

We are looking for a solution of the form

$$y(t) = a \cos\Psi + \varepsilon y_1(a\Psi), \tag{2.2.36}$$
$$\dot{a} = \varepsilon A_1(a), \tag{2.2.37}$$
$$\dot{\Psi} = \alpha_0 + \varepsilon B_1(a), \tag{2.2.38}$$

and according to (2.2.13) we get

$$\frac{\partial^2 y_1}{\partial \Psi^2} + y_1 = \frac{1}{\alpha_0^2}(f_0 + 2A_1\alpha_0 \sin\Psi + 2B_1\alpha_0 a \cos\Psi), \tag{2.2.39}$$

where

$$f_0 = Q(a\cos\Psi - a\alpha_0 \sin\Psi)$$
$$= a\alpha_0 \left(\frac{3}{4}ga^2\alpha_0^2 - 2h \right) \sin\Psi - \frac{1}{4}ga^3\alpha_0^3 \sin 3\Psi. \tag{2.2.40}$$

Substituting (2.2.40) in (2.2.39), we get

$$\frac{\partial^2 y_1}{\partial \Psi^2} + y_1 = \frac{1}{\alpha_0^2} \left\{ \left[2A_1\alpha_0 + a\alpha_0 \left(\frac{3}{4}ga^2\alpha_0^2 - 2h \right) \right] \sin\Psi \right.$$
$$\left. + 2B_1\alpha_0 a \cos\Psi - \frac{1}{4}ga^3\alpha_0^3 \sin 3\Psi \right\}. \tag{2.2.41}$$

From the condition of avoiding secular terms, we determine the unknown coefficients

$$A_1 = \frac{1}{2}a\left(2h - \frac{3}{4}ga^2\alpha_0^2\right),$$ (2.2.42)

$$B_1 = 0.$$ (2.2.43)

The solution to (2.2.41) because of (2.2.42) and (2.2.43) takes the form

$$y_1 = \frac{1}{32}ga^3\alpha_0 \sin 3\Psi.$$ (2.2.44)

From (2.2.37), taking into account (2.2.42), we get

$$\dot{a} = \varepsilon ah(1 - K^2a^2),$$ (2.2.45)

where

$$K^2 = \frac{3}{8}\frac{g\alpha_0^2}{h}.$$ (2.2.46)

If $K^2 > 0$, then we obtain

$$\frac{da}{a(1 - Ka)(1 + Ka)} = \varepsilon h\,dt.$$ (2.2.47)

The above equation can be presented in the form

$$\int \frac{da}{a} + \frac{K}{2}\int \frac{da}{1 - Ka} - \frac{K}{2}\int \frac{da}{1 + Ka} + \ln L = \varepsilon ht,$$ (2.2.48)

where L is the integration constant. After integration we get

$$\ln \frac{aL}{\sqrt{1 - K^2a^2}} = \varepsilon ht,$$ (2.2.49)

and then

$$a = \frac{a_0 e^{\varepsilon ht}}{\sqrt{1 + a_0^2 K^2 e^{2\varepsilon ht}}},$$ (2.2.50)

where $a_0 = L^{-1}$ is a constant. The phase Ψ can be approximated by

$$\psi = \alpha_0 t + \Theta_0,$$ (2.2.51)

and both a_0 and Θ_0 are defined by the initial conditions.

2.3 Equivalent Linearization

Leaving only the first term in the series (2.2.2), we have

$$y(t) = a\cos\Psi,$$ (2.3.1)

and in the series (2.2.5) and (2.2.6) we take only the terms

$$\dot{a} = \varepsilon A_1(a),$$ (2.3.2)

$$\dot{\Psi} = \alpha_0 + \varepsilon B_1(a).$$ (2.3.3)

Then the solution (2.3.1) will be the first simplified approximation solution of (2.2.1). We prove below that the solution (2.3.1) of (2.2.1) fulfils the equation

$$\ddot{y} + 2h_e(a)\dot{y} + \alpha_e^2(a)y = O(\varepsilon). \tag{2.3.4}$$

We call the above equation the equivalent linear approximation to the non-linear equation. Two parameters appearing in (2.3.4), the equivalent unit damping coefficient h_e and the frequency α_e, are defined as follows

$$h_e(a) = \frac{\varepsilon}{2\pi\alpha_0 a} \int\limits_0^{2\pi} Q(a\cos\Psi, -a\alpha_0\sin\Psi)\sin\Psi\,d\Psi, \tag{2.3.5}$$

$$\alpha_e(a) = \alpha_0 - \frac{\varepsilon}{2\pi\alpha_0 a} \int\limits_0^{2\pi} Q(a\cos\Psi, -a\alpha_0\sin\Psi)\cos\Psi\,d\Psi. \tag{2.3.6}$$

From (2.3.1) we have

$$\dot{y} = \dot{a}\cos\Psi - a\dot{\Psi}\sin\Psi. \tag{2.3.7}$$

Taking into account (2.2.21) and (2.3.5), the equation (2.3.2) will take the form

$$\dot{a} = -ah_e(a). \tag{2.3.8}$$

Taking into account (2.2.22) and (2.3.6), the equation (2.3.3) will take the form

$$\dot{\Psi} = \alpha_e(a). \tag{2.3.9}$$

Taking into account (2.3.5) and (2.3.6) for (2.3.1), we obtain

$$\dot{y} = -ah_e\cos\Psi - a\alpha_e\sin\Psi. \tag{2.3.10}$$

Differentiating (2.3.7) we get

$$\ddot{y} = -\dot{a}h_e\cos\Psi - a\frac{dh_e}{da}\cos\Psi + ah_e\dot{\Psi}\sin\Psi - \dot{a}\alpha_e\sin\Psi$$
$$- a\frac{d\alpha_e}{da}\dot{a}\sin\Psi - a\alpha_e\dot{\Psi}\cos\Psi, \tag{2.3.11}$$

and after taking into account (2.3.8) and (2.3.9), we get

$$\ddot{y} = ah_e^2\cos\Psi + 2h_e a\alpha_e\sin\Psi - a\alpha_e^2\cos\Psi + a^2 h_e\frac{dh_e}{da}\cos\Psi$$
$$+ a^2 h_e\frac{d\alpha_e}{da}\sin\Psi - 2h_e^2 a\cos\Psi - 2h_e\alpha_e\sin\Psi + \alpha_e^2 a\cos\Psi. \tag{2.3.12}$$

Finally, we obtain

$$\ddot{y} + 2h_e\dot{y} + \alpha_e^2 y = -ah_e^2\cos\Psi + a^2 h_e\frac{dh_e}{da}\cos\Psi + a^2 h_e\frac{dh_e}{da}\sin\Psi. \tag{2.3.13}$$

The right-hand side of (2.3.13) is of order ε^2, because according to (2.3.5) and (2.3.6) the expressions h_e, dh_e/da, $d\alpha_e/da$ are of order ε.

To summarize, we have illustrated that on the basis of the approximations introduced by (2.3.2) and (2.3.3) we reduce the problem governed by the nonlinear equation (2.2.1) to an equivalent linear one. Equation (2.3.13) possesses a periodic solution if

$$h_e(a) = 0, \tag{2.3.14}$$

and the solution to the above algebraic nonlinear equation is an amplitude a_0 of the periodic solutions of the form

$$y_0 = a_0 \cos[\alpha_e(a_0)t]. \tag{2.3.15}$$

In the case when $h_e(a) \equiv 0$, the solution is given by

$$y = A\cos(\alpha_e(a_0)t + \theta), \tag{2.3.16}$$

where A and θ are constants dependent on the initial conditions.

Example 2.3.1. Using the method of equivalent linearization, determine the amplitude of oscillations of the Rayleigh equation (2.2.34).

According to (2.3.5) and (2.3.6), the equivalent damping coefficient and the equivalent frequency are equal to

$$h_e(a) = \frac{1}{2\pi\alpha_0 a} \int_0^{2\pi} \left[2h(-a\alpha_0 \sin \Psi) - g(-a\alpha_0 \sin \Psi)^3\right] \sin \Psi \, d\Psi$$

$$= -h + \frac{3}{8}ga^2\alpha_0^2, \tag{2.3.17}$$

$$\alpha_e(a) = \alpha_0 - \frac{1}{2\pi\alpha_0 a} \int_0^{2\pi} \left[2h(-a\alpha_0 \sin \Psi) - g(-a\alpha_0 \sin \Psi)^3\right] \cos \Psi \, d\Psi$$

$$= \alpha_0. \tag{2.3.18}$$

Therefore, the equivalent linear equation takes the form

$$\ddot{y} + \left(-h + \frac{3}{8}ga^2\alpha_0^2\right)\dot{y} + \alpha_0^2 y = 0, \tag{2.3.19}$$

and the amplitude of the periodic oscillation is equal to

$$a_0 = \frac{1}{\alpha_0}\sqrt{\frac{8h}{3g}}. \tag{2.3.20}$$

2.4 Analysis of Nonconservative Nonautonomous Systems

2.4.1 Introduction

We consider one-degree-of-freedom systems of the form

$$\ddot{y} + \alpha_0^2 y = \varepsilon\phi(y, \dot{y}, \omega t), \tag{2.4.1}$$

and the exciting force fulfils the periodicity conditions $\phi(\omega t + 2\pi) = \phi(\omega t)$. Therefore, it can be developed into the Fourier series

$$\phi(y, \dot{y}, \omega t) = Q_{10}(y, \dot{y}) + \sum_{m=1}^{M} \{Q_{1m}(y, \dot{y}) \cos m\omega t$$
$$+ Q_{2m}(y, \dot{y}) \sin m\omega t\}. \tag{2.4.2}$$

We also assume that the functions Q_{10}, Q_{1m} and Q_{2m} are analytical because of their variables. This means that they can be expanded into a Taylor series in the vicinity of the equilibrium (here taken as the trivial one $y = \dot{y} = 0$).

In the case of linear systems with a periodic excitation, the resonant oscillations can appear for m-harmonics, when $m\omega = \alpha_0$, whereas in the case of nonlinear systems, the resonance occurs when

$$m\omega = n\alpha_0, \tag{2.4.3}$$

where $m, n = 1, 2, 3, \dots$. In dissipative systems during resonant oscillations, an increase in the amplitude of oscillations is observed.

The resonance occurring in nonlinear systems can be classified as follows:

1. Main resonance ($m = n = 1$).
2. Subharmonic resonance ($m = 1$, $n > 1$).
3. Ultraharmonic resonance ($m > 1$, $n = 1$).
4. Ultrasubharmonic resonance ($m > 1$, $n > 1$).

2.4.2 Nonresonance Oscillations

Consider an oscillator governed by the equation

$$\ddot{y} + \alpha_0^2 y = \varepsilon \left[Q(y, \dot{y}) + P(\eta) \right], \tag{2.4.4}$$

where ε is the perturbation parameter and the exciting force $P(\eta) = P(\eta + 2\pi)$, where $\eta = \omega t$. We use the KBM method described earlier to analyse (2.4.4).

We are looking for the solution

$$y = a \cos \Psi + \varepsilon y_1(a, \Psi, \eta), \tag{2.4.5}$$

where

$$\dot{a} = \varepsilon A_1(a), \tag{2.4.6}$$
$$\dot{\Psi} = \alpha_0 + \varepsilon B_1(a), \tag{2.4.7}$$

Here we restrict ourselves to the $O(\varepsilon^2)$ approximation.

From (2.4.5) one obtains

$$\dot{y} = \dot{a} \cos \Psi + a\dot{\Psi} \sin \Psi + \varepsilon \left(\frac{\partial y_1}{\partial a} \dot{a} + \frac{\partial y_1}{\partial \Psi} \dot{\Psi} + \frac{\partial y_1}{\partial \eta} \omega \right), \tag{2.4.8}$$

$$\ddot{y} = \ddot{a}\cos\Psi - 2\dot{a}\dot{\Psi}\sin\Psi - a\ddot{\Psi}\sin\Psi - a\dot{\Psi}^2\cos\Psi$$

$$+\varepsilon\left(\frac{\partial^2 y_1}{\partial a^2}\dot{a}^2 + 2\frac{\partial^2 y_1}{\partial a\partial\Psi}\dot{a}\dot{\Psi} + 2\frac{\partial^2 y_1}{\partial a\partial\eta}\dot{a}\omega + \frac{\partial y_1}{\partial a}\ddot{a}\right.$$

$$\left.+\frac{\partial^2 y_1}{\partial\Psi^2}\dot{\Psi}^2 + 2\frac{\partial^2 y_1}{\partial\Psi\partial\eta}\dot{\Psi}\omega + \frac{\partial y_1}{\partial\Psi}\ddot{\Psi} + \frac{\partial^2 y_1}{\partial\eta^2}\omega^2\right). \tag{2.4.9}$$

Taking into account (2.4.6) and (2.4.7) in (2.4.8) and (2.4.9), we get

$$\dot{y} = -a\alpha_0\sin\Psi + \varepsilon\left(A_1\cos\Psi - aB_1\sin\Psi + \frac{\partial y_1}{\partial\Psi}\alpha_0 + \frac{\partial y_1}{\partial\eta}\omega\right), \tag{2.4.10}$$

$$\ddot{y} = -a\alpha_0^2\cos\Psi + \varepsilon\left(-2A_1\alpha_0\sin\Psi - 2B_1\alpha_0 a\cos\Psi + \frac{\partial^2 y_1}{\partial\Psi^2}\alpha_0^2\right.$$

$$\left.+2\frac{\partial^2 y_1}{\partial\Psi\partial\eta}\alpha_0\omega + \frac{\partial^2 y_1}{\partial\eta^2}\omega^2\right). \tag{2.4.11}$$

Because

$$\ddot{a} = \frac{d}{dt}(\dot{a}) = \varepsilon^2 A_1\frac{dA_1}{da},$$

$$\ddot{\Psi} = \frac{d}{dt}(\dot{\Psi}) = \varepsilon^2\frac{dB_1}{da}A_1, \tag{2.4.12}$$

$$\dot{\Psi}^2 = \alpha_0^2 + \varepsilon a\alpha_0 B_1 + \varepsilon^2 B_1,$$

therefore, the left-hand side L of (2.4.4) takes the form

$$L = \varepsilon\left(2A_1\alpha_0\sin\Psi - 2B_1\alpha_0 a\cos\Psi + \frac{\partial^2 y_1}{\partial\Psi^2}\alpha_0^2 + 2\frac{\partial^2 y_1}{\partial\Psi\partial\eta}\alpha_0\omega\right.$$

$$\left.+\frac{\partial^2 y_1}{\partial\eta^2}\omega^2\right) + \varepsilon\alpha_0^2 y_1. \tag{2.4.13}$$

Developing the right-hand side R of (2.4.4) into a power series in ε, we have

$$P = \varepsilon\left[Q(y,\dot{y}) + P(\eta)\right]$$

$$= \varepsilon\left\{\left[Q(y,\dot{y}) + P(\eta)\right]|_{\varepsilon=0} + \varepsilon\frac{1}{1!}\left[\frac{\partial Q}{\partial y}\frac{dy}{d\varepsilon} + \frac{\partial Q}{\partial y}\frac{d\dot{y}}{d\varepsilon}\right]|_{\varepsilon=0} + \cdots\right\}$$

$$= \varepsilon\left[Q(a\cos\Psi, -a\alpha_0\sin\Psi) + P(\eta)\right] + O(\varepsilon). \tag{2.4.14}$$

Because the functions $- Q(\Psi) = Q(\Psi + 2\pi)$ and $P(y) = P(\omega(t+T))$, where $T = 2\pi/\omega$, then we develop the right-hand side of (2.4.14) into the Fourier series of the form

$$P = \varepsilon\left[\frac{1}{2}b_0(a) + \sum_{n=1}^{\infty}(b_n(a)\cos n\Psi + c_n(a)\sin\Psi) + \frac{1}{2}p_0\right.$$

$$\left.+\sum_{m=1}^{\infty}(p_m\cos m\eta + q_m\sin m\eta)\right] \tag{2.4.15}$$

where

$$b_n = \frac{1}{\pi} \int_0^{2\pi} Q(a \cos \Psi, -a\alpha_0 \sin \Psi) \cos n\Psi \, d\Psi,$$

$$c_n = \frac{1}{\pi} \int_0^{2\pi} Q(a \cos \Psi, -a\alpha_0 \sin \Psi) \sin n\Psi \, d\Psi,$$

$$p_m = \frac{2}{T} \int_0^T P(\eta) \cos m\eta \, d\eta,$$

$$q_m = \frac{2}{T} \int_0^T P(\eta) \sin m\eta \, d\eta, \quad m, n = 0, 1, 2, \ldots \quad (2.4.16)$$

Equating the terms of the same powers of ε in (2.4.13) and (2.4.15), we obtain

$$\frac{\partial^2 y_1}{\partial \Psi^2} + 2\frac{\omega}{\alpha_0}\frac{\partial^2 y_1}{\partial \Psi \partial \eta} + \frac{\omega^2}{\alpha_0^2}\frac{\partial^2 y_1}{\partial \eta^2} + y_1 = \frac{1}{\alpha_0^2}\left\{ [2A_1\alpha_0 - c_1(a)] \sin \Psi \right.$$

$$+ [2B_1\alpha_0 a - b_1(a)] \cos \Psi + \frac{1}{2}b_0(a) + \sum_{n=2}^{\infty} (b_n(a) \cos n\Psi$$

$$\left. + c_n(a) \sin \Psi) + \frac{1}{2}p_0 + \sum_{m=1}^{\infty} (p_m \cos m\eta + q_m \sin m\eta) \right\}. \quad (2.4.17)$$

The first two terms standing on the right-hand side of that equation give the secular terms and we get

$$A_1 = -\frac{c_1(a)}{2\alpha_0} = -\frac{1}{2\pi\alpha_0} \int_0^{2\pi} Q(a \cos \Psi, -a\alpha_0 \sin \Psi) \sin \Psi \, d\Psi, \quad (2.4.18)$$

$$B_1 = -\frac{b_1(a)}{2\alpha_0} = -\frac{1}{2a\pi\alpha_0} \int_0^{2\pi} Q(a \cos \Psi, -a\alpha_0 \sin \Psi) \cos \Psi \, d\Psi. \quad (2.4.19)$$

From (2.4.17) we have

$$\frac{\partial^2 y_1}{\partial \Psi^2} + 2\frac{\omega}{\alpha_0}\frac{\partial^2 y_1}{\partial \Psi \partial \eta} + \frac{\omega^2}{\alpha_0^2}\frac{\partial^2 y_1}{\partial \eta^2} + y_1 = \frac{1}{\alpha_0^2}\left[\frac{1}{2}b_0(a) + \sum_{n=2}^{\infty} (b_n(a) \cos n\Psi \right.$$

$$\left. + c_n(a) \sin \Psi) + \frac{1}{2}p_0 + \sum_{m=1}^{\infty} (p_m \cos m\eta + q_m \sin m\eta) \right]. \quad (2.4.20)$$

From the condition that the solution to (2.4.20) will not contain the first harmonic of the force oscillations, we find

$$y_1 = \frac{1}{\alpha_0^2}\left[\frac{b_0(a)}{2} + \sum_{n=2}^{\infty}\left(\frac{b_n(a)}{1-n^2}\cos n\Psi + \frac{c_n(a)}{1-n^2}\sin\Psi\right)\right.$$ (2.4.21)

$$\left. + \frac{p_0}{2} + \sum_{m=1}^{\infty}\left(\frac{p_m}{1-\left(\frac{\omega}{\alpha_0}m\right)^2}\cos m\eta + \frac{q_m}{1-\left(\frac{\omega}{\alpha_0}m\right)^2}\sin m\eta\right)\right].$$

Taking into account (2.4.21) in (2.4.5), we get the general solution obtained with an accuracy of $O(\varepsilon^2)$. According to (2.4.6), (2.4.7) and (2.4.18), (2.4.19), we get

$$\dot{a} = -\frac{\varepsilon}{2\pi\alpha_0}\int_0^{2\pi} Q(a\cos\Psi, -a\alpha_0\sin\Psi)\sin\Psi\,d\Psi,$$ (2.4.22)

$$\dot{\Psi} = -\frac{\varepsilon}{2a\pi\alpha_0}\int_0^{2\pi} Q(a\cos\Psi, -a\alpha_0\sin\Psi)\cos\Psi\,d\Psi.$$ (2.4.23)

The above equations govern the transitional dynamical state of the systems investigated. When we consider the steady state, the problem is reduced to the solution of the nonlinear algebraic equation

$$\int_0^{2\pi} Q(a\cos\Psi, -a\alpha_0\sin\Psi)\sin\Psi\,d\Psi = 0,$$ (2.4.24)

from which we determine the amplitude a. It can possess several solutions. The phase corresponding to each of them can be found from

$$\Psi = \int\left(\alpha_0 - \frac{\varepsilon}{2a\pi\alpha_0}\int_0^{2\pi} Q(a\cos\Psi, -a\alpha_0\sin\Psi)\cos\Psi\,d\Psi\right)dt + \Theta_0.$$ (2.4.25)

The solution y includes two parts. The first one governs the free oscillations (harmonics of Ψ), whereas the second one governs the excited oscillations (harmonics of η). In the general case a separation of those two types of oscillations is impossible. The solution is not defined for the resonance case, i.e. when $\alpha_0 = m\omega$. This problem will be solved in the next section.

Example 2.4.1. Investigate the nonresonant motion of the oscillator

$$\ddot{y} + \alpha_0^2 y = \varepsilon\left[(2h - g\dot{y}^2)\dot{y} + p\cos\omega t\right].$$ (2.4.26)

According to (2.4.15) and (2.4.16), we get $p_m = 0$ for $m \neq 1$, $p_1 = p$, $q_m = 0$, $b_n = 0$, $c_2 = 0$, $c_3 = \frac{1}{4}ga^3\alpha_0^2$ and $c_n = 0$ for $n > 3$. On the basis of (2.4.5), we obtain

$$y = a\cos\Psi + \varepsilon\left[\frac{1}{32}ga^3\alpha_0^2\sin 3\Psi + \frac{p}{\alpha_0^2 - \omega^2}\cos\omega t\right].$$ (2.4.27)

For this case we have from (2.4.22) and (2.4.23)

$$\dot{a} = -\frac{\varepsilon}{2\pi\alpha_0} \int_0^{2\pi} \left[2h - g(-a\alpha_0 \sin \Psi^2)\right] (-a\alpha_0 \sin \Psi) \sin \Psi \, d\Psi$$

$$= ah(1 - K^2 a^2), \qquad (2.4.28)$$
$$\dot{\Psi} = \alpha_0, \qquad (2.4.29)$$

where

$$K^2 = \frac{3}{8} \frac{g\alpha_0^2}{h}. \qquad (2.4.30)$$

For $K^2 > 0$ the solutions to (2.4.22) and (2.4.23) are as follows

$$a = \frac{a_0 e^{\varepsilon h t}}{\sqrt{1 + a_0^2 k^2 e^{2\varepsilon h t}}}, \quad \Psi = \alpha_0 t + \Theta, \qquad (2.4.31)$$

where a_0 and Θ are defined on the basis of the initial conditions. Taking into account (2.4.31) in (2.4.27), we obtain

$$y = \frac{a_0 e^{\varepsilon h t}}{\sqrt{1 + a_0^2 k^2 e^{2\varepsilon h t}}} \cos(\alpha_0 t + \Theta) + \frac{\varepsilon}{32} g\alpha_0 \left[\frac{a_0 e^{\varepsilon h t}}{\sqrt{1 + a_0^2 K^2 e^{2\varepsilon h t}}}\right]^3$$

$$\cdot \sin(3\alpha_0 t + 3\Theta) + \frac{\varepsilon p}{\alpha_0^2 + \omega^2} \cos \omega t. \qquad (2.4.32)$$

In the steady state $t \to +\infty$ and from (2.4.32) we obtain

$$\lim_{t \to +\infty} y = \frac{1}{K} \cos(\alpha_0 t + \Theta) + \frac{\varepsilon}{32} \frac{g\alpha_0}{K^3} \sin(3\alpha_0 t + 3\Theta) + \frac{\varepsilon p}{\alpha_0^2 + \omega^2} \cos \omega t. (2.4.33)$$

Finally, we have to remind the reader that the solution is valid for small amplitude of the excited force and when it is far enough from resonance.

2.4.3 Oscillations in the Neighbourhood of Resonance

Let us consider again the oscillator (2.4.4) and now we use the equivalent linearization method described earlier to solve the problem stated in the title of the subchapter. The solution sought is of the form

$$y = a \cos \Psi + \sum_{k=1}^{K} \varepsilon^k y_k(a, \Psi, \eta), \qquad (2.4.34)$$

where

$$\dot{a} = \sum_{k=1}^{K} \varepsilon^k A_k(a, \vartheta). \qquad (2.4.35)$$

$$\dot{\Psi} = \alpha_0 + \sum_{k=1}^{K} \varepsilon^k \overline{B}_k(a, \vartheta), \qquad (2.4.36)$$

depend additionally on the phase shift ϑ. This phase is defined by

$$n\vartheta(t) = n\Psi(t) - m\eta(t),\qquad\qquad(2.4.37)$$

which allows us to eliminate $n\Psi$ and obtain the following equations

$$y = a\cos\left(\frac{m}{n}\eta + \vartheta\right) + \sum_{k=1}^{K}\varepsilon^{k}y_{k}(a,\vartheta,\eta),\qquad\qquad(2.4.38)$$

$$\dot{a} = \sum_{k=1}^{K}\varepsilon^{k}A_{k}(a,\vartheta),\qquad\qquad(2.4.39)$$

$$\dot{\vartheta} = \alpha_{0} - \frac{m}{n}\omega + \sum_{k=1}^{K}\varepsilon^{k}\overline{B}_{k}(a,\vartheta).\qquad\qquad(2.4.40)$$

Further considerations will be focused on the oscillations near the ultra-subharmonic resonance, which emphasizes the equation

$$\alpha_{0}^{2} - \left(\frac{m}{n}\omega\right)^{2} = \varepsilon\Delta.\qquad\qquad(2.4.41)$$

Taking into account (2.4.41) in (2.4.4), we obtain

$$\ddot{y} + \left(\frac{m}{n}\omega\right)^{2}y = \varepsilon\left[Q(y,\dot{y}) + p(\eta) - \Delta y\right].\qquad\qquad(2.4.42)$$

From (2.4.37) we get

$$\alpha_{0} = \sqrt{\left(\frac{m}{n}\omega\right)^{2} + \varepsilon\Delta} \cong \frac{m}{n}\omega + \frac{\Delta}{2\frac{m}{n}}\varepsilon.\qquad\qquad(2.4.43)$$

From (2.4.34)–(2.4.36) we obtain (with an accuracy of $O(\varepsilon^{2})$)

$$y = a\cos\left(\frac{m}{n}\eta + \vartheta\right) + \varepsilon y_{1}(a,\vartheta,\eta),\qquad\qquad(2.4.44)$$

$$\dot{a} = \varepsilon A_{1}(a,\vartheta),\qquad\qquad(2.4.45)$$

$$\dot{\vartheta} = \varepsilon\left[\overline{B}_{1}(a,\vartheta) + \frac{\Delta}{2\frac{m}{n}\omega}\right] = \varepsilon B_{1}(a,\vartheta).\qquad\qquad(2.4.46)$$

Differentiating (2.4.44) and taking into account (2.4.45) and (2.4.46), we get

$$\dot{y} = -a\frac{m}{n}\omega\sin\left(\frac{m}{n}\eta + \vartheta\right) + \varepsilon\left[A_{1}\cos\left(\frac{m}{n}\eta + \vartheta\right)\right.$$
$$\left. -aB_{1}\sin\left(\frac{m}{n}\eta + \vartheta\right) + \frac{\partial y_{1}}{\partial\eta}\omega\right],\qquad\qquad(2.4.47)$$

$$\ddot{y} = -a\left(\frac{m}{n}\omega\right)^{2}\cos\left(\frac{m}{n}\eta + \vartheta\right) + \varepsilon\left[-2a\frac{m}{n}\omega B_{1}\cos\left(\frac{m}{n}\eta + \vartheta\right)\right.$$
$$\left. -2\frac{m}{n}\omega A_{1}\sin\left(\frac{m}{n}\eta + \vartheta\right) + \frac{\partial^{2}y_{1}}{\partial\eta^{2}}\omega^{2}\right].\qquad\qquad(2.4.48)$$

Taking into account the above equations in (2.4.4), we obtain

$$-a\left(\frac{m}{n}\eta\right)^2\cos\left(\frac{m}{n}\eta+\vartheta\right)+a\left(\frac{m}{n}\eta\right)^2\cos\left(\frac{m}{n}\eta+\vartheta\right)$$
$$+\varepsilon\left[-2a\frac{m}{n}\omega B_1\cos\left(\frac{m}{n}\eta+\vartheta\right)-2\frac{m}{n}\omega A_1\sin\left(\frac{m}{n}\eta+\vartheta\right)\right.$$
$$\left.+\frac{\partial^2 y_1}{\partial\eta^2}\omega^2+\left(\frac{m}{n}\eta\right)^2 y_1\right]$$
$$=\varepsilon\left\{Q\left[a\cos\left(\frac{m}{n}\eta+\vartheta\right),-a\frac{m}{n}\omega\sin\left(\frac{m}{n}\eta+\vartheta\right)\right]\right.$$
$$\left.+P(\eta)-\varDelta a\cos\left(\frac{m}{n}\eta+\vartheta\right)\right\}. \tag{2.4.49}$$

Comparing the terms in ε, we get

$$\frac{\partial^2 y_1}{\partial\eta^2}+\left(\frac{m}{n}\right)^2 y_1=2\frac{m}{n}\frac{1}{\omega}A_1\sin\left(\frac{m}{n}\eta+\vartheta\right) \tag{2.4.50}$$
$$+\left(2a\frac{m}{n}\frac{1}{\omega}B_1\frac{\varDelta a}{\omega^2}\right)\cos\left(\frac{m}{n}\eta+\vartheta\right)+\frac{1}{\omega^2}\left[\frac{1}{2}b_0(a)\right.$$
$$+\sum_{n'}^{\infty}b_{n'}(a)\cos n'\left(\frac{m}{n}\eta+\vartheta\right)+c_{n'}(a)\sin n'\left(\frac{m}{n}\eta+\vartheta\right)$$
$$\left.+\frac{1}{2}p_0+\sum_{m'}^{\infty}p_{m'}\cos m'\eta+q_{m'}\sin m'\eta\right],$$

where

$$b_n=\frac{1}{\pi}\int_0^{2\pi}Q\left(a\cos\varPsi,-a\frac{m}{n}\omega\sin\varPsi\right)\cos n'\varPsi\,\mathrm{d}\varPsi,$$

$$c_n=\frac{1}{\pi}\int_0^{2\pi}Q\left(a\cos\varPsi,-a\frac{m}{n}\omega\sin\varPsi\right)\sin n'\varPsi\,\mathrm{d}\varPsi,$$

$$p_{m'}=\frac{2}{T}\int_0^T P(\eta)\cos m'\eta\,\mathrm{d}\eta,$$

$$q_{m'}=\frac{2}{T}\int_0^T P(\eta)\sin m'\eta\,\mathrm{d}\eta,\quad m',n'=0,1,2,\ldots \tag{2.4.51}$$

Further calculations will be made for $n=1$. From (2.4.50) we obtain

$$\frac{\partial^2 y_1}{\partial\eta^2}+m^2 y_1=\left(2m\frac{1}{\omega}A_1+\frac{1}{\omega^2}c_1\right)\sin(m\eta+\vartheta)+\frac{1}{\omega^2}g_m\sin m\eta$$
$$+\left(2am\frac{1}{\omega}B_1\frac{\varDelta a}{\omega^2}+\frac{1}{\omega^2}b_1\right)\cos(m\eta+\vartheta)+\frac{1}{\omega^2}p_m\cos m\eta$$

$$+ \frac{1}{\omega^2} \left[\frac{1}{2} b_0(a) + \sum_{n'}^{\infty} b_{n'} \cos n' \left(m\eta + \vartheta \right) + c_{n'} \sin n' \left(m\eta + \vartheta \right) \right.$$

$$+ \frac{1}{2} p_0 + \sum_{\substack{m' = 1 \\ m' \neq m}}^{\infty} \left(p_{m'} \cos m'\eta + q_{m'} \sin m'\eta \right) \Bigg]. \tag{2.4.52}$$

After introducing the following quantities

$$2m \frac{1}{\omega} A_1 + \frac{1}{\omega^2} c_1 = \alpha, \qquad\qquad \frac{1}{\omega^2} q_m = \beta,$$

$$2am \frac{1}{\omega} B_1 - \frac{\Delta a}{\omega^2} + \frac{1}{\omega^2} b_1 = \gamma, \qquad \frac{1}{\omega^2} p_m = \delta, \tag{2.4.53}$$

we get

$$\alpha \sin \left(m\eta + \vartheta \right) + \beta \sin m\eta + \gamma \cos \left(m\eta + \vartheta \right) + \delta \cos m\eta = \left(\alpha \cos \vartheta \right.$$
$$+ \beta - \gamma \sin \vartheta) \sin m\eta + \left(\alpha \sin \vartheta + \gamma \cos \vartheta + \delta \right) \cos m\eta. \tag{2.4.54}$$

The secular terms equal zero when

$$\alpha \cos \vartheta + \beta \sin \vartheta = 0,$$
$$\alpha \sin \vartheta + \gamma \cos \vartheta = 0. \tag{2.4.55}$$

Multiplying the first equation of (2.4.55) by $\cos \vartheta$ and the second one by $\sin \vartheta$, and adding both of them, we have

$$\alpha + \beta \cos \vartheta + \delta \sin \vartheta = 2m \frac{1}{\omega} A_1 + \frac{1}{\omega^2} c_1 + \frac{1}{\omega^2} q_m \cos \vartheta$$

$$+ \frac{1}{\omega^2} p_m \sin \vartheta = 0. \tag{2.4.56}$$

Multiplying the first equation of (2.4.55) by $\sin \vartheta$ and the second one by $\cos \vartheta$, and adding both of them, we have

$$\beta \sin \vartheta - \delta \cos \vartheta - \gamma = \frac{1}{\omega^2} q_m \sin \vartheta - \frac{1}{\omega^2} p_m \cos \vartheta$$

$$- 2am \frac{1}{\omega} B_1 + \frac{\Delta a}{\omega^2} - \frac{b_1}{\omega^2} = 0. \tag{2.4.57}$$

From the above equations we obtain

$$A_1 = -\frac{c_1}{2m\omega} - \frac{q_m}{2m\omega} \cos \vartheta - \frac{p_m}{2m\omega} \sin \vartheta,$$

$$B_1 = -\frac{\Delta}{2m\omega} - \frac{b_1}{2ma\omega} - \frac{q_m}{2ma\omega} \sin \vartheta - \frac{p_m}{2ma\omega} \cos \vartheta. \tag{2.4.58}$$

Taking into account (2.4.58) in (2.4.45) and (2.4.46) we get

$$\dot{a} = \varepsilon \left(-\frac{c_1}{2m\omega} - \frac{q_m}{2m\omega} \cos \vartheta - \frac{p_m}{2m\omega} \sin \vartheta \right), \tag{2.4.59}$$

$$\dot{\vartheta} = \alpha_0 - m\omega + \varepsilon \left(-\frac{b_1}{2ma\omega} - \frac{q_m}{2ma\omega} \sin \vartheta - \frac{p_m}{2ma\omega} \cos \vartheta \right). \tag{2.4.60}$$

In order to simplify this procedure, we take $m = 1$, i.e. we are looking for a solution of the form

$$y = a \cos(\omega t + \vartheta). \tag{2.4.61}$$

From (2.4.43) we get

$$\varepsilon \Delta = 2\omega(\alpha_0 - \omega), \tag{2.4.62}$$

whereas from (2.4.59) and (2.4.60) we have

$$\dot{a} = \varepsilon \left(-\frac{c_1}{2\omega} - \frac{q_m}{2\omega} \sin \vartheta \right)$$

$$= -\frac{\varepsilon}{2\pi\omega} \int_0^{2\pi} Q(a \cos \Psi, -a\omega \sin \Psi) \sin \Psi \, d\Psi - \frac{\varepsilon p}{2\omega} \sin \vartheta, \tag{2.4.63}$$

$$\dot{\vartheta} = \alpha_0 - \omega + \varepsilon \left(-\frac{b_1}{2a\omega} - \frac{q_m}{2a\omega} \cos \vartheta \right) = \alpha_0 - \omega$$

$$- \frac{\varepsilon}{2\pi\omega a} \int_0^{2\pi} Q(a \cos \Psi, -a\omega \sin \Psi) \cos \Psi \, d\Psi - \frac{\varepsilon p}{2a\omega} \cos \vartheta. \tag{2.4.64}$$

Now we introduce the following quantities

$$h_e(a) = -\frac{\varepsilon}{2\pi\alpha_0 a} \int_0^{2\pi} Q(a \cos \Psi, -a\alpha_0 \sin \Psi) \sin \Psi \, d\Psi, \tag{2.4.65}$$

$$\alpha_e(a) = \alpha_0 - \frac{\varepsilon}{2\pi\alpha_0 a} \int_0^{2\pi} Q(a \cos \Psi, -a\alpha_0 \sin \Psi) \cos \Psi \, d\Psi. \tag{2.4.66}$$

We now show that (2.4.61) fulfils the equivalent linear equations of the form

$$\ddot{y} + 2h_e(a)\dot{y} + \alpha_e^2(a)y = \varepsilon p \cos \omega t. \tag{2.4.67}$$

The equation (2.4.63) can be transformed into the form

$$\dot{a} = -ah_e + ah_e - \frac{\varepsilon}{2\pi\omega} \int_0^{2\pi} Q(a \cos \Psi, -a\omega \sin \Psi) \sin \Psi \, d\Psi - \frac{\varepsilon p}{2\omega} \sin \vartheta$$

$$= -ah_e + \frac{\varepsilon}{2\pi\alpha_0} \int_0^{2\pi} Q(a \cos \Psi, -a\alpha_0 \sin \Psi) \sin \Psi \, d\Psi$$

$$- \frac{\varepsilon}{2\pi\omega} \int_0^{2\pi} Q(a \cos \Psi, -a\alpha_0 \sin \Psi) \sin \Psi \, d\Psi - \frac{\varepsilon p}{2\omega} \sin \vartheta, \tag{2.4.68}$$

and then taking into account (2.4.62), into the form

$$\dot{a} = -ah_e + \frac{\varepsilon}{2\pi\left(\omega + \frac{\varepsilon\Delta}{2\omega}\right)} \int_0^{2\pi} Q\left(a\cos\Psi, -a\left(\omega + \frac{\varepsilon\Delta}{2\omega}\right)\sin\Psi\right)\sin\Psi\,d\Psi$$

$$-\frac{\varepsilon}{2\pi\omega} \int_0^{2\pi} Q(a\cos\Psi, -a\omega\sin\Psi)\sin\Psi\,d\Psi - \frac{\varepsilon p}{2\omega}\sin\vartheta. \tag{2.4.69}$$

After expanding the second term of the right-hand side of (2.4.69) in a power series because of ε, we obtain

$$\dot{a} = -ah_e + \frac{\varepsilon}{2\pi\left(\omega + \frac{\varepsilon\Delta}{2\omega}\right)} \int_0^{2\pi} Q\left(a\cos\Psi, -a\left(\omega + \frac{\varepsilon\Delta}{2\omega}\right)\sin\Psi\right)\sin\Psi\,d\Psi$$

$$-\frac{\varepsilon}{2\pi\omega} \int_0^{2\pi} Q(a\cos\Psi, -a\omega\sin\Psi)\sin\Psi\,d\Psi - \frac{\varepsilon p}{2\omega}\sin\vartheta$$

$$= -ah_e - \frac{\varepsilon p}{2\omega}\sin\vartheta. \tag{2.4.70}$$

Similar considerations lead to

$$\dot{\vartheta} = \alpha_e - \omega - \frac{\varepsilon p}{2a\omega}\cos\vartheta. \tag{2.4.71}$$

From (2.4.61) we get

$$\dot{y} = \dot{a}\cos(\omega t + \vartheta) - a(\omega + \dot{\vartheta})\sin(\omega t + \vartheta), \tag{2.4.72}$$

$$\ddot{y} = \ddot{a}\cos(\omega t + \vartheta) - 2a(\omega + \dot{\vartheta})\sin(\omega t + \vartheta) - a\ddot{\vartheta}\sin(\omega t + \vartheta)$$
$$-a(\omega + \dot{\vartheta})^2\cos(\omega t + \vartheta). \tag{2.4.73}$$

Taking into account (2.4.70) and (2.4.71) in the above equation, we find (with an accuracy of $O(\varepsilon^2)$)

$$\dot{y} = -ah_e\cos(\omega t + \vartheta) + \frac{\varepsilon p}{2\omega}\sin\omega t - a\alpha_e\sin(\omega t + \vartheta), \tag{2.4.74}$$

$$\ddot{y} = 2ah_e\alpha_e\sin(\omega t + \vartheta) + \varepsilon p\cos\omega t - a\alpha_e\cos(\omega t + \vartheta). \tag{2.4.75}$$

The left-hand side L of (2.4.67), taking into account (2.4.61), (2.4.72) and (2.4.73), can be transformed into the form

$$L = \varepsilon p\cos\omega t - 2ah_e^2\cos(\omega t + \vartheta) + h_e\frac{\varepsilon p}{\omega}\sin\omega t \cong \varepsilon p\cos\omega t. \tag{2.4.76}$$

Taking into account (2.4.76) and the right-hand side of (2.4.67), we see that the solution (2.4.61) fulfills (2.4.67) with an accuracy of ε. Thus, the method of equivalent linearization allows us to replace (2.4.4) by (2.4.67), which is valid near resonance. The unit equivalent coefficient of damping $h_e(a)$ and the equivalent frequency $\alpha_e(a)$ are functions of the amplitude a. This amplitude can be found from the formula

$$a = \frac{\varepsilon p}{\sqrt{(\alpha_e^2(a) - \omega^2)^2 + 4h_e^2(a)\omega^2}},$$ (2.4.77)

which allows us to obtain

$$\omega = \sqrt{[\alpha_e^2(a) - 2h_e^2(a)] \pm \sqrt{4h_e^2(a)\,[h_e^2(a) - \alpha_e^2(a)] + \frac{\varepsilon^2 p^2}{\alpha^2}}}.$$ (2.4.78)

For a given amplitude, we can have one, two or no value of frequency according to (2.4.78). According to linear oscillation theory, we have

$$\vartheta = \arctan \frac{-2h_e(a)\omega}{\alpha_e^2(a) - \omega^2}.$$ (2.4.79)

Therefore, for each amplitude a and ω defined by (2.4.78), it is possible to find the corresponding phase ϑ. The exemplary results are shown in Figure 2.2. However, not all parts of the resonance curves are stable. In order to check stability, let us consider the steady state defined by

$$y = a_0 \cos(\omega t + \vartheta_0),$$ (2.4.80)

where a_0 and ϑ_0 fulfil equations (2.4.70) and (2.4.71). Therefore, we have

$$-a_0 h_e(a_0) - \frac{\varepsilon p}{2\omega} \sin \vartheta_0 = 0,$$ (2.4.81)

$$\alpha_e(a_0) - \omega - \frac{\varepsilon p}{2a_0\omega} \cos \vartheta_0 = 0.$$ (2.4.82)

In order to investigate the stability of (2.4.80), we have to consider the near by solution

$$y = a \cos(\omega t + \vartheta),$$ (2.4.83)

where $a(t)$ and $\vartheta(t)$ are the solutions of (2.4.70) and (2.4.71)

$$\dot{a} = -a h_e(a) - \frac{\varepsilon p}{2\omega} \sin \vartheta = \varepsilon A\,[a(t), \vartheta(t), \omega],$$ (2.4.84)

$$\dot{\vartheta} = \alpha_e(a) - \omega - \frac{\varepsilon p}{2a\omega} \cos \vartheta = \varepsilon B\,[a(t), \vartheta(t), \omega].$$ (2.4.85)

We will consider the solutions close to the investigated solutions

$$a(t) = a_0 + \delta_a(t),$$
$$\vartheta(t) = \vartheta_0 + \delta_\vartheta(t),$$ (2.4.86)

where $\delta(t)$ are small enough. Taking into account (2.4.85) in (2.4.82), we obtain

$$\dot{\delta}_a = \varepsilon A\,[(a_0 + \delta_a(t)), (\vartheta_0 + \delta_\vartheta(t)), \omega],$$
$$\dot{\delta}_\vartheta = \varepsilon B\,[(a_0 + \delta_a(t)), (\vartheta_0 + \delta_\vartheta(t)), \omega],$$ (2.4.87)

and next, we develop the right-hand sides of (2.4.87) into a Taylor series because of δ_a and δ_ϑ near the point $(a_0\,,\,\vartheta_0)$, and finally we obtain

a)

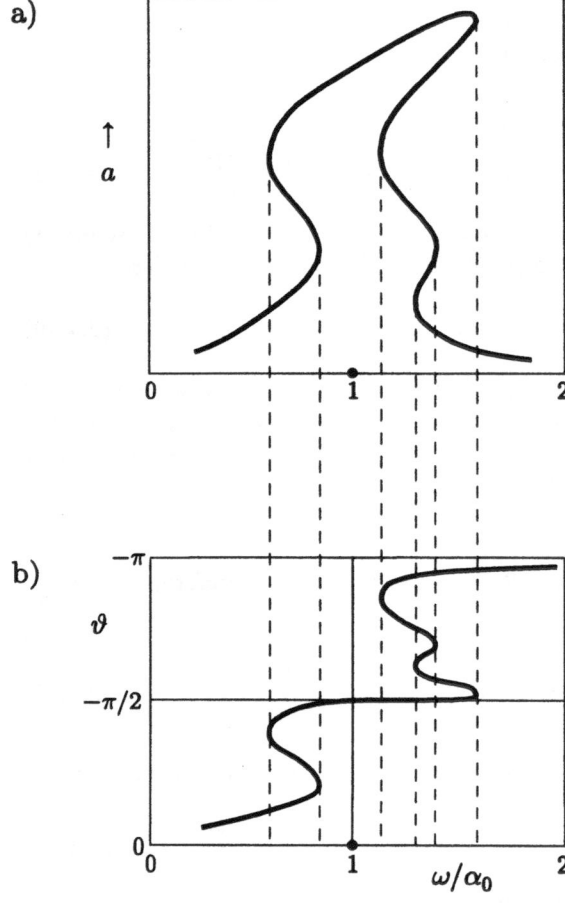

b)

Fig. 2.2. Amplitude of oscillations (**a**) and phase shift (**b**) versus ω/α_0

$$\dot{\delta}_{\mathrm{a}} = \varepsilon \left[A(a_0, \vartheta_0, \omega) + \frac{\partial A}{\partial a}(a_0, \vartheta_0)\delta_{\mathrm{a}} + \frac{\partial A}{\partial a}(a_0, \vartheta_0)\delta_\vartheta \right],$$

$$\dot{\delta}_\vartheta = \varepsilon \left[B(a_0, \vartheta_0, \omega) + \frac{\partial B}{\partial a}(a_0, \vartheta_0)\delta_{\mathrm{a}} + \frac{\partial B}{\partial a}(a_0, \vartheta_0)\delta_\vartheta \right]. \tag{2.4.88}$$

According to (2.4.80) and (2.4.82), we have

$$A(a_0, \vartheta_0, \omega) = 0,$$
$$B(a_0, \vartheta_0, \omega) = 0. \tag{2.4.89}$$

Solutions to the linear differential equations (2.4.88) are sought in the form

$$\delta_{\mathrm{a}} = D_{\mathrm{a}} \mathrm{e}^{rt},$$
$$\delta_\vartheta = D_\vartheta \mathrm{e}^{rt}. \tag{2.4.90}$$

Taking into account (2.4.90) in (2.4.88), we obtain the following characteristic equations

$$r^2 - \varepsilon r \left[\frac{\partial A}{\partial a}(a_0, \vartheta_0) + \frac{\partial B}{\partial \vartheta}(a_0, \vartheta_0) \right]$$

$$+ \varepsilon^2 \left[\frac{\partial A}{\partial a}(a_0, \vartheta_0) \frac{\partial B}{\partial \vartheta}(a_0, \vartheta_0) - \frac{\partial A}{\partial \vartheta}(a_0, \vartheta_0) \frac{\partial B}{\partial a}(a_0, \vartheta_0) \right] = 0. \quad (2.4.91)$$

The solution will be stable, if $\delta_a(t)$ and $\delta_\vartheta(t)$ approach zero with $t \to +\infty$. This happens when the real parts of the roots of (2.4.91) are less than zero. According to Vieta's formulas we have

$$\frac{\partial A}{\partial a} + \frac{\partial A}{\partial \vartheta} < 0, \quad (2.4.92)$$

$$\frac{\partial A}{\partial a} \frac{\partial B}{\partial \vartheta} - \frac{\partial A}{\partial \vartheta} \frac{\partial B}{\partial a} > 0. \quad (2.4.93)$$

These conditions will be transformed into a form allowing us to estimate the stability of the solution on the basis of the resonance curve given in Fig. 2.3.

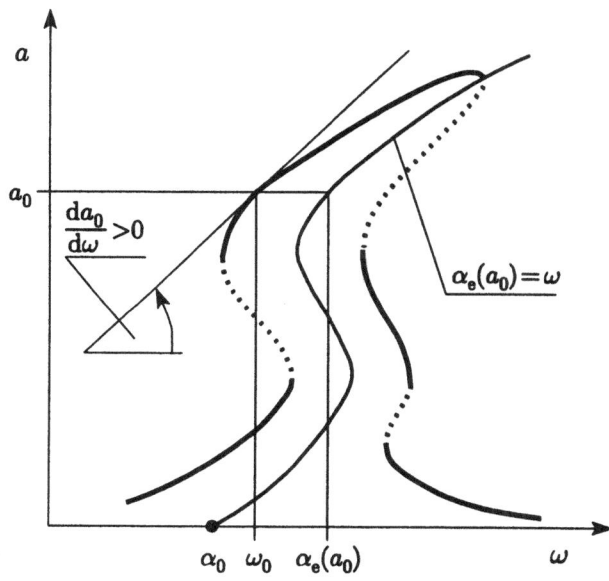

Fig. 2.3. Resonance curve with stable (*continuous line*) and unstable (*dashed line*) parts

According to (2.4.84) and (2.4.85), we obtain

$$\frac{\partial A}{\partial a} = -h_e(a_0) - a_0 \frac{\partial h_e}{\partial a}(a_0), \qquad \frac{\partial A}{\partial \vartheta} = -\frac{\varepsilon p}{2\omega} \cos \vartheta_0,$$

$$\frac{\partial B}{\partial a} = \frac{\partial h_e}{\partial a}(a_0) + \frac{\varepsilon p}{2\alpha_0} \cos \vartheta_0, \qquad \frac{\partial B}{\partial \vartheta} = -\frac{\varepsilon p}{2\alpha_0 \omega} \sin \vartheta_0. \quad (2.4.94)$$

On the basis of the above results, the stability conditions will take the form

$$- h_e(a_0) - a_0 \frac{\partial h_e}{\partial a}(a_0) + \frac{\varepsilon p}{2\alpha_0 \omega} \sin \vartheta_0 < 0. \tag{2.4.95}$$

Taking into account equation (2.4.81), we have

$$- h_e(a_0) - a_0 \frac{\partial h_e}{\partial a}(a_0) < 0. \tag{2.4.96}$$

This condition is transformed into the form

$$\frac{d}{da_0} \left[a_0^2 h_e(a_0) \right] > 0, \quad \text{for } a_0 > 0. \tag{2.4.97}$$

According to (2.4.82), we obtain

$$A \left[a_0(\omega), \vartheta_0(\omega), \omega \right] = 0,$$
$$B \left[a_0(\omega), \vartheta_0(\omega), \omega \right] = 0. \tag{2.4.98}$$

Differentiating the above equations with respect to ω, we have

$$\frac{\partial A}{\partial a} \frac{\partial a_0}{\partial \omega} + \frac{\partial A}{\partial \vartheta} \frac{\partial \vartheta_0}{\partial \omega} = - \frac{\partial A}{\partial \omega},$$
$$\frac{\partial B}{\partial a} \frac{\partial a_0}{\partial \omega} + \frac{\partial B}{\partial \vartheta} \frac{\partial \vartheta_0}{\partial \omega} = - \frac{\partial B}{\partial \omega}. \tag{2.4.99}$$

Multiplying the first equation of (2.4.99) by $\partial B / \partial \vartheta$, and the second one by $\partial A / \partial \vartheta$ and adding up both of them, we get

$$\frac{\partial A}{\partial a} \frac{dB}{d\vartheta} - \frac{\partial A}{\partial \vartheta} \frac{dB}{da} = \left(\frac{da_0}{d\omega} \right)^{-1} \left(\frac{\partial A}{\partial \vartheta} \frac{dB}{d\omega} - \frac{\partial B}{\partial \vartheta} \frac{dA}{d\omega} \right). \tag{2.4.100}$$

Because

$$\frac{\partial A}{\partial \omega} = \frac{\varepsilon p}{2\omega^2} \sin \vartheta_0,$$
$$\frac{\partial B}{\partial \omega} = -1 + \frac{\varepsilon p}{2\alpha_0 \omega^2} \cos \vartheta_0 \tag{2.4.101}$$

and taking into account (2.4.94), we obtain according to (2.4.93),

$$- \left(\frac{da_0}{d\omega} \right)^{-1} \left[\frac{\varepsilon p}{2\omega} \cos \vartheta_0 \left(-1 + \frac{\varepsilon p}{2\alpha_0 \omega^2} \cos \vartheta_0 \right) \right.$$
$$\left. + \frac{\varepsilon^2 p^2}{4\alpha_0 \omega^3} \sin^2 \vartheta_0 \right] > 0, \tag{2.4.102}$$

which, after limiting considerations to the terms in the first power of ε, leads to the condition

$$\frac{\frac{\varepsilon p}{2\omega} \cos \vartheta_0}{\frac{da_0}{d\omega}} > 0. \tag{2.4.103}$$

According to (2.4.85), we have

$$\frac{\varepsilon p}{2\omega} \cos \vartheta_0 = a_0 \left[\alpha_0(a_0) - \omega \right],$$ (2.4.104)

then, for $a_0 > 0$ we have

$$\frac{\alpha_e(a_0) - \omega}{\frac{da_0}{d\omega}} > 0.$$ (2.4.105)

On the basis of this inequality we can formulate the following conclusions: the solution is stable if

$$\frac{da_0}{d\omega} > 0 \quad \text{and} \quad \alpha_e(a_0) > \omega,$$ (2.4.106)

or if

$$\frac{da_0}{d\omega} < 0 \quad \text{and} \quad \alpha_e(a_0) < \omega.$$ (2.4.107)

This analysis allowed us to determine the stability of the solutions on the basis of the consideration of the resonance curve, which is illustrated in Fig. 2.3. In this figure the "skeleton line" is defined by the equation $\alpha_e(a_0) = \omega$.

Now we will analyse the slow transition through the resonance taking into consideration Fig. 2.4.

The amplitude of driven oscillations is increased along sector AB of the resonance curve (Fig. 2.4a). In point B a sudden jump into a new branch on the resonance curve has occurred (point D) and a further increase in the frequency ω is accompanied by an increase in the amplitude of oscillations up to point E. In this point a sudden amplitude change to the value defined by point J has appeared. A further increase in ω causes a slight decrease in the amplitude of oscillations.

In a similar way we are able to analyse the dynamics with an increase in the frequency ω (Fig. 2.4b). We have to emphasize that the process of nonlinear and discontinuous changes of the amplitude corresponding to the increase in the frequency differs from a similar process accompanying the decrease in the frequency.

As has been mentioned earlier, for the considered parameters of the system different kinds of oscillations can occur (they depend on the initial conditions).

Example 2.4.2. Analyse the dynamics of the system

$$\ddot{y} + \alpha_0^2 y = \varepsilon(-2h - \beta y^3 + p \cos \omega t)$$ (2.4.108)

in the neighbourhood of the resonance using the method of equivalent linearization.

According to (2.4.65) and (2.4.66), we obtain

$$h_e(a) = \frac{\varepsilon}{2\pi\alpha_0 a} \int_0^{2\pi} \left[-\beta(a\cos\Psi)^3 - 2h(-a\alpha_0\sin\Psi) \right] \sin\Psi \, d\Psi$$

$$= \varepsilon h, \tag{2.4.109}$$

$$\alpha_e(a) = \alpha_0 - \frac{\varepsilon}{2\pi\alpha_0 a} \int_0^{2\pi} \left[-\beta(a\cos\Psi)^3 - 2h(-a\alpha_0\sin\Psi) \right] \cos\Psi \, d\Psi$$

$$= \alpha_0 + \frac{3\varepsilon}{3\alpha_0} a^2\beta. \tag{2.4.110}$$

According to (2.4.78), the resonance curve is given by

$$\omega = \left(\left[\alpha_0 + \frac{3\varepsilon}{8\alpha_0} a^2\beta \right]^2 - 2\varepsilon^2 h^2 \right.$$

$$\left. \pm \sqrt{4\varepsilon^2 h^2 \left[\varepsilon^2 h^2 - \left(\alpha_0 + \frac{3\varepsilon}{3\alpha_0} a^2\beta \right)^2 \right] + \frac{\varepsilon^2 p^2}{a^2}} \right)^{\frac{1}{2}}, \tag{2.4.111}$$

and following (2.4.79), we find the shift of the phase equal to

$$\vartheta = \text{arctg} \, \frac{-2\varepsilon h\omega}{\left(\alpha_0 + \frac{3}{8}\frac{\varepsilon}{\alpha_0} a^2\beta \right)^2 - \omega^2}. \tag{2.4.112}$$

On the basis of the last two equations and for the parameters $\alpha_0 = 1$, $\varepsilon = 0.1$, $p = 2$, $\beta = 2$, resonance curves are obtained, which are shown in Fig. 2.5. The skeleton line is defined by

$$\alpha_0(a) = \alpha_0 + \frac{3\varepsilon}{8\alpha_0} a^2\beta = \omega. \tag{2.4.113}$$

In this figure the continuous line denotes the stable solutions, whereas the dashed line corresponds to the unstable solutions. The conditions (2.4.97) have the form $2\varepsilon h < 0$, which is true for $h > 0$.

2.5 Nonstationary Nonlinear Systems

In what follows, we consider nonstationary linear and nonlinear mechanical systems. We begin with a simple example of the parametric oscillations of the mathematical pendulum with a periodic change of its length (Fig. 2.6).

For small φ the governing equation has the form

$$\frac{d^2\varphi}{dt^2} + \frac{2}{L}\frac{dL}{dt}\frac{d\varphi}{dt} + \frac{g}{L}\varphi = 0, \tag{2.5.1}$$

where $L(t)$ is the time-dependent length and g is the gravitational acceleration. We consider small changes of the length according to the formula

a)

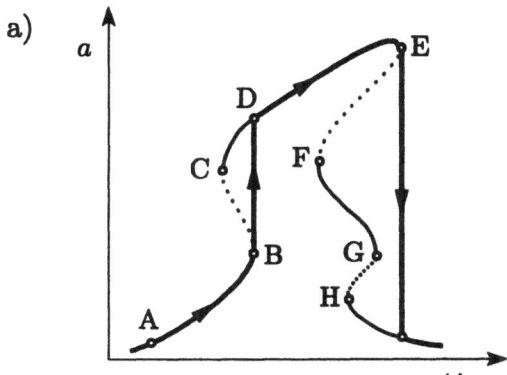

b)

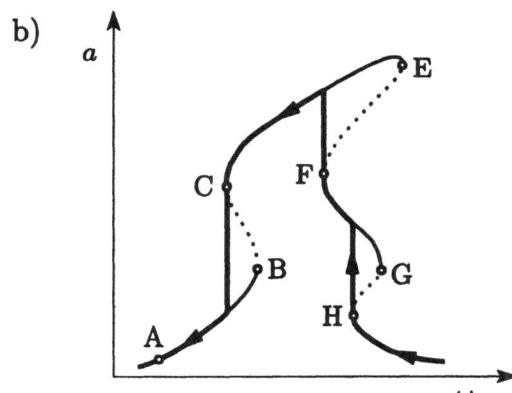

Fig. 2.4. Nonlinear jumps of amplitudes accompanying an increase (a) and a decrease (b) in frequency

$$L = L_0(1 + \mu \cos \omega t). \tag{2.5.2}$$

Then, by introducing the new variable $x = l\varphi$ and after some transformations, we obtain

$$\frac{\mathrm{d}^2 x}{\mathrm{d}t^2} + \left(\frac{g}{L_0} + \mu \omega^2 \cos \omega t \right) x = 0, \tag{2.5.3}$$

where $0 < \mu \ll 1$ is a small perturbation parameter.

When the fixed point zero (trivial solution) undergoes periodic movement (Fig. 2.6b), we get the following governing equation

$$mL_0 \ddot{\varphi} = -mg \sin \varphi + m\ddot{y} \sin \varphi, \tag{2.5.4}$$

which leads to the Mathieu equation

$$\ddot{x} + (a + 16q \cos 2\tau) x = 0, \tag{2.5.5}$$

where

$$2\tau = \omega t, \quad x = \varphi, \quad a = \frac{4g}{L_0 \omega^2}, \quad q = \frac{y_0}{4L_0}. \tag{2.5.6}$$

a)

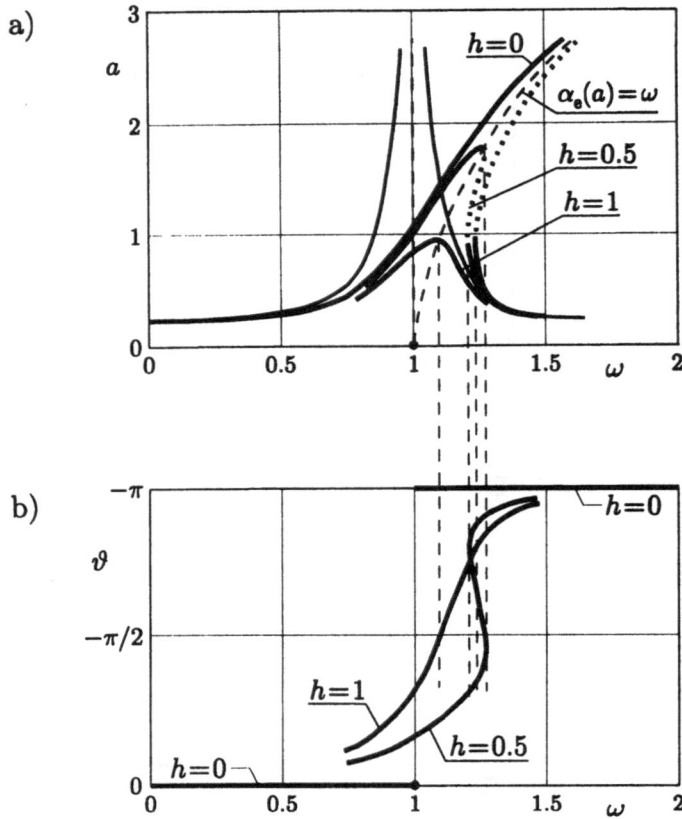

b)

Fig. 2.5. Amplitude (a) and phase (b) versus ω for different values of h

In the third example (Fig. 2.6c) the excitation $y(t)$ can be defined as follows

$$y(t) = \frac{r^2}{4L} + r\cos\omega t - \frac{r^2}{4L}\cos 2\omega t, \tag{2.5.7}$$

and the problem is reduced to the Hill equation

$$\ddot{x} + [\sigma + \mu\Phi(\tau)]\,x = 0, \tag{2.5.8}$$

where

$$\sigma = \frac{g}{i_0}, \quad 2\tau = \omega t, \quad \mu = \frac{r}{L_0}, \quad \Phi(\tau) = \omega^2\cos 2\tau - r\omega^2\cos 4\tau. \tag{2.5.9}$$

If $\Phi(\tau) = \cos\tau$, then we have the following equation

$$\frac{\mathrm{d}^2 x}{\mathrm{d}\tau^2} + (\sigma + \mu\cos\tau)x = 0. \tag{2.5.10}$$

The periodic solutions with periods 2π and 4π appear on the stability limits of the system governed by the Mathieu equation (2.5.10). We would like to

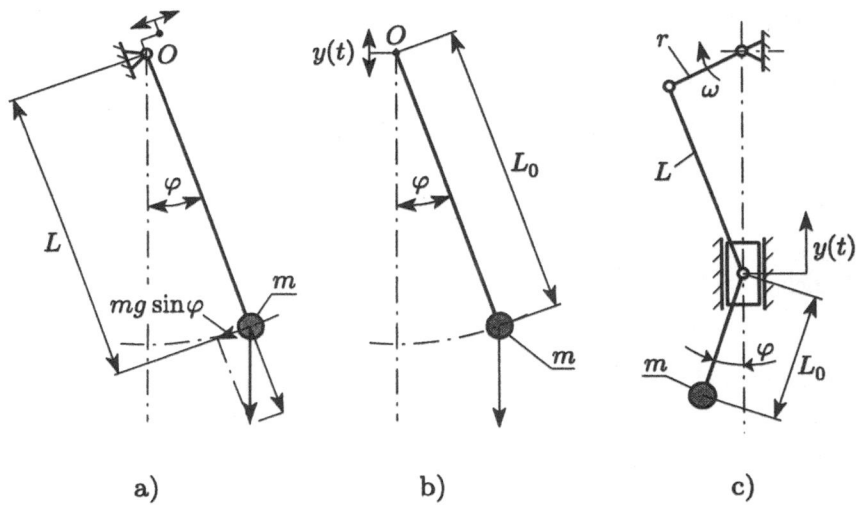

Fig. 2.6. Mathematical pendulum with a periodically changed length $L(t)$ (a), with the periodic movement of the fixed point O (b), and with anharmonic periodic excitation (c)

estimate analytically the (σ, γ) values for which the system possesses a 4π-periodic solution. We are looking for the following solution

$$x(\tau, \mu) = x_0(\tau) + \mu x_1(\tau) + \mu^2 x_2(\tau) + \ldots,$$
$$\sigma(\mu) = \sigma_0 + \mu\sigma_1 + \mu^2\sigma_2 + \ldots. \qquad (2.5.11)$$

Substituting (2.5.11) into (2.5.10) and comparing the terms with the same power of the small parameter $\mu \ll 1$, we get

$$\frac{d^2 x_0}{d\tau^2} + \sigma_0 x_0 = 0,$$
$$\frac{d^2 x_1}{d\tau^2} + \sigma_0 x_1 = -(\sigma_1 + \cos\tau)x_0, \qquad (2.5.12)$$
$$\frac{d^2 x_2}{d\tau^2} + \sigma_0 x_2 = -(\sigma_1 + \cos\tau)x_1 - \sigma_2 x_0,$$
$$\cdots\cdots\cdots$$

The first equation of (2.5.12) gives

$$x_0(\tau) = A_0 \cos\sqrt{\sigma_0}\tau + B_0 \sin\sqrt{\sigma_0}\tau. \qquad (2.5.13)$$

Because we are looking for the 4π-periodic solution, therefore

$$\sigma_0 = \frac{k^2}{4}, \qquad k = 0, 1, 2, \ldots \qquad (2.5.14)$$

and from (2.5.13) we obtain

$$x_0(\tau) = A_0 \cos \frac{k\tau}{2} + B_0 \sin \frac{k\tau}{2}. \tag{2.5.15}$$

We consider first the case when $k = 0$. From (2.5.12) we find

$$x_1(\tau) = C_1 + C_2\tau - \sigma_1 A_0 \frac{\tau^2}{2} + A_0 \cos \tau, \tag{2.5.16}$$

where C_i $(i = 1, 2)$ are the integration constants. Taking into account the periodicity, we have

$$x_1(\tau) = C_1 + A_0 \cos \tau. \tag{2.5.17}$$

From the second equation of (2.5.12) we get

$$x_2(\tau) = C_3 + C_4\tau - \left(\sigma_2 + \frac{1}{2}\right) A_0 \frac{\tau^2}{2} + C_1 \cos \tau + \frac{A_0}{8} \cos 2\tau. \tag{2.5.18}$$

Again, the 4π-periodicity gives $\sigma_0 = \sigma_1 = 0$, $\sigma_2 = -\frac{1}{2}$ and from the second equation of (2.5.11), we obtain

$$\sigma \approx -\frac{\mu^2}{2}. \tag{2.5.19}$$

For $n = 1$ we have $\sigma_0 = \frac{1}{4}$, and

$$x_0(\tau) = A_0 \cos \frac{\tau}{2} + B_0 \sin \frac{\tau}{2}. \tag{2.5.20}$$

The second equation of (2.5.12) has the new form

$$\frac{d^2 x}{d\tau^2} + \frac{1}{4} x_1 = -\left(\sigma_1 + \frac{1}{2}\right) A_0 \cos \frac{\tau}{2} + \left(-\sigma_1 + \frac{1}{2}\right) B_0 \sin \frac{\tau}{2}$$
$$- \frac{A_0}{2} \cos \frac{3}{2}\tau - \frac{B_0}{2} \sin \frac{3}{2}\tau. \tag{2.5.21}$$

In order to have a periodic solution, secular terms should vanish, which leads to the conditions

$$-\left(\sigma_1 + \frac{1}{2}\right) A_0 = 0,$$

$$\left(-\sigma_1 + \frac{1}{2}\right) B_0 = 0. \tag{2.5.22}$$

There are three different solutions of (2.5.22), which are given below:

1. $A_0 \neq 0$; $\sigma_1^{(1)} = -\frac{1}{2}$; $B_0 = 0$.

2. $A_0 = 0$; $\sigma_1^{(2)} = +\frac{1}{2}$; $B_0 \neq 0$. $\qquad(2.5.23)$

3. $A_0 = B_0 = 0$ with $\sigma_1^{(3)}$ temporarily unknown.

The first two solutions of (2.5.23) give the following estimation of the stability limits

$$\sigma^{(1)} \approx \frac{1}{4} - \frac{\mu}{2},$$

$$\sigma^{(2)} \approx \frac{1}{4} + \frac{\mu}{2}. \tag{2.5.24}$$

For $n = 2$ we have the following stability limits

$$\sigma^{(1)} \approx 1 + \frac{5}{12}\mu^2,$$

$$\sigma^{(2)} \approx 1 - \frac{1}{2}\mu^2. \tag{2.5.25}$$

In all these cases we introduced the small positive perturbation parameter μ, which characterizes the modulation depth of the parametric excitation. However, linear systems are an idealization of real systems which are nonlinear. Therefore, we consider the general form of the system of equations

$$\{\dot{x}\} = [[\Phi_0] + \mu[\Phi_1(t)] + \mu^2[\Phi_2(t)] + \ldots]\{x\} + \varepsilon[F(x_1, \ldots, x_n)]. \tag{2.5.26}$$

Above, we have two independent perturbation parameters μ and ε. The first one defines the parametric modulation, whereas the second one characterizes the nonlinearity. Additionally, $[\Phi_0]$ is the constant matrix, and $[\Phi_i(t)] = [\Phi_i(t + T)]$, where T is the period.

The solution of (2.5.26) is sought in the form

$$\{x(t, \varepsilon, \mu)\} = x^{(0,0)} + \mu x^{(1,0)} + \mu^2 x^{(2,0)} + \ldots$$
$$+ \varepsilon\left(x^{(0,1)} + \mu x^{(1,1)} + \mu^2 x^{(2,1)} + \ldots\right) \tag{2.5.27}$$
$$+ \varepsilon^2\left(x^{(0,2)} + \mu x^{(1,2)} + \mu^2 x^{(2,2)} + \ldots\right)$$
$$+ \ldots$$

Substituting (2.5.27) into (2.5.26), and after comparing the expressions with the same power of $\mu^k \varepsilon^l$, we get

$$\varepsilon^0, \mu^0 \; : \; \left\{\dot{x}^{(0,0)}\right\} - [\Phi_0]\left\{x^{(0,0)}\right\} = \{0\},$$

$$\varepsilon^1 \; : \; \left\{\dot{x}^{(0,1)}\right\} - [\Phi_0]\left\{x^{(0,1)}\right\} = \left[\frac{\partial F}{\partial \varepsilon}\right]\left\{x^{(0,0)}\right\},$$

$$\mu^1 \; : \; \left\{\dot{x}^{(1,0)}\right\} - [\Phi_0]\left\{x^{(1,0)}\right\} = [\Phi_1]\left\{x^{(0,0)}\right\}, \tag{2.5.28}$$

$$\varepsilon^2 \; : \; \left\{\dot{x}^{(0,2)}\right\} - [\Phi_0]\left\{x^{(0,2)}\right\} = \left[\frac{\partial F}{\partial \varepsilon}\right]\left\{x^{(0,1)}\right\} + \left[\frac{\partial^2 F}{\partial \varepsilon^2}\right]\left\{x^{(0,0)}\right\},$$

$$\mu^2 \; : \; \left\{\dot{x}^{(2,0)}\right\} - [\Phi_0]\left\{x^{(2,0)}\right\} = [\Phi_2]\left\{x^{(0,0)}\right\} + [\Phi_1]\left\{x^{(1,0)}\right\},$$

$$\varepsilon\mu \; : \; \left\{\dot{x}^{(1,1)}\right\} - [\Phi_0]\left\{x^{(1,1)}\right\} = [\Phi_1]\left\{x^{(0,1)}\right\} + \left[\frac{\partial F}{\partial \varepsilon}\right]\left\{x^{(1,0)}\right\},$$

$$\vdots$$

The first equation of the set (2.5.28) gives the fundamental part of the sought solution (2.5.27), which can be succesfully approximated by solving the recurrent set of the rest of equations (2.5.28). To achieve a complete ordering of all of the recurrent equations, we take the following additional condition $\varepsilon < \mu$. The tringle below gives the ordering from the smallest to the largest values asymptotically on each horizontal row, i.e. defines the sequence of recurrent equations

$$
\begin{array}{c}
\varepsilon \quad \mu \\
\varepsilon^2 \quad \varepsilon\mu \quad \mu^2 \\
\varepsilon^3 \quad \varepsilon^2\mu \quad \mu^2\varepsilon \quad \mu^3 \\
\varepsilon^4 \quad \varepsilon^3\mu \quad \varepsilon^2\mu^2 \quad \mu^3\varepsilon \quad \mu^4
\end{array}
\qquad (2.5.29)
$$

. . . .

The question arises: which initial conditions should be taken for $x^{(0,0)}$, $x^{(1,0)}$, $x^{(0,1)}$ and so on. The following condition should be fulfilled

$$
\{x(t_0, \varepsilon, \mu)\} = \left\{x^{(0,0)}(t_0)\right\} + \varepsilon \left\{x^{(0,1)}(t_0)\right\} + \mu \left\{x^{(1,0)}(t_0)\right\} + \dots, \quad (2.5.30)
$$

and usually it is useful to take

$$
\left\{x^{(0,0)}(t_0)\right\} = \{x(t_0, \varepsilon, \mu)\}, \qquad (2.5.31)
$$

which leads to the equations

$$
x^{(1,0)}(t_0) = x^{(0,1)}(t_0) = x^{(2,0)}(t_0) = \dots = 0. \qquad (2.5.32)
$$

Examples of two calculation to support the above consideration will be considered now [28].

Example 2.5.1. Consider the standard example of vibrations of a system with one degree of freedom, of the form

$$
m\frac{d^2 x}{dt} + (k_0 + k_1 \cos 2t)x + k_2 x^3 = 0, \qquad (2.5.33)
$$

where m is the mass of the vibrating body, and k_0, k_1 and k_2 are the stiffness coefficients.

After assuming the parameters $\lambda^2 = k_0/m$, $\mu = k_1/k_0$, and $\varepsilon = k_2/m$, we obtain the equation

$$
\frac{d^2 x}{dt} + \lambda^2(1 + \mu \cos 2t)x + \varepsilon x^3 = 0. \qquad (2.5.34)
$$

For $\mu = 0$, we obtain the Duffing equation, and for $\varepsilon = 0$, the Mathieu equation.

Let us develop the quantities λ^2 and x into a power series in the small parameters μ and ε:

$$\lambda^2 = n^2 + a^{(1,0)}\mu + \mu^2 a^{(2,0)} + \ldots + \varepsilon \left(a^{(0,1)} + \mu a^{(1,1)} + \ldots \right)$$

$$+ \varepsilon^2 \left(a^{(0,2)} + \mu a^{(1,2)} + \mu^2 a^{(2,2)} + \ldots \right) + \ldots \tag{2.5.35}$$

$$x = x^{(0,0)} + \mu x^{(1,0)} + \mu^2 x^{(2,0)} + \ldots$$

$$+ \varepsilon \left(x^{(0,1)} + \mu x^{(1,1)} + \mu^2 x^{(2,1)} + \ldots \right)$$

$$+ \varepsilon^2 \left(x^{(0,2)} + \mu x^{(1,2)} + \mu^2 x^{(2,2)} + \ldots \right) + \ldots \tag{2.5.36}$$

Now, consider the first unstable zone ($n = 1$). After substituting (2.5.35) and (2.5.36) into (2.5.34), and equating the expressions representing the same power μ and ε and their combinations, we obtain the following recurrent system of linear differential equations

$$\ddot{x}^{(0,0)} + x^{(0,0)} = 0,$$

$$\ddot{x}^{(1,0)} + x^{(1,0)} = -x^{(0,0)} \cos 2t - a^{(1,0)} x^{(0,0)},$$

$$\ddot{x}^{(2,0)} + x^{(2,0)} = -a^{(2,0)} x^{(0,0)} - a^{(1,0)} x^{(1,0)} - a^{(1,0)} \cos 2t x^{(0,0)} - x^{(1,0)} \cos 2t,$$

$$\vdots$$

$$\ddot{x}^{(0,1)} + x^{(0,1)} = \left(x^{(0,0)} \right)^3 - a^{(0,1)} x^{(0,0)}, \tag{2.5.37}$$

$$\ddot{x}^{(1,1)} + x^{(1,1)} = -3 \left(x^{(0,0)} \right)^2 x^{(1,0)} - a^{(1,0)} x^{(0,1)} - a^{(0,1)} x^{(0,0)} \cos 2t$$

$$- a^{(1,1)} x^{(0,0)} - x^{(0,1)} \cos 2t,$$

$$\ddot{x}^{(2,1)} + x^{(2,1)} = -3 x^{(0,0)} \left(x^{(1,0)} \right)^2 - 3 \left(x^{(0,0)} \right)^2 x^{(2,0)} - a^{(2,1)} x^{(0,0)}$$

$$- a^{(1,1)} x^{(1,0)} - a^{(0,1)} x^{(2,0)} - a^{(0,1)} x^{(1,0)} \cos 2t - a^{(2,0)} x^{(0,1)}$$

$$- x^{(0,1)} a^{(1,0)} \cos 2t - x^{(1,1)} a^{(1,0)} - x^{(1,1)} \cos 2t,$$

$$\vdots$$

$$\ddot{x}^{(0,2)} + x^{(0,2)} = -3 \left(x^{(0,0)} \right)^2 x^{(0,1)} - a^{(0,2)} x^{(0,0)} - a^{(0,1)} x^{(0,1)},$$

$$\vdots$$

Let us assume the initial conditions $x(t = 0) = A_0$, $\dot{x}(t = 0) = B_0$. From the first equation of system (2.5.37) we obtain

$$x^{(0,0)} = A_0 \cos t + B_0 \sin t. \tag{2.5.38}$$

Substituting (2.5.38) into the second equation of system (2.5.37), from the condition of avoiding the secular terms, we get: $1°$ $a^{(1,0)} = -\frac{1}{2}$ and $B_0 = 0$ or $2°$ $a^{(1,0)} = \frac{1}{2}$ and $A_0 = 0$. Let us consider the first case: then $x^{(1,0)} = \frac{1}{16} A_0 \cos 3t$ and, taking into account the third equation, we have

$$a^{(2,0)} = \frac{7}{32}, \quad x^{(2,0)} = -\frac{9}{256} A_0 \cos 3t + \frac{A_0}{768} \cos 5t. \tag{2.5.39}$$

Acting analogously, we obtain

$$a^{(0,1)} = -\frac{3}{4}A_0^2, \quad x^{(0,1)} = \frac{1}{32}A_0^3 \cos 3t,$$

$$a^{(1,1)} = \frac{5}{16}A_0^2, \quad x^{(1,1)} = -\frac{11A_0^3}{256}\cos 3t + \frac{A_0^3}{384}\cos 5t, \tag{2.5.40}$$

$$a^{(2,1)} = \frac{75}{1024}A_0^2, \quad x^{(2,1)} = \frac{73}{24576}A_0^3 \cos 3t - \frac{235}{73728}A_0^3 \cos 5t$$

$$+\frac{16A_0^4}{147456}\cos 7t, \tag{2.5.41}$$

$$a^{(0,2)} = -\frac{3}{128}A_0^4, \quad x^{(0,2)} = \frac{3}{1024}A_0^5 \cos 3t + \frac{3}{3072}A_0^5 \cos 5t.$$

Taking the above calculation into account in (2.5.35), we obtain

$$\lambda^2 = 1 - \frac{1}{2}\mu + \frac{7}{32}\mu^2 + \ldots + \varepsilon A_0^2 \left(-\frac{3}{4} + \frac{5}{16}\mu + \frac{75}{1024}\mu^2 + \ldots \right)$$

$$+\varepsilon^2 \left(-\frac{3}{128}A_0^4 + \ldots \right) + \ldots . \tag{2.5.42}$$

For $\varepsilon = 0$ we obtain one branch of the first limit of the stability loss of the Mathieu equation, and for $\mu = 0$, the dependence of the frequency on the amplitude for a conservative system with the Duffing characteristics.

Example 2.5.2. Consider an unsteady nonlinear system with two-degree-of-freedom, the motion of which is governed by the following differential equations

$$M\ddot{z}_1 + c(\dot{z}_1 - \dot{z}_2) + k(z_1 - z_2) + k_1(z_1 - z_2)^3 = 0, \tag{2.5.43}$$

$$m\ddot{z}_2 - c(\dot{z}_1 - \dot{z}_2) - k(z_1 - z_2) - k_1(z_1 - z_2)^3 = -(k_3 - k_0 \cos 2\omega t)z_2,$$

where m and M are the masses, k, k_1, k_3 and k_0 are the rigidities, and c is the damping coefficient.

The frequencies of free vibrations of the conservative system are calculated after assuming $c = k_1 = k_0 = 0$ in (2.5.43). They amount to

$$\alpha_{1,2}^2 = \frac{1}{2}\left[\frac{k}{M} + \frac{k + k_3}{m} \mp \sqrt{\left(\frac{k}{M} + \frac{k + k_3}{m} \right)^2 - \frac{4kk_3}{Mm}} \right]. \tag{2.5.44}$$

Our consideration is limited to calculating the first simple parametric resonance around the frequency α_1.

Equation (2.5.43) will be rearranged to the form

$$\ddot{y}_1 + \lambda_1^2 y_1 = -\varepsilon \overline{M} \lambda_1^2 (\varepsilon_1 y_2 - \varepsilon_2 y_1)^3 - \mu \delta_1 \lambda_1 (\varepsilon_1 \dot{y}_2 - \varepsilon_2 \dot{y}_1)$$

$$+\mu \rho_1 \lambda_1^2 (y_1 - y_2) \cos 2\tau,$$

$$\ddot{y}_2 + \lambda_1^2 \nu^2 y_2 = -\varepsilon \overline{M} \lambda_1^2 \nu^2 (\varepsilon_1 y_2 - \varepsilon_2 y_1)^3 - \mu \delta_1 \lambda_1 \nu (\varepsilon_1 \dot{y}_2 - \varepsilon_2 \dot{y}_1) \tag{2.5.45}$$

$$+\mu \rho_2 \lambda_1^2 \nu^2 (y_1 - y_2) \cos 2\tau,$$

where

$$\mu = \frac{k_0}{k_1}, \quad \varepsilon = \frac{k_1}{k}, \quad \tau = \omega t, \quad \lambda_1^2 = \frac{\alpha_1^2}{\omega^2}, \quad \overline{M} = \frac{M}{m},$$

$$z_1 = \beta_1 y_2 - \beta_2 y_1, \quad z_2 = \psi(y_1 - y_2), \quad \nu = \frac{\alpha_2}{\alpha_1}, \quad \delta_1 = \frac{M\alpha_1 c}{mk},$$

$$\rho_1 = \frac{\gamma_1 k_1 \psi}{m\alpha_1^2}, \quad \rho_2 = \frac{\gamma_2 k_1 \psi}{m\alpha_2^2}, \quad \gamma_1 = \frac{k - M\alpha_1^2}{k}, \quad \gamma_2 = \frac{k - M\alpha_2^2}{k}, \quad (2.5.46)$$

$$\varepsilon_1 = \beta_1 + \psi, \quad \varepsilon_2 = \beta_2 + \psi, \quad \beta_1 = \frac{m\gamma_1}{M(\gamma_1 - \gamma_2)}, \quad \beta_2 = \frac{m\gamma_2}{M(\gamma_1 - \gamma_2)},$$

$$\psi = \frac{1}{\gamma_1 - \gamma_2},$$

and the dot now denotes the differentiation with respect to τ.

We seek the solution of (2.5.45) of the form

$$y_1 = y_1^{(0,0)} + \mu y_1^{(1,0)} + \mu^2 y_1^{(2,0)} + \ldots + \varepsilon y_1^{(0,1)}$$
$$+ \varepsilon^2 y_1^{(0,2)} + \ldots + \mu\varepsilon y_1^{(1,1)} + \ldots$$

$$y_2 = y_2^{(0,0)} + \mu y_2^{(1,0)} + \mu^2 y_2^{(2,0)} + \ldots + \varepsilon y_2^{(0,1)}$$
$$+ \varepsilon^2 y_2^{(0,2)} + \ldots + \mu\varepsilon y_2^{(1,1)} + \ldots \qquad (2.5.47)$$

$$\lambda_1^2 = 1 + \mu a_1^{(1,0)} + \mu^2 a_1^{(2,0)} + \ldots + \varepsilon a_1^{(0,1)} + \varepsilon^2 a_1^{(0,2)} + \ldots + \mu\varepsilon a_1^{(1,1)} + \ldots$$

$$\lambda_1 = 1 + \frac{a_1^{(1,0)}}{2}\mu + \frac{a_1^{(0,1)}}{2}\varepsilon + \frac{a_1^{(1,1)}}{2}\mu\varepsilon + \mu^2 \frac{a_1^{(2,0)}}{4} + \varepsilon^4 \frac{a_1^{(0,2)}}{4} + \ldots$$

After substituting (2.5.47) into (2.5.45), we obtain the following system of linear differential equations

$$\ddot{y}_1^{(0,0)} + y_1^{(0,0)} = 0,$$

$$\ddot{y}_1^{(1,0)} + y_1^{(1,0)} = -a_1^{(1,0)} y_1^{(0,0)} - \delta_1 \left(\varepsilon_1 \dot{y}_2^{(0,0)} - \varepsilon_2 \dot{y}_1^{(0,0)} \right)$$
$$+ \rho_1 \left(y_1^{(0,0)} - y_2^{(0,0)} \right) \cos 2\tau,$$

$$\ddot{y}_1^{(2,0)} + y_1^{(2,0)} = -a_1^{(1,0)} y_1^{(1,0)} - a_1^{(2,0)} y_1^{(0,0)} - \delta_1 \frac{a_1^{(1,0)}}{2} \left(\varepsilon_1 \dot{y}_2^{(0,0)} - \varepsilon_2 \dot{y}_1^{(0,0)} \right)$$
$$- \delta_1 \left(\varepsilon_1 \dot{y}_2^{(1,0)} - \varepsilon_2 \dot{y}_1^{(1,0)} \right) + \rho_1 a_1^{(1,0)} \left(y_1^{(0,0)} - y_2^{(0,0)} \right) \cos 2\tau$$
$$+ \rho_1 \left(y_1^{(1,0)} - y_2^{(1,0)} \right) \cos 2\tau,$$

$$\ddot{y}_1^{(0,1)} + y_1^{(0,1)} = -a_1^{(0,1)} y_1^{(0,0)} - \overline{M} \left(\varepsilon_1 y_2^{(0,0)} - \varepsilon_2 y_1^{(0,0)} \right)^3, \qquad (2.5.48)$$

$$\ddot{y}_1^{(0,2)} + y_1^{(0,2)} = -a_1^{(0,1)} y_1^{(0,1)} - a_1^{(0,2)} y_1^{(0,0)} - \overline{M} a_1^{(0,1)} \left(\varepsilon_1 y_2^{(0,0)} - \varepsilon_2 y_1^{(0,0)} \right)^3$$
$$- \overline{M} \left\{ 3\varepsilon_1^3 (y_2^{(0,0)})^2 y_2^{(0,1)} - 6\varepsilon_1^2 \varepsilon_2 y_2^{(0,1)} y_1^{(0,0)} \right.$$

$$+ 6\varepsilon_1\varepsilon_2^2 y_2^{(0,0)} y_1^{(0,0)} y_1^{(0,1)} + 3\varepsilon_1\varepsilon_2^2 y_2^{(0,1)} (y_1^{(0,0)})^2$$
$$-3\varepsilon_2^3 (y_1^{(0,0)})^2 y_1^{(0,1)} - 3\varepsilon_1^2\varepsilon_2 (y_2^{(0,0)})^2 y_1^{(0,1)} \Big\},$$

$$\ddot{y}_1^{(1,1)} + y_1^{(1,1)} = -a_1^{(1,1)} y_1^{(0,0)} - a_1^{(0,1)} y_1^{(1,0)} - a_1^{(1,0)} y_1^{(0,1)}$$
$$-\overline{M} a_1^{(1,0)} \left(\varepsilon_1 y_2^{(0,0)} - \varepsilon_2 y_1^{(0,0)} \right)^3 - \overline{M} \Big\{ 3\varepsilon_1^3 (y_2^{(0,0)})^2 y_2^{(1,0)}$$
$$-6\varepsilon_1^2 y_2^{(0,0)} y_2^{(1,0)} \varepsilon_2 y_1^{(0,0)} + 6\varepsilon_1\varepsilon_2^2 y_2^{(0,0)} y_1^{(1,0)} y_1^{(0,0)}$$
$$-\delta_1 \frac{a_1^{(0,1)}}{2} \left(\varepsilon_1 y_2^{(0,0)} - \varepsilon_2 y_1^{(0,0)} \right) - \delta_1 \left(\varepsilon_1 \dot{y}_2^{(0,1)} - \varepsilon_2 \dot{y}_1^{(0,1)} \right)$$
$$+\rho_1 \left(y_1^{(0,1)} - y_2^{(0,1)} \right) \cos 2\tau + \rho_1 a_1^{(0,1)} \left(y_1^{(0,0)} \right.$$
$$\left. - y_2^{(0,0)} \right) \cos 2\tau \Big\},$$

$$\ddot{y}_2^{(0,0)} + \nu^2 y_2^{(0,0)} = 0,$$
$$\ddot{y}_2^{(1,0)} + \nu^2 y_2^{(1,0)} = -\nu^2 a_1^{(1,0)} y_2^{(0,0)} - \delta_2\nu \left(\varepsilon_1 \dot{y}_2^{(0,0)} - \varepsilon_2 \dot{y}_1^{(0,0)} \right)$$
$$+\rho_2\nu^2 \left(y_1^{(0,0)} - y_2^{(0,0)} \right) \cos 2\tau,$$

$$\ddot{y}_2^{(2,0)} + \nu^2 y_2^{(2,0)} = -\nu^2 a_1^{(1,0)} y_2^{(1,0)} - \nu^2 a_1^{(2,0)} y_2^{(0,0)} - \delta_2\nu \frac{a_1^{(1,0)}}{2} \left(\varepsilon_1 \dot{y}_2^{(0,0)} \right.$$
$$\left. - \varepsilon_2 \dot{y}_1^{(0,0)} \right) - \delta_2\nu \left(\varepsilon_1 \dot{y}_2^{(1,0)} - \varepsilon_2 \dot{y}_1^{(1,0)} \right)$$
$$\rho_2\nu^2 a_1^{(1,0)} \left(y_1^{(0,0)} - y_2^{(0,0)} \right) \cos 2\tau$$
$$+\rho_2\nu^2 \left(y_1^{(1,0)} - y_2^{(1,0)} \right) \cos 2\tau,$$

$$\ddot{y}_2^{(0,1)} + \nu^2 y_2^{(0,1)} = -\nu^2 a_1^{(0,1)} y_1^{(0,0)} - \overline{M}\nu^2 \left(\varepsilon_1 y_2^{(0,0)} - \varepsilon_2 y_1^{(0,0)} \right)^3, \quad (2.5.49)$$
$$\ddot{y}_2^{(0,2)} + \nu^2 y_2^{(0,2)} = -\nu^2 y_2^{(0,1)} a_1^{(0,1)} - a_1^{(0,2)} \nu^2 y_2^{(0,0)} - \nu^2 \overline{M} a_1^{(0,1)} \left(\varepsilon_1 y_2^{(0,0)} \right.$$
$$\left. - \varepsilon_2 y_1^{(0,0)} \right)^3 - \overline{M}\nu^2 \Big(3\varepsilon_1^3 (y_2^{(0,0)})^2 y_2^{(0,1)}$$
$$-6\varepsilon_1^2 \varepsilon_2 y_2^{(0,0)} y_2^{(0,1)} y_1^{(0,0)} + 6\varepsilon_1\varepsilon_2^2 y_2^{(0,0)} y_1^{(0,0)} y_1^{(0,1)}$$
$$+3\varepsilon_1\varepsilon_2^2 y_2^{(0,1)} (y_1^{(0,0)})^2 - 3\varepsilon_2^3 (y_1^{(0,0)})^2 y_1^{(0,1)}$$
$$-3\varepsilon_1^2 (y_2^{(0,0)})^2 \varepsilon_2 y_1^{(0,1)} \Big),$$

$$\ddot{y}_2^{(1,1)} + y_2^{(1,1)} = -\nu^2 a_1^{(1,1)} y_2^{(0,0)} - \nu^2 a_1^{(0,1)} y_2^{(1,0)} - \nu_1^2 a_1^{(1,0)} y_2^{(0,1)} - \nu^2 \overline{M} a_1^{(1,0)}$$
$$\cdot \left(\varepsilon_1 y_2^{(0,0)} - \varepsilon_2 y_1^{(0,0)} \right)^3 - \nu^2 \overline{M} \Big(3\varepsilon_1^3 (y_2^{(0,0)})^2 y_2^{(1,0)}$$

$$-6\varepsilon_1^2\varepsilon_2 y_2^{(0,0)}y_2^{(1,0)}y_1^{(0,0)} + 6\varepsilon_1\varepsilon_2^2 y_2^{(0,0)}y_1^{(1,0)}y_1^{(0,0)}$$

$$-3\varepsilon_1^2\varepsilon_2(y_2^{(0,0)})^2 y_1^{(1,0)} + 3\varepsilon_1\varepsilon_2^2 y_2^{(1,0)}(y_1^{(0,0)})^2$$

$$-3\varepsilon_2^3(y_1^{(0,0)})^2 y_1^{(1,0)}\Big) - \delta_2\nu\frac{a_1^{(0,1)}}{2}\left(\varepsilon_1\dot{y}_2^{(0,0)} - \varepsilon_2\dot{y}_1^{(0,0)}\right)$$

$$-\delta_2\nu\left(\varepsilon_1\dot{y}_2^{(0,1)} - \varepsilon_2\dot{y}_1^{(0,1)}\right) + \rho_2\nu^2\left(y_1^{(0,1)} - y_2^{(0,1)}\right)\cos 2\tau$$

$$+\rho_2\nu^2 a_1^{(0,1)}\left(y_1^{(0,0)} - y_2^{(0,0)}\right)\cos 2\tau.$$

Let us assume the following initial conditions: $y_1^{(0,0)}(0) = A_0$, $\dot{y}_1^{(0,0)}(0) = B_0$ and $y_2^{(0,0)} = 0$. Let us limit ourselves to considering the equations occurring near μ, ε, and $\mu\varepsilon$. After calculation we obtain

$$y_1^{(0,0)} = A_0\cos\tau + B_0\sin\tau,$$

$$a_1^{(1,0)} = \pm\sqrt{\frac{\rho_1^2}{4} - \varepsilon_2^2\delta_1^2},$$

$$y_1^{(1,0)} = -\frac{\rho_1 A_0}{16}\cos 3\tau - \frac{\rho_1 B_0}{16}\sin 3\tau,$$

$$y_2^{(1,0)} = -\frac{\left(A_0\delta_0\nu\varepsilon_2 + \frac{\rho_2\nu^2 B_0}{2}\right)}{\nu^2 - 1}\sin\tau + \frac{\left(B_0\delta_2\nu\varepsilon_2 + \frac{\rho_2\nu^2 A_0}{2}\right)}{\nu^2 - 1}\cos\tau$$

$$+\frac{\rho_2\nu^2 A_0}{2(\nu^2 - 9)}\cos 3\tau + \frac{\rho_2\nu^2 B_0}{2(\nu^2 - 9)}\sin 3\tau,$$

$$a_1^{(0,1)} = \frac{3}{4}\overline{M}\varepsilon_2^3(A_0^2 + B_0^2), \tag{2.5.50}$$

$$y_1^{(0,1)} = -\frac{1}{8}\overline{M}\varepsilon_2^3 A_0\left(\frac{A_0^2}{4} - \frac{3}{4}B_0^4\right)\cos 3\tau$$

$$-\frac{1}{8}\overline{M}\varepsilon_2^3 B_0\left(\frac{3}{4}A_0^2 - \frac{1}{4}B_0^2\right)\sin 3\tau,$$

$$y_2^{(0,1)} = \frac{\left(-\nu^2 a_1^{(0,1)}A_0 + \frac{3}{4}\overline{M}\nu^2\varepsilon_2^3 A_0^3 + \frac{3}{4}\overline{M}\nu^2\varepsilon_2^3 A_0 B_0^2\right)}{\nu^2 - 1}\cos\tau$$

$$+\frac{\left(-\nu^2 a_1^{(0,1)}B_0 + \frac{3}{4}\overline{M}\nu^2\varepsilon_2^3 A_0^2 B_0 + \frac{3}{4}\overline{M}\nu^2\varepsilon_2^3 B_0^3\right)}{\nu^2 - 1}\sin\tau$$

$$+\frac{\left(\frac{1}{4}\overline{M}\nu^2\varepsilon_2^3 A_0^3 - \frac{3}{4}\overline{M}\nu^2\varepsilon_2^3 A_0 B_0^2\right)}{\nu^2 - 9}\cos 3\tau$$

$$+\frac{\left(\frac{3}{4}\overline{M}\nu^2\varepsilon_2^3 A_0^2 B_0 - \frac{1}{4}\overline{M}\nu^2\varepsilon_2^3 B_0^3\right)}{\nu^2 - 9}\sin 3\tau.$$

From the conditions of avoiding the terms unrestrictedly growing in time, we obtain

$$
-a_1^{(1,1)} A_0 + \frac{3}{4}\overline{M}a_1^{(1,0)} \varepsilon_2^3 A_0^3 + \overline{M}a_1^{(1,0)} \varepsilon_2^3 A_0 B_0^2
$$

$$
-3\overline{M}\varepsilon_1\varepsilon_2^2 \left[-\frac{\left(A_0\delta_2\nu\varepsilon_2 + \frac{\rho_2\nu^2 B_0}{2}\right)}{\nu^2 - 1}\frac{A_0 B_0}{2} \right.
$$

$$
+\frac{\left(\sigma_2\nu\varepsilon_2 B_0 + \frac{\rho_2\nu^2 A_0}{2}\right)}{\nu^2 - 1}\left(\frac{3}{4}A_0^2 + \frac{1}{4}B_0^2\right)
$$

$$
+\frac{\rho_2\nu^2 A_0}{2(\nu^2 - 9)}\left(\frac{1}{4}A_0^2 - \frac{1}{4}B_0^2\right) + \left. \frac{\rho_2\nu^2 B_0}{2(\nu^2 - 9)}\frac{A_0 B_0}{2}\right]
$$

$$
+3\overline{M}\varepsilon_2^3\left[-\frac{A_0^3}{64}\rho_1 - \frac{A_0\rho_1 B_0^2}{32} + \frac{1}{64}B_0^2\rho_1 A_0\right] + \delta_1\frac{a_1^{(0,1)}}{2}\varepsilon_2 B_0
$$

$$
-\delta_1\varepsilon_1\frac{\left(-\nu^2 a_1^{(0,1)} B_0 + \frac{3}{4}\overline{M}\nu^2\varepsilon_2^3 A_0^2 B_0 + \frac{3}{4}\overline{M}\nu^2\varepsilon_2^3 B_0^3\right)}{\nu^2 - 1}
$$

$$
-\frac{\rho_1}{16}A_0\overline{M}\varepsilon_2^3\left[\frac{1}{4}A_0^2 - \frac{3}{4}B_0^2\right.
$$

$$
-\frac{\rho_1}{2}\frac{\left(-\nu^2 a_1^{(0,1)} A_0 + \frac{3}{4}\overline{M}\nu^2\varepsilon_2^3 A_0^3 + \frac{3}{4}\overline{M}\varepsilon_2^3 A_0 B_0^2\right)}{\nu^2 - 1}
$$

$$
+\frac{1}{2}\frac{\left(\frac{1}{4}\overline{M}\nu^2\varepsilon_2^3 A_0^3 - \frac{3}{4}\overline{M}\nu^2\varepsilon_2^3 A_0 B_0^2\right)}{\nu^2 - 9} + \left. \frac{1}{2}\rho_1 a_1^{(0,1)} A_0\right] = 0, \qquad (2.5.51)
$$

$$
-a_1^{(1,1)} B_0 + \frac{1}{4}\overline{M}a_1^{(1,0)} \varepsilon_2^3 3A_0 B_0 + \frac{3}{4}\overline{M}a_1^{(1,0)} \varepsilon_2^3 B_0^3
$$

$$
-3\overline{M}\varepsilon_1\varepsilon_2^2 \left[-\frac{\left(A_0\delta_2\nu\varepsilon_2 + \frac{\rho_2\nu^2 B_0}{2}\right)}{\nu^2 - 1}\left(\frac{1}{4}A_0^2 + \frac{3}{4}B_0^2\right) \right.
$$

$$
+\frac{\left(\delta_2\nu\varepsilon_2 B_0 + \frac{\rho_2\nu^2 A_0}{2}\right)}{\nu^2 - 1}\frac{A_0 B_0}{2}
$$

$$
-\frac{1}{4}\frac{\rho_2\nu^2 A_0^2 B_0}{(\nu^2 - 9)} + \left. \frac{\rho_2\nu^2 B_0}{2(\nu^2 - 9)}\left(-\frac{1}{4}B_0^2 + \frac{1}{4}A_0^2\right)\right]
$$

$$
+3\overline{M}\varepsilon_2^3\left[-\frac{\rho_1 A_0^2 B_0}{64} + \frac{A_0^2 B_0\rho_1}{32} + \frac{1}{64}B_0^3\rho_1\right] - \frac{\delta_1 a_1^{(0,1)}}{2}\varepsilon_2 A_0 +
$$

$$
-\delta_1\varepsilon_1\left(-\frac{\left(-\nu^2 a_1^{(0,1)} A_0 + \frac{3}{4}\overline{M}\nu^2\varepsilon_2^3 A_0^3 + \frac{3}{4}\overline{M}\nu^2\varepsilon_2^3 A_0 B_0^2\right)}{\nu^2 - 1}\right)
$$

$$
-\frac{\rho_1}{16}\overline{M}\varepsilon_2^3 B_0\left(\frac{3}{4}A_0^2 - \frac{1}{4}B_0^2\right)
$$

$$-\frac{\rho_1}{2}\left(\frac{-\nu^2 a_1^{(0,1)} B_0 + \frac{3}{4}\overline{M}\nu^2\varepsilon_2^3 A_0^2 B_0 + \frac{3}{4}\overline{M}\nu^2\varepsilon_2^3 B_0^3}{\nu^2 - 1}\right)$$

$$+\left(\frac{\frac{3}{4}\overline{M}\nu^2\varepsilon_2^3 A_0^2 B_0 - \frac{1}{4}\overline{M}\nu^2\varepsilon_2^3 B_0^3}{2(\nu^2 - 9)}\right) - \frac{\rho_1 a_1^{(0,1)} B_0}{2}\Bigg] = 0. \qquad (2.5.52)$$

In order to avoid time-consuming calculations, we assume that $B_0 = 0$. Then

$$a_1^{(1,1)} = \frac{3}{4}\overline{M}a_1^{(1,0)}\varepsilon_2^3 A_0^3 - \overline{M}\varepsilon_1\varepsilon_2^2\frac{\rho_2\nu^2}{2(\nu^2 - 1)}\cdot\frac{9}{8}A_0^2 - \frac{3}{8}\overline{M}\varepsilon_1\varepsilon_2^2\frac{\rho_2\nu^2}{(\nu^2 - 9)}A_0^2$$

$$-\frac{3}{64}\overline{M}\varepsilon_2^3\rho_1 A_0^2 - \frac{1}{64}\overline{M}\varepsilon_2^3 A_0^2 + \frac{\rho_1\nu^2 a_1^{(0,1)}}{2(\nu^2 - 1)} - \frac{3\overline{M}\nu^2\varepsilon_2^3 A_0^2}{8(\nu^2 - 1)}$$

$$+\frac{1}{8}\frac{\overline{M}\nu^2\varepsilon_2^3 A_0^2}{(\nu^2 - 9)} + \frac{\rho_1 a_1^{(0,1)}}{2}. \qquad (2.5.53)$$

Substituting the calculated values $a_1^{(1,0)}$, $a_1^{(0,1)}$, $a_1^{(1,1)}$ into the third equation of (2.5.47), we obtain the analytic form of the limit of the loss of stability, which depends on μ, ε and the other parameters.

2.6 Parametric and Self-Excited Oscillation in a Three-Degree-of-Freedom Mechanical System

2.6.1 Analysed System and Equation of Motion

We consider a mechanical system with friction induced self-excited oscillations and parametric oscillations [14d, 17d]. The latter come from a rotor with rectangular cross-sections, which has a mass concentrated eccentrically at its centre. The rotor is fixed on the base placed on the belt moving with a constant velocity. At a certain value of the belt velocity and the frequency of rotor revolutions, parametric and self-excited oscillations appear in addition to the forced oscillations caused by the unbalanced rotor. The diagram of the system is presented in Fig. 2.7.

A weightless shaft with a rectangular cross-section and with a cylinder-like mass concentrated at its centre is supported by the base placed on the belt moving with constant velocity v_0. The friction coefficient between the belt and the base depends on their relative velocity. The character of this dependence (Fig. 2.8) causes self-excited vibrations. The effect is described in basic works on nonlinear vibrations. On the other hand, considering the nonidentical cross-section of the rotor at various values of its rotational speed, parametric vibrations occur. The vibrations cause a change to the normal force holding down the base to the belt in the vertical direction, and hence they cause changes to the friction force. It is assumed that despite the vibration of the rotor, the contact between the base and the belt is maintained.

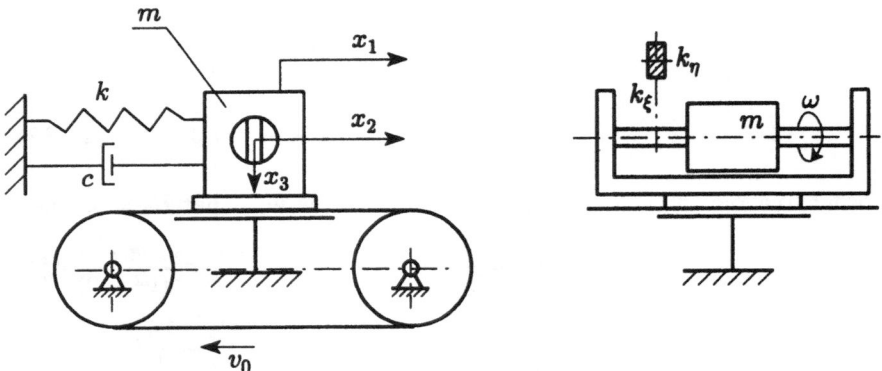

Fig. 2.7. Diagram of the system

The calculation model of the system is presented in Fig. 2.9. The equations of motion of the system have the form

$$m\ddot{x}_c = -\xi_w k_\xi \cos\varphi - \eta_w k_\eta \sin\varphi,$$
$$m\ddot{y}_c = \xi_w k_\xi \sin\varphi - \eta_w k_\eta \cos\varphi + mg, \qquad (2.6.1)$$
$$I_{z''}\ddot{\varphi} = -M_0 + a(-\xi_w k_\xi \cos\varphi_0 + \eta_w k_\eta \sin\varphi_0),$$

where

x_c, y_c	:	coordinates of the centre of mass of the cylinder,
$I_{z''}$	:	mass moment of inertia of the cylider with mass m in relation to the axis z'' of the system $O''x''y''z''$ moving with translatory motion in relation to $Oxyz$,
ξ_w, η_w	:	coordinates of the point of puncture by the shaft in the coordinate system $O'\xi\eta$,
$O'\xi\eta$	:	coordinate system whose axes are parallel to the main, central inertia axes of the cross-section of the shaft,
k_ξ, k_η	:	shaft rigidities in the direction of the axes ξ and η,
M_0	:	driving torque reduced by all the resistance torques,
a, φ_0	:	parameters characterizing the position of the centre of mass of the disk C in relation to the point of puncture by the shaft.

For the states near the steady states, the torque M_0 is very small. Let

$$I_{z''} = mi_s^2, \qquad (2.6.2)$$

where i_s is the inertia radius. Then the third equation of (2.6.1) will assume the form

$$\ddot{\varphi} = \frac{1}{m}\frac{a}{i_s^2}(-\xi_w k_\xi \cos\varphi_0 + \eta_w k_\eta \sin\varphi_0). \qquad (2.6.3)$$

As the eccentricity a and the shaft deflection ξ_w and η_w are small if compared to the inertia radius, the following can be assumed

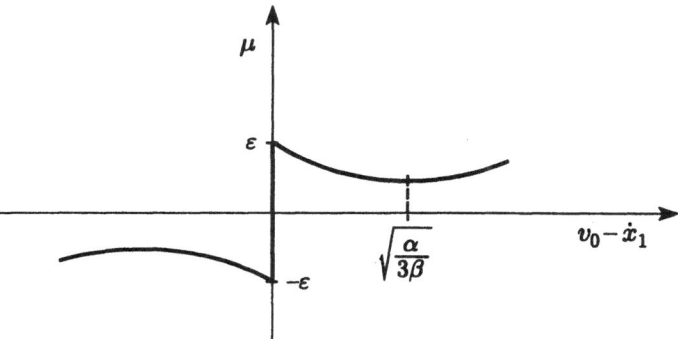

Fig. 2.8. Dependence of the friction coefficient on the relative velocity

$$\ddot{\varphi} = 0, \qquad \dot{\varphi} = \omega = \text{const}, \qquad \varphi = \omega t. \tag{2.6.4}$$

The following geometric dependences result from Fig. 2.9:

$$\xi_w = (x_w - x) \cos \varphi - y_w \sin \varphi,$$
$$\eta_w = (x_w - x) \sin \varphi + y_w \cos \varphi,$$
$$y_c = y_w + a \cos(\varphi + \varphi_0), \tag{2.6.5}$$
$$x_c = x_w + a \sin(\varphi + \varphi_0),$$

where x_w, y_w are the coordinates of the point of puncture by the shaft W in the system $Oxyz$.

In order to write down the equation of motion of the mass M, it is necessary to determine the dynamic reactions on the shaft at its points of support. They are determined from the equations

$$X_1 + X_2 + \xi_w k_\xi \cos \omega t + \eta_w k_\eta \sin \omega t = 0,$$
$$Y_1 + Y_2 - \xi_w k_\xi \sin \omega t + \eta_w k_\eta \cos \omega t = 0, \tag{2.6.6}$$

where X_1, Y_1 and X_2, Y_2 denote the support reactions on the left and right end of the shaft, respectively. The rotor reactions on the supports are

$$R_x = -X_1 - X_2,$$
$$R_y = -Y_1 - Y_2. \tag{2.6.7}$$

The equation of motion of a body with mass M, on the assumption that $Mg + R_y > 0$, has the form

$$M\ddot{x} = -kx - c\dot{x} + R_x + (Mg + R_y)\mu(w),$$
$$w = v_0 - \dot{x}. \tag{2.6.8}$$

The dependence of the friction coefficient on the relative velocity w can be described by the polynomial

$$\mu(w) = \varepsilon \, \text{sgn} \, w - \alpha w + \beta w^3. \tag{2.6.9}$$

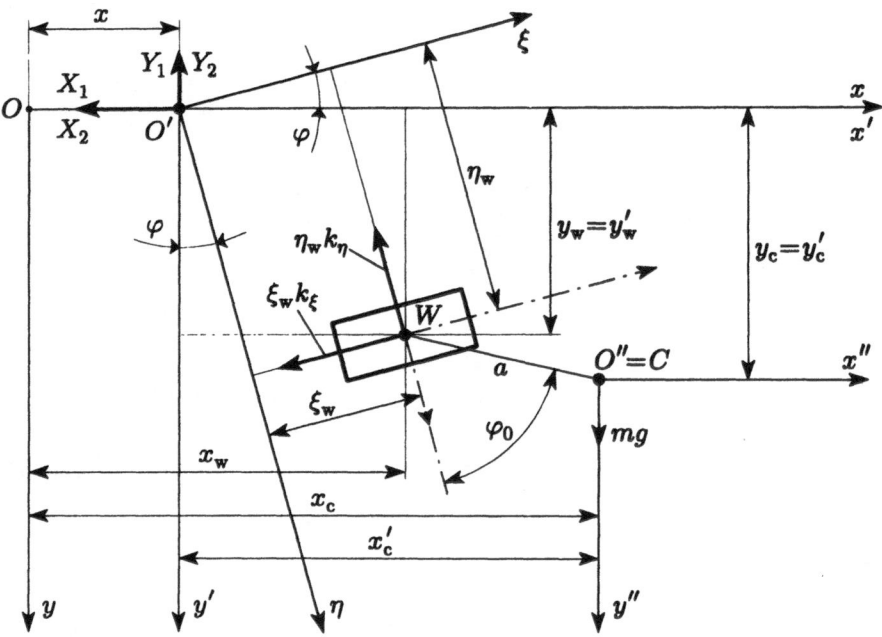

Fig. 2.9. Calculation model of the system

Finally, the equations of motion of the system, after assuming that $x = x_1$, $x_w = x_2$, $y_w = x_3$, have the form

$$\ddot{x}_1 = -x_1 \left(\Omega^2 + \Omega_\xi^2 + \Omega_\eta^2 + (\Omega_\xi^2 - \Omega_\eta^2) \cos 2\omega t \right) - H\dot{x}_1$$
$$-x_2 \left(-(\Omega_\xi^2 + \Omega_\eta^2) - (\Omega_\xi^2 - \Omega_\eta^2) \cos 2\omega t \right) - x_3(\Omega_\xi^2 - \Omega_\eta^2) \sin 2\omega t$$
$$+ \left[g - x_2(\Omega_\xi^2 - \Omega_\eta^2) \sin 2\omega t - x_3 \left(-(\Omega_\xi^2 + \Omega_\eta^2) \right. \right.$$
$$\left. + (\Omega_\xi^2 - \Omega_\eta^2) \cos 2\omega t) + x_1(\Omega_\xi^2 - \Omega_\eta^2) \sin 2\omega t \right] \left(\varepsilon \mathrm{sgn}\, (v_0 - \dot{x}_1) \right.$$
$$\left. - \alpha(v_0 - \dot{x}_1) + \beta(v_0 - \dot{x}_1)^3 \right),$$

$$\ddot{x}_2 = x_1 \left(\omega_\xi^2 + \omega_\eta^2 + (\omega_\xi^2 - \omega_\eta^2) \cos 2\omega t \right) \qquad (2.6.10)$$
$$-x_2 \left(\omega_\xi^2 + \omega_\eta^2 + (\omega_\xi^2 - \omega_\eta^2) \cos 2\omega t \right) + x_3(\omega_\xi^2 - \omega_\eta^2) \sin 2\omega t$$
$$+ \alpha^2 \omega \sin(\omega t + \varphi_0),$$

$$\ddot{x}_3 = -x_1(\omega_\xi^2 - \omega_\eta^2) \sin 2\omega t + x_2(\omega_\xi^2 - \omega_\eta^2) \sin 2\omega t$$
$$+ x_3 \left(-(\omega_\xi^2 + \omega_\eta^2) + (\omega_\xi^2 - \omega_\eta^2) \cdot \cos 2\omega t \right) + \alpha \omega^2 \cos(\omega t + \varphi_0) + g,$$

where: $\Omega^2 = k/M$, $\Omega_\xi^2 = k_\xi/(2M)$, $\Omega_\eta^2 = k_\eta/(2M)$, $H = C/M$, $\omega_\xi^2 = k_\xi/(2m)$, $\omega_\eta^2 = k_\eta/(2m)$.

2.6.2 Transformation of the Equations of Motion to the Main Coordinates

Let us introduce the following notation

$$\Omega_1^2 = \Omega_\xi^2 + \Omega_\eta^2, \quad \Omega_2^2 = \Omega_\xi^2 - \Omega_\eta^2, \quad H = \mu H_1,$$
$$\omega_1^2 = \omega_\xi^2 + \omega_\eta^2, \quad \omega_2^2 = \omega_\xi^2 - \omega_\eta^2, \quad \chi = \frac{\alpha}{\varepsilon}, \quad \rho = \frac{\beta}{\varepsilon}, \tag{2.6.11}$$
$$a \cos\varphi_0 = \mu P, \quad a \sin\varphi_0 = \mu Q, \quad \mu G = g,$$

where $\mu = \omega_2^2/\omega_1^2 = \Omega_2^2/\Omega_1^2 = (k_\xi - k_\eta)/(k_\xi + k_\eta)$ is the perturbation parameter.

After accounting for (2.6.11) in the equation system (2.6.10), it will assume the form

$$\ddot{x}_1 = -x_1\Omega^2 - x_1\Omega_1^2(3 + \mu\cos 2\omega t) - \mu H_1\dot{x}_1$$
$$+x_2\Omega_1^2(1 + \mu\cos 2\omega t) + x_3\Omega_1^2\mu\sin 2\omega t$$
$$+\varepsilon(g - x_2\Omega_1^2\mu\sin 2\omega t + x_1\Omega_1^2\mu\sin 2\omega t$$
$$+x_3\Omega_1^2(-\mu\cos 2\omega t))(\text{sgn}(v_0 - \dot{x}_1) - \chi(v_0 - \dot{x}_1) + \rho(v_0 - \dot{x}_1)^3),$$
$$\ddot{x}_2 = -x_1\omega_1^2(1 + \mu\cos 2\omega t) - x_2\omega_1^2(1 + \mu\cos 2\omega t) \tag{2.6.12}$$
$$+x_3\omega_1^2\mu\sin 2\omega t + \mu(P\sin\omega t + Q\sin\omega t)\omega^2,$$
$$\ddot{x}_3 = -x_1\omega_1^2\mu\sin 2\omega t + x_2\omega_1^2\mu\sin 2\omega t - x_3\omega_1^2(1 - \mu\cos 2\omega t)$$
$$+\mu(P\cos\omega t - Q\sin\omega t)\omega^2 + \mu G.$$

If we introduce $\mu = \varepsilon = 0$ into the equation system (2.6.12), we obtain the homogeneous linear differential equation system

$$\ddot{x}_1 + x_1(\Omega^2 + \Omega_1^2) - x_2\Omega_1^2 = 0,$$
$$\ddot{x}_2 + \omega_1^2(x_2 - x_1) = 0, \tag{2.6.13}$$
$$\ddot{x}_3 + \omega_1^2 x_3 = 0.$$

When we assume the solution of (2.6.13) in the form $x_i = A_i \cos pt$, $i = 1, 2, 3$, we find the following frequencies

$$p_{1,2}^2 = \frac{1}{2}\left(\Omega_2 + \Omega_1^2 + \omega_1^2 \pm \sqrt{(\Omega^2 + \Omega_1^2 + \omega_1^2)^2 - 4\Omega^2\omega_1^2}\right),$$
$$p_3^2 = \omega_1^2. \tag{2.6.14}$$

Let us introduce the main coordinates ξ_1, for which at $\mu = \varepsilon = 0$ the uncoupling of the linear part of the first two equations of the system (2.6.12) will occur. Let us now multiply these equations by ξ_1 and ξ_2, respectively, and add the sides. The result will be as follows

$$\ddot{x}_1\xi_1 + x_1(\Omega^2 + \Omega_1^2)\xi_1 - x_2\Omega_1^2\xi_1 + \ddot{x}_2\xi_2 + x_2\omega_1^2\xi_2 - x_1\omega_1^2\xi_2$$
$$= \mu\big[-x_1\Omega_1^2\xi_1\cos 2\omega t - H_1\dot{x}_1\xi_1 + \xi_1 x_2\Omega_1^2\cos 2\omega t$$
$$-\xi_1 x_3\Omega_1^2\sin 2\omega t + \xi_2 x_1\omega_1^2\cos 2\omega t - \xi_2 x_2\omega_1^2\cos 2\omega t \tag{2.6.15}$$

$$+\xi_2 x_3 \omega_1^2 \sin 2\omega t + \omega^2 \xi_2 (P \sin \omega t + Q \cos \omega t)]$$
$$+\varepsilon \xi_1 \left[g - x_2 \Omega_1^2 \mu \sin 2\omega t + x_3 \omega_1^2 (1 - \mu \cos 2\omega t) \right.$$
$$+ x_1 \Omega_1^2 \mu \sin 2\omega t \left] \left[\operatorname{sgn}(v_0 - \dot{x}_1) - \chi(v_0 - \dot{x}_1) + \rho(v_0 - \dot{x}_1)^3 \right].$$

By denoting

$$\xi_1 \Theta^2 = (\Omega^2 + \Omega_1^2)\xi_1 - \omega_1^2 \xi_2,$$
$$\xi_2 \Theta^2 = -\Omega_1^2 \xi_1 + \omega_1^2 \xi_2 \qquad (2.6.16)$$

we find

$$(\Omega^2 + \Omega_1^2 - \theta^2)\xi_1 - \omega_1^2 \xi_2 = 0,$$
$$-\Omega_1^2 \xi_1 + (\omega_1^2 - \theta^2)\xi_2 = 0. \qquad (2.6.17)$$

In order for (2.6.17) to be fulfilled for ξ_1 and ξ_2 different from zero, the following dependence must occur

$$\begin{vmatrix} \Omega^2 + \Omega_1^2 - \Theta^2 & -\omega_1^2 \\ -\Omega_1^2 & \omega_1^2 - \Theta^2 \end{vmatrix} = 0. \qquad (2.6.18)$$

Hence, $\Theta_1^2 = p_1^2$ and $\Theta_2^2 = p_2^2$.

Let $\xi_1 = \xi_1'$ and $\xi_2 = \xi_2'$ be denoted for $\theta_1 = p_1$. From the second equation of the system (2.6.16) we find

$$\xi_2' = \gamma_1 \xi_1', \qquad (2.6.19)$$

where

$$\gamma_1 = \frac{\Omega_1^2}{\omega_1^2 - p_1^2}.$$

Making use of the dependences (2.6.16) and (2.6.19), (2.6.15) is transformed to

$$\ddot{x}_1 + x_1 p_1^2 + \ddot{x}_2 \gamma_1 + x_2 p_1^2 \gamma_1 = \mu \left[x_1 p_1^2 \gamma_1 \cos 2\omega t - H_1 \dot{x}_1 \right.$$
$$- p_1^2 \gamma_1 x_2 \cos 2\omega t + p_1^2 \gamma_1 x_3 \sin 2\omega t + \gamma_1 \omega^2 (P \sin \omega t + Q \cos \omega t) \right]$$
$$+ \varepsilon \left[g - x_2 \Omega_1^2 \mu \sin 2\omega t + x_3 \Omega_1^2 (1 - \mu \cos 2\omega t) + x_1 \Omega_1^2 \mu \sin 2\omega t \right]$$
$$\cdot \left[\operatorname{sgn}(v_0 - \dot{x}_1) - \chi(v_0 - \dot{x}_1) + \beta(v_0 - \dot{x}_1)^3 \right]. \qquad (2.6.20)$$

Analogously, for $\Theta_2 = p_2$ the following notation is used: $\xi_1 = \xi_1''$ and $\xi_2 = \xi_2''$, while

$$\xi_2'' = \gamma_2 \xi_1'', \qquad (2.6.21)$$

where

$$\gamma_2 = \frac{\Omega_1^2}{\omega_1^2 - p_2^2}.$$

If we take (2.6.16) and (2.6.21) into account in (2.6.15), the equation will assume the form

$$\ddot{x}_1 + x_1 p_2^2 + \ddot{x}_2 \gamma_2 + x_2 p_2^2 \gamma_2 = \mu\left[x_1 p_2^2 \gamma_2 \cos 2\omega t - H_1 \dot{x}_1\right.$$
$$-p_2^2 \gamma_2 x_2 \cos 2\omega t + p_2^2 \gamma_2 x_3 \sin 2\omega t + \gamma_2 \omega^2 (P \sin \omega t + Q \cos \omega t)\right]$$
$$+\varepsilon\left[g - x_2 \Omega_1^2 \mu \sin 2\omega t + x_3 \Omega_1^2 (1 - \mu \cos 2\omega t) + x_1 \Omega_1^2 \sin 2\omega t\right]$$
$$\cdot\left[\mathrm{sgn}\,(v_0 - \dot{x}_1) - \chi(v_0 - \dot{x}_1) + \rho(v_0 - \dot{x}_1)^3\right]. \tag{2.6.22}$$

Let us denote

$$y_1 = x_1 + \gamma_1 x_2,$$
$$y_2 = x_1 + \gamma_2 x_2. \tag{2.6.23}$$

The reverse dependences can be determined from (2.6.23)

$$x_1 = \beta_1 y_1 - \beta_2 y_2,$$
$$x_2 = \psi(y_1 - y_2), \tag{2.6.24}$$

where

$$\beta_1 = \frac{\gamma_2}{\gamma_2 - \gamma_1}, \quad \beta_2 = \frac{\gamma_1}{\gamma_2 - \gamma_1}, \quad \psi = \frac{1}{\gamma_1 - \gamma_2}.$$

Let us additionally assume that $x_3 = y_3$.

Taking (2.6.23) and (2.6.24) into account in (2.6.22), (2.6.20) and (2.6.12), we obtain the following differential equation system

$$\ddot{y}_1 + p_1^2 y_1 = \mu\left[p_1^2 \gamma_1 (\beta_1 y_1 - \beta_2 y_2) \cos 2\omega t - H_1 (\beta_1 \dot{y}_1 - \beta_2 \dot{y}_2)\right.$$
$$-p_1^2 \gamma_1 \psi \cdot (y_1 - y_2) \cos 2\omega t + p_1^2 \gamma_1 y_3 \sin 2\omega t$$
$$+\omega^2 \gamma_1 (P \sin \omega t + Q \cos \omega t)\right]$$
$$+\varepsilon\left[g - \mu \Omega_1^2 \psi(y_1 - y_2) \sin 2\omega t + \mu \Omega_1^2 (\beta_1 y_1 - \beta_2 y_2) \sin 2\omega t\right.$$
$$+\Omega_1^2 y_3 (1 - \mu \cos 2\omega t)\right]\left[\mathrm{sgn}(v_0 - \beta_1 \dot{y}_1 + \beta_2 \dot{y}_2)\right.$$
$$-\chi(v_0 - \beta_2 \dot{y}_1 + \beta_2 \dot{y}_2) + \rho(v_0 - \beta_1 \dot{y}_1 + \beta_2 \dot{y}_2)^3\right];$$

$$\ddot{y}_2 + p_2^2 y_2 = \mu\left[p_2^2 \gamma_2 (\beta_1 y_1 - \beta_2 y_2) \cos 2\omega t - H_1 (\beta_1 \dot{y}_1 - \beta_2 \dot{y}_2)\right. \tag{2.6.25}$$
$$-p_2^2 \gamma_2 \psi(y_1 - y_2) \cos 2\omega t + p_2^2 \gamma_2 y_3 \sin 2\omega t$$
$$+\omega^2 \gamma_2 (P \sin \omega t + Q \cos \omega t)\right]$$
$$+\varepsilon\left[g - \mu \Omega_1^2 \psi(y_1 - y_2) \sin 2\omega t + \mu \Omega_1^2 (\beta_1 y_1 - \beta_2 y_2) \sin 2\omega t\right.$$
$$+\Omega_1^2 y_3 (1 - \mu \cos 2\omega t)\right]\left[\mathrm{sgn}(v_0 - \beta_1 \dot{y}_1 + \beta_2 \dot{y}_2)\right.$$
$$-\chi(v_0 - \beta_1 \dot{y}_1 + \beta_2 \dot{y}_2) + \rho(v_0 - \beta_1 \dot{y}_1 + \beta_2 \dot{y}_2)^3\right];$$

$$\ddot{y}_3 + p_3^2 y_3 = \mu\left[-p_3^2 (\beta_1 y_1 - \beta_2 y_2) \sin 2\omega t + p_3^2 \psi(y_1 - y_2) \sin 2\omega t\right.$$
$$+p_3^2 y_3 \cos 2\omega t + \omega^2 (P \cos \omega t - Q \sin \omega t) + G\right].$$

After introducing the dimensionless time $\tau = \omega t$, we obtain

$$\ddot{y}_1 + \lambda_1^2 y_1 = \mu\big[\lambda_1^2\gamma_1(\varepsilon_1 y_1 - \varepsilon_2 y_2)\cos 2\tau - \lambda_1\bar{H}_1(\beta_1 y_1' - \beta_2 y_2')$$
$$+\lambda_1^2\gamma_1 y_3\sin 2\tau + \gamma_1(P\sin\tau + Q\cos\tau)\big]$$
$$+\varepsilon\big[g + \mu\Omega_1^2(\varepsilon_1 y_1 - \varepsilon_2 y_2)\sin 2\tau + \Omega_1^2 y_3(1 - \mu\cos 2\tau)\big]$$
$$\cdot\big[\frac{\lambda_1^2}{p_1^2}\mathrm{sgn}(v_0 - \beta_1\omega y_1' - \beta_2\omega y_2') - \lambda_1\bar{\chi}_1(\lambda_1 v_0' - \beta_1 y_1' + \beta_2 y_2')$$
$$+\omega\rho(\lambda_1 v_0' - \beta_1 y_1' + \beta_2 y_2')^3\big];$$

$$\ddot{y}_2 + \lambda_2^2 y_2 = \mu\big[\lambda_2^2\gamma_2(\varepsilon_1 y_1 - \varepsilon_2 y_2)\cos 2\tau - \lambda_2\bar{H}_1(\beta_1 y_1' - \beta_2 y_2') \qquad (2.6.26)$$
$$+\lambda_2^2\gamma_2 y_3\sin 2\tau + \gamma_2(P\sin\tau + Q\cos\tau)\big]$$
$$+\varepsilon\big[g + \mu\Omega_1^2(\varepsilon_1 y_1 - \varepsilon_2 y_2)\sin 2\tau + \Omega_1^2 y_3(1 - \mu\cos 2\tau)\big]$$
$$\cdot\big[\frac{\lambda_2^2}{p_1^2}\mathrm{sgn}(v_0 - \beta_1\omega y_1' - \beta_2\omega y_2') - \lambda_2\bar{\chi}_2(\lambda_2 v_0'' - \beta_1 y_1' + \beta_2 y_2')$$
$$+\omega\rho(\lambda_2 v_0'' - \beta_1 y_1' + \beta_2 y_2')^3\big];$$

$$\ddot{y}_3 + \lambda_3^2 y_3 = \mu\big[-\lambda_3^2(\varepsilon_1 y_1 - \varepsilon_2 y_2)\sin 2\tau$$
$$+\lambda_3^2 y_3\cos 2\tau + P\cos\tau + Q\sin\tau + \lambda_3^2\bar{G}\big],$$

where

$$y_i = \frac{dy_i}{d\tau}, \quad \lambda_i^2 = \frac{p_i^2}{\omega^2}, \quad i = 1, 2, 3,$$

$$\varepsilon_k = \beta_k - \psi, \quad \bar{\chi}_k = \frac{\chi}{p_k}, \quad k = 1, 2,$$

$$\bar{G} = \frac{G}{p_3^2}, \quad \bar{H}_1 = \frac{H_1}{p_1}, \quad v_0' = \frac{v_0}{p_1}, \quad v_0'' = \frac{v_0}{p_2}, \quad \bar{H}_2 = \frac{H_2}{p_2}.$$

2.6.3 Zones of Instability of the First Order

The procedure of solving the system of equations (2.6.26) consists in assuming two perturbation parameters μ and ε connected with the parametric excitation and friction, respectively.

The sought periodic solutions of $y_i(\tau)$ are presented in the form of the double power series

$$y_i(\tau) = y_{0,0}^{(i)} + \mu y_{0,1}^{(i)} + \mu^2 y_{0,2}^{(i)} + \ldots + \varepsilon\left(y_{1,0}^{(i)} + \mu y_{1,1}^{(i)} + \mu^2 y_{1,2}^{(i)} + \ldots\right) + \ldots (2.6.27)$$

where $y_{k,l}^{(i)}$, $k, l = 0, 1, 2, \ldots$ must fulfil the condition of periodicity. Periodic solutions are only possible for certain values of the parameters λ_i^2 presented in the form of the analogous series

$$\lambda_i^2 = n^2 + \mu\alpha_{0,1} + \mu^2\alpha_{0,2} + \ldots + \varepsilon\left(\alpha_{1,0} + \mu\alpha_{1,1} + \mu^2\alpha_{2,2} + \ldots\right) + \ldots (2.6.28)$$

where $\alpha_{k,l}$ $k, l = 0, 1, 2, \ldots$ are unknown coefficients, which are determined from the condition of periodicity, avoiding in the solution terms unrestrictedly growing in time. For a first-order resonance $n^2 = 1$ we shall determine the parametric unstability zones, for which the frequency of parameter modulation fulfils, consecutively, the dependences $\omega \cong p_1$, $\omega \cong p_2$, and $\omega \cong p_3$. In series (2.6.27) and (2.6.28) for $\omega \cong p_1$ and $\omega \cong p_2$ we shall limit our considerations to the first powers of the small parameters μ and ε. On the other hand, for $\omega \cong p_3$ we shall limit ourselves in the calculations to the second approximation. In all three cases we shall assume that $\text{sgn}(v_0 - \beta_1\omega y_1' - \beta_2\omega y_2') = 1$.

Let

$$\lambda_2^2 = \nu_{2,1}^2\lambda_1^2,$$
$$\lambda_3^2 = \nu_{3,1}^2\lambda_1^2, \tag{2.6.29}$$

where

$$\nu_{2,1} = \frac{p_2}{p_1},$$
$$\nu_{3,1} = \frac{p_3}{p_1},$$

and let us assume that $\nu_{2,1}$ and $\nu_{3,1}$ are not integers.

Let us first consider the case $\omega \cong p_1$, assuming that

$$y_{0,0}^{(2)}(\tau) = y_{0,0}^{(3)}(\tau) = 0. \tag{2.6.30}$$

The assumption is accounted for by a weak conjugation of (2.6.26) for $\varepsilon \ll 1$ and $\mu \ll 1$. For $\mu = \varepsilon = 0$ we shall obtain an unlinked system of three linear differential equations. For the resonance coordinate $y_i^{(\tau)}$, the magnitude of the oscillation of the other two main coordinates should be of the order of the small parameters μ and ε.

Let us substitute series (2.6.27) and (2.6.27) into the differential equation (2.6.26), taking into consideration the dependences (2.6.19) and (2.6.30) and the expansion

$$\lambda_1 \cong 1 + \mu\frac{\alpha_{0,1}}{2} + \varepsilon\frac{\alpha_{1,0}}{2} + \dots. \tag{2.6.31}$$

After equating to zero the coefficients of the same powers ε and μ, we obtain the system of recurrent differential equations

$$y_{0,0}^{(1)\prime\prime} + y_{0,0}^{(1)} = 0;$$

$$y_{1,0}^{(1)\prime\prime} + y_{1,0}^{(1)} = -\alpha_{1,0}y_{0,0}^{(1)} + \frac{g}{p_1^2} - g\bar{\chi}_1 v_0^1 + g\bar{\chi}_1\beta_1 y_{0,0}^{(1)\prime} + g\omega\rho(v_0^1)^3$$
$$-3g\omega\rho(v_0^1)^2\beta_1 y_{0,0}^{(1)\prime} + 3g\omega\rho v_0^1\beta_1^2\left(y_{0,0}^{(1)\prime}\right)^2 + g\omega\rho\beta_1^3\left(y_{0,0}^{(1)\prime}\right)^3;$$

$$y_{0,1}^{(1)\prime\prime} + y_{0,1}^{(1)} = -\alpha_{0,1}y_{0,0}^{(1)} + \gamma_1\varepsilon_1 y_{0,0}^{(1)}\cos 2\tau$$
$$-\bar{H}_1\beta_2 y_{0,0}^{(1)\prime} + \gamma_1 P\sin\tau + \gamma_1 Q\cos\tau;$$

$$y_{1,0}^{(2)''} + \nu_{2,1}^{(2)} y_{1,0}^{(2)} = g \frac{v_{2,1}^2}{P_2^2 g v_{2,1}^2 \bar{X}_2 v_0''} + g\nu_{2,1}\bar{X}_2\beta_1 y_{0,0}^{(1)'} \tag{2.6.32}$$

$$+ g\omega\rho v_{2,1}^3 (v_0'')^3 + 3g\omega v_{2,1}^2 (v_0'')^2 \beta_1 y_{0,0}^{(1)'}$$

$$+ 3g\omega\rho v_{2,1} v_0'' \beta_1^2 \left(y_{0,0}^{(1)'}\right)^2 + g\omega\rho\beta_1^3 \left(y_{0,0}^{(1)'}\right)^3;$$

$$y_{0,1}^{(2)''} + \nu_{2,1}^{(2)} y_{0,1}^{(2)} = \gamma_2 v_{2,1}^2 \varepsilon_1 y_{0,0}^{(1)} \cos 2\tau - \nu_{2,1}\bar{H}_2\beta_1 y_{0,0}^{(1)'}$$

$$+ \gamma_2 P \sin \tau + \gamma_2 Q \cos \tau;$$

$$y_{1,0}^{(3)''} + \nu_{3,1}^{(2)} y_{1,0}^{(3)} = 0;$$

$$y_{0,1}^{(3)''} + \nu_{3,1}^2 y_{0,1}^{(3)} = -\nu_{3,1}^{(1)} \varepsilon_1 y_{0,0}^{(1)} \sin 2\tau + P \cos \tau - Q \sin \tau + \nu_{3,1}^2 \bar{G}.$$

Assuming the solution of the first equation of the system (2.6.32) in the form

$$y_{0,0}^{(1)} = a_1 \cos \tau + b_1 \sin \tau \tag{2.6.33}$$

we obtain from the second equation:

$$y_{1,0}^{(1)''} + y_{1,0}^{(1)} = \frac{p_1^2}{g} - g\bar{X}_1 v_0^1 + g\omega\rho(v_0') + \frac{3}{2}g\omega\rho(v_0')\beta_1^2(a_1^2 + b_1^2)$$

$$+ \cos \tau \left[-\alpha_{1,0}a_1 + g\bar{X}\beta_1 b_1 - 3g\omega\rho(v_0')^2\beta_1 b_1 - \frac{3}{4}g\omega\rho\beta_1^3 b_1^3 \right.$$

$$\left. + \frac{3}{4}g\omega\rho\beta_1^3 b_1 a_1^2 \right] + \sin \tau \left[-\alpha_{1,0}b_1 - g\bar{X}_1\beta_1 a_1 \right.$$

$$\left. + 3g\omega\rho(v_0')^2\beta_1 a_1 + \frac{3}{4}g\omega\rho\beta_1^3 b_1^2 a_1 + \frac{3}{4}g\omega\rho\beta_1^3 a_1^3 \right] \tag{2.6.34}$$

$$+ \frac{3}{2}g\omega\beta v_0'\beta_1^2(b_1^2 - a_1^2) \cos 2\tau + 3g\omega\rho v_0'\beta_1^2 a_1 b_1 \sin 2\tau$$

$$+ \cos 3\tau \left[-\frac{1}{4}g\omega\rho\beta_1^3 b_1^3 + \frac{3}{4}g\omega\rho\beta_1^3 b_1 a_1^2 \right]$$

$$+ \sin 3\tau \left[-\frac{1}{4}g\omega\rho\beta_1^3 a_1^3 + \frac{3}{4}g\omega\rho\beta_1^3 b_1^2 a_1 \right].$$

From the condition of periodicity we obtain two algebraic equations

$$-\alpha_{1,0}a_1 + \left(g\bar{X}_1\beta_1 - 3g\omega\rho(v_0')^2\beta_1 - \frac{3}{4}g\omega\rho\beta_1^3 A_1^2 \right) b_1 = 0,$$

$$-\left(g\bar{X}_1\beta_1 - 3g\omega\rho(v_0')^2\beta_1 - \frac{3}{4}g\omega\rho\beta_1^3 A_1^2 \right) a_1 - \alpha_{1,0}b_1 = 0, \tag{2.6.35}$$

where $A_1^2 = a_1^2 + b_1^2$.

For non-zero a_1 and b_1, the following relation must occur

$$\begin{vmatrix} -\alpha_{1,0} & g\bar{X}_1\beta_1 - 3g\omega\rho(v_0')^2\beta_1 - \frac{3}{4}g\omega\rho\beta_1^3 A_1^2 \\ g\bar{X}_1\beta_1 - 3g\omega\rho(v_0')^2\beta_1 - \frac{3}{4}g\omega\rho\beta_1^3 A_1^2 & -\alpha_{1,0} \end{vmatrix} = 0. \tag{2.6.36}$$

Hence,

$$\alpha_{1,0}^2 + \left(g\bar{X}_1\beta_1 - 3g\omega\rho(v_0')^2\beta_1 - \frac{3}{4}g\omega\rho\beta_1^3 A_1^2 \right)^2 = 0. \tag{2.6.37}$$

The only real solution of (2.6.37) is

$$\alpha_{1,0} = 0,$$

$$A_1^2 = \frac{\bar{\chi} - 3\omega\rho(v_0')^2}{\frac{3}{4}\omega\rho\beta_1^2}. \tag{2.6.38}$$

The following function is the solution of (2.6.34)

$$
\begin{aligned}
y_{1,0}^{(1)} = & \frac{g}{p_1^2} - g\bar{\chi}_1 v_0' + g\omega\rho(v_0')^3 + \frac{3}{2}g\omega\rho v_0'\beta_1^2 A_1^2 \\
& + \frac{1}{2}g\omega\rho v_0'\beta_1^2(b_1^2 - a_1^2)\cos 2\tau + g\omega\rho v_0'\beta_1^2 a_1 b_1 \sin 2\tau \\
& + \left(\frac{1}{32}g\omega\rho\beta_1^3 b_1^3 + \frac{3}{32}g\omega\rho\beta_1^3 b_1 a_1^2\right)\cdot\cos 3\tau \\
& + \left(\frac{1}{32}g\omega\rho\beta_1^3 a_1^3 - \frac{3}{32}g\omega\rho\beta_1^3 b_1^2 a_1\right)\sin 3\tau.
\end{aligned} \tag{2.6.39}
$$

The solution omits the general integral of the homogeneous equation by associating it with $y_{0,0}^{(1)}$.

When we substitute (2.6.33) into the fourth and sixth equation of the equation system (2.6.32), after transformations, we obtain

$$
\begin{aligned}
y_{1,0}^{(2)\prime\prime} + \nu_{2,1}^2 y_{1,0}^{(2)} = & \; \nu_{2,1}^2\frac{g}{p_2^2} - \nu_{2,1}^2 g\bar{\chi}_2 v_0'' + g\omega\rho_{2,1}^3(v_0'')^3 \\
& + \frac{3}{2}\nu_{2,1}g\omega\rho v_0''\beta_1^2 A_1^2 + \cos\tau\left[\nu_{2,1}g\bar{\chi}_2\beta_1 b_1\right. \\
& \left. -3g\omega\rho\nu_{2,1}^2(v_0'')^2\beta_1 b_1 + \frac{3}{4}g\omega\rho\beta_1^3 A_1^2 b_1\right] \\
& + \sin\tau\left[-\nu_{2,1}g\bar{\chi}_2\beta_1\alpha_1 + 3g\omega\rho\nu_{2,1}^2(v_0'')^2\beta_1 a_1\right. \\
& \left. + \frac{3}{4}g\omega\rho\beta_1^3 A_1^2 a_1\right] + \frac{3}{2}\nu_{2,1}g\omega\rho v_0''\beta_1^2(b_1^2 - a_1^2)\cos 2\tau \\
& + 3\nu_{2,1}g\omega\rho v_0''\beta_1^2 a_1 b_1\sin 2\tau \\
& + \left(-\frac{1}{4}g\omega\rho\beta_1^3 b_1^3 + \frac{1}{2}g\omega\rho\beta_1^3 b_1 a_1^2\right)\cos 3\tau \\
& + \left(-\frac{1}{4}g\omega\rho\beta_1^3 a_1^3 + \frac{3}{4}g\omega\rho\beta_1^3 b_1^2 a_1\right)\sin 3\tau;
\end{aligned} \tag{2.6.40}
$$

$$y_{1,0}^{(3)\prime\prime} + \nu_{3,1}^2 y_{1,0}^{(3)} = 0. \tag{2.6.41}$$

The following functions are the particular solutions of the above equations

$$
\begin{aligned}
y_{1,0}^{(2)} = & \; \frac{g}{p_2^2}\bar{\chi}_2 v_0'' + g\omega\rho\nu_{2,1}(v_0'')^3 + \frac{3}{2\nu_{2,1}}g\omega\rho v_0''\beta_1^2 A_1^2 \\
& + \frac{1}{\nu_{2,1}^2 - 1}\left[\nu_{2,1}g\bar{\chi}_2\beta_1 - 3g\omega\rho\nu_{2,1}^2(v_0'')^3\beta_1 - \frac{3}{4}g\omega\rho\beta_1^3 A_1^3\right]b_1\cos\tau \\
& + \frac{1}{\nu_{2,1}^2 - 1}\left[-\nu_{2,1}g\bar{\chi}_2\beta_1 a_1 + 3g\omega\rho\nu_{2,1}^2(v_0'')^2\beta_1 a_1\right.
\end{aligned} \tag{2.6.42}
$$

$$\left. +\frac{3}{4}g\omega\rho\beta_1^3 A_1^2 a_1 \right] \sin\tau + \frac{3}{2(\nu_{2,1}^2 - 4)}g\omega\rho v_0'' \beta_1^2 a_1 b_1 \sin 2\tau$$

$$+\frac{1}{\nu_{2,1}^2 - g}\left(-\frac{1}{4}g\omega\rho\beta_1^3 a_1^3 + \frac{3}{4}g\omega\rho\beta_1^3 b_1 a_1^2\right)\cos 3\tau$$

$$+\frac{1}{\nu_{2,1}^2 - g}\left(-\frac{1}{4}g\omega\rho\beta_1^3 a_1^3 + \frac{3}{4}g\omega\rho\beta_1^3 b_1^2 a_1\right)\sin 3\tau$$

and

$$y_{1,0}^{(3)} = 0. \tag{2.6.43}$$

By substituting (2.6.39) into the third equation of the system (2.6.32), we obtain

$$y_{0,1}^{(1)\prime\prime} + y_{0,1}^{(1)} = \left(-\alpha_{0,1}a_1 + \frac{1}{2}\gamma_1\varepsilon_1 a_1 - \bar{H}_1\beta_1 b_1 + \gamma_1 Q\right)\cos\tau$$

$$+\left(-\alpha_{0,1}b_1 + \frac{1}{2}\gamma_1\varepsilon_1 b_1 - \bar{H}_1\beta_1 a_1 + \gamma_1 P\right)\sin\tau \tag{2.6.44}$$

$$+\frac{1}{2}\gamma_1\varepsilon_1 a_1 \cos 3\tau + \frac{1}{2}\gamma_1\varepsilon_1 b_1 \sin 3\tau.$$

We avoid terms unrestrictedly growing in time in its solution if the following equations are fulfilled

$$\left(\alpha_{0,1} - \frac{1}{2}\gamma_1\varepsilon_1\right)a_1 + \bar{H}_1\beta_1 b_1 = \gamma_1 Q,$$

$$-\bar{H}_1\beta_1 a_1 + \left(\alpha_{0,1} + \frac{1}{2}\gamma_1\varepsilon_1\right)b_1 = \gamma_1 P. \tag{2.6.45}$$

For the case $P = Q$, after transformations, we obtain from (2.6.45):

$$\alpha_{0,1} = \pm\left(\frac{P^2}{A_1^2}\gamma_1^2 + \frac{1}{4}\gamma_1^2\varepsilon_1^2 - \bar{H}_1^2\beta_1^2\right.$$

$$\left.\pm\sqrt{(\frac{P}{A_1}\gamma_1)^4 + \gamma_1^4\varepsilon_1^2\frac{P^2}{A_1^2} - 2\bar{H}_1\beta_1\frac{P^2}{A_1^2}\gamma_1^3\varepsilon_1}\right)^{\frac{1}{2}}. \tag{2.6.46}$$

The particular solution of (2.6.44) is

$$y_{0,1}^{(1)} = -\frac{1}{16}\gamma_1\varepsilon_1(a_1 \cos 3\tau + b_1 \sin 3\tau). \tag{2.6.47}$$

Taking (2.6.33) into consideration in the fifth and seventh equation of the system (2.6.32), we find the particular solutions

$$y_{0,1}^{(2)} = \frac{1}{\nu_{2,1}^2 - 1}\left(\frac{1}{2}\nu_{2,1}^2\gamma_2\varepsilon_1 a_1 - \nu_{2,1}\bar{H}_2\beta_1 b_1 + \gamma_2 Q\right)\cos\tau$$

$$+\frac{1}{\nu_{2,1}^2 - 1}\left(-\frac{1}{2}\nu_{2,1}^2\gamma_2\varepsilon_1 b_1 + \nu_{2,1}\bar{H}_2\beta_1 a_1 + \gamma_2 P\right)\sin\tau$$

$$+\frac{\nu_{2,1}^2\gamma_2\varepsilon_1}{2(\nu_{2,1}^2-9)}a_1\cos 3\tau+\frac{\nu_{2,1}^2\gamma_2\varepsilon_1}{2(\nu_{2,1}^2-9)}b_1\sin 3\tau; \tag{2.6.48}$$

$$y_{0,1}^{(3)}=\bar{G}+\frac{1}{\nu_{3,1}^2-1}\left(-\frac{1}{2}\nu_{3,1}^2\varepsilon_1 b_1+P\right)\cos\tau$$

$$+\frac{1}{\nu_{3,1}^2-1}\left(\frac{1}{2}\nu_{3,1}^2\varepsilon_1 a_1+Q\right)\sin\tau$$

$$+\frac{\nu_{3,1}^2\varepsilon_1}{2(\nu_{3,1}^2-9)}b_1\cos 3\tau+\frac{\nu_{3,1}^2\varepsilon_1}{2(\nu_{3,1}^2-9)}a_1\sin 3\tau. \tag{2.6.49}$$

We have thus determined the particular terms of the series (2.6.27) and (2.6.28), limiting the calculations to the first approximation.

Let us now concentrate on the analysis of the case $\omega\cong p_2$. The solutions will be sought, as has been done previously, in the form of the series (2.6.27) and (2.6.28) for $i=2$. From (2.6.28) we obtain

$$\lambda_2=1+\varepsilon\frac{\alpha_{1,0}}{2}+\mu\frac{\alpha_{0,1}}{2}+\dots. \tag{2.6.50}$$

Let us denote

$$\lambda_1^2=\nu_{1,2}^2\lambda_2^2,$$
$$\lambda_3^2=\nu_{3,2}^2\lambda_2^2, \tag{2.6.51}$$

where $\nu_{1,2}=p_1/p_2$, $\nu_{3,2}=p_3/p_2$, and $\nu_{1,2}$ and $\nu_{3,2}$ are not integers.

Analogously to (2.6.30), we have

$$y_{0,0}^{(1)}(\tau)=y_{0,0}^{(3)}=0. \tag{2.6.52}$$

Substituting (2.6.27) into (2.6.26), with (2.6.50), (2.6.3) and (2.6.52) taken into account, after equating to zero the coefficients of the same powers of μ and ε, we obtain

$$y_{1,0}^{(1)\prime\prime}+\nu_{1,2}^2 y_{1,0}^{(1)}=\frac{g}{p_2^2}-g\bar{\chi}\nu_{1,2}^2 v_0'-g\bar{\chi}\nu_{1,2}\beta_2 y_{0,0}^{(2)\prime}+g\omega\rho\nu_{1,2}^3(v_0')^3$$

$$+3g\omega\rho\beta_2\nu_{1,2}^2(v_0')^2 y_{0,0}^{(2)\prime}+3g\omega\rho\nu_{1,2}\beta_2^2 v_0'\left(y_{0,0}^{(2)\prime}\right)^2$$

$$+g\omega\rho\beta_2^3\left(y_{0,0}^{(2)\prime}\right)^3;$$

$$y_{0,1}^{(1)\prime\prime}+\nu_{1,2}^2 y_{0,1}^{(1)}=-\gamma_1\nu_{1,2}^2\varepsilon_2 y_{0,0}^{(2)}\cos 2\tau+\nu_{1,2}\bar{H}_1\beta_2 y_{0,0}^{(2)\prime}$$

$$+\beta_1 P\sin\tau+\gamma_1 Q\cos\tau;$$

$$y_{0,0}^{(2)\prime\prime}+y_{0,0}^{(2)}=0; \tag{2.6.53}$$

$$y_{1,0}^{(2)\prime\prime}+y_{1,0}^{(2)}=-\alpha_{1,0}y_{0,0}^{(2)}+\frac{g}{p_2^2}-g\bar{\chi}_2 v_0''-g\bar{\chi}_2\beta_2 y_{0,0}^{(2)\prime}+g\omega\rho(v_0'')^3$$

$$+3g\omega\rho(v_0'')^2\beta_2 y_{0,0}^{(2)\prime}+3g\omega\rho v_0''\beta_2^2\left(y_{0,0}^{(2)\prime}\right)^2+g\omega\rho\beta_2^3\left(y_{0,0}^{(2)\prime}\right)^3;$$

$$y_{0,1}^{(2)\prime\prime}+y_{0,1}^{(2)}=-\alpha_{0,1}y_{0,0}^{(2)}-\gamma_2\varepsilon_2 y_{0,0}^{(2)}\cos 2\tau+\bar{H}_2\beta_2 y_{0,0}^{(2)\prime}$$

$$+\gamma_2 P\sin\tau+\gamma_2 Q\cos\tau;$$

$$y_{1,0}^{(3)''} + \nu_{3,2}^2 y_{1,0}^{(3)} = 0;$$

$$y_{0,1}^{(3)''} + \nu_{3,2}^3 y_{0,1}^{(3)} = \nu_{3,2}^2 \varepsilon_2 y_{0,0}^{(2)} \sin 2\tau + P \cos \tau - Q \sin \tau + \nu_{3,2}^2 \bar{G}.$$

After substituting the following expression in the fourth equation of the system (2.6.53)

$$y_{0,0}^{(2)} = a_2 \cos \tau + b_2 \sin \tau \tag{2.6.54}$$

and using the trigonometric relations, we obtain

$$y_{1,0}^{(2)''} + y_{1,0}^{(2)} = \frac{g}{p_2^2} - g\bar{\chi}_2 v_0'' + gw\rho(v_0'')^3 + \frac{3}{2}gw\rho(v_0'')^2(a_2^2 + b_2^2)$$

$$+\left(-\alpha_{1,0}a_2 - g\bar{\chi}_2\beta_2 b_2 + 3gw\rho(v_0'')^2\beta_1 b_1 + \frac{3}{4}gw\rho\beta_2^3 b_2^3\right.$$

$$\left.+\frac{3}{4}gw\rho\beta_2^3 b_2 a_2^2\right)\cos\tau + \left(-\alpha_{1,0}b_2 + g\bar{\chi}_2\beta_2 a_2\right.$$

$$\left.-3gw\rho(v_0'')^2\beta_2 a_2 + \frac{3}{4}gw\rho\beta_2^3 b_2^2 a_2 - \frac{3}{4}gw\rho\beta_2^3 a_2^3\right)\sin\tau$$

$$+\frac{3}{2}gw\rho v_0''\beta_2^2(b_2^2 - a_2^2)\cos 2\tau$$

$$+3gw\rho v_0''\beta_2^2 a_2 b_2 \sin 2\tau + \frac{1}{4}(b_2^2 - 3a_2^2)gw\rho\beta_2^3 b_2 \cos 3\tau$$

$$+\frac{1}{4}(a_2^2 - 3b_2^2)gw\rho\beta_2^3 a_2 \sin 3\tau. \tag{2.6.55}$$

From the condition of periodicity of the solution, we get

$$-\alpha_{1,0}a_2 + \left(-g\bar{\chi}_2\beta_2 + 3gw\rho(v_0'')^2\beta_2 + \frac{3}{4}gw\rho\beta_2^3 A_2^2\right)b_2 = 0,$$

$$\left(g\bar{\chi}_2\beta_2 - 3gw\rho(v_0'')^2\beta_2 - \frac{3}{4}gw\rho\beta_2^3 A_2^2\right)a_2 - \alpha_{1,0}b_2 = 0, \tag{2.6.56}$$

where $A_2^2 = a_2^2 + b_2^2$.

For the non-zero a_2 and b_2 the main determinant of equation system (2.6.56) must equal zero. From this condition we obtain

$$\alpha_{1,0} = 0,$$

$$A_2^2 = \frac{\bar{\chi}_2 - 3w\rho(v_0'')^2}{\frac{3}{4}w\rho\beta_2^2}. \tag{2.6.57}$$

The particular solution of (2.6.55) is

$$y_{1,0}^{(2)} = \frac{g}{p_2^2} - g\bar{\chi}_2 v_0'' + gw\rho(v_0'')^3 + \frac{3}{2}gw\rho v_0''\beta_2^2 A_2^2$$

$$-\frac{1}{2}gw\rho v_0''\beta_2^2(b_2^2 - a_2^2)\cos 2\tau + gw\rho v_0''\beta_2^2 a_2 b_2 \sin 2\tau \tag{2.6.58}$$

$$+\frac{1}{32}(3a_2^2 - b_2^2)gw\rho\beta_2^3 b_2 \cos 3\tau + \frac{1}{32}(3b_2^2 - a_2^2)gw\rho\beta_2^3 a_2 \sin 3\tau.$$

Making use of (2.6.54) in the first and sixth equation of system (2.6.53), we obtain their particular integrals

$$y_{1,0}^{(1)} = \frac{g}{p_1^2} - g\bar{\chi}_1 v_0' + g\omega\rho\nu_{1,2}(v_0')^3 + \frac{3}{2\nu_{1,2}}g\omega\rho v_0'\beta_2^2 A_2^2$$

$$+ \frac{1}{\nu_{1,2}^2 - 1}\left(-g\bar{\chi}_1\nu_{1,2}\beta_2 + 3g\omega\rho\beta_2\nu_{1,2}^2(v_0')^2 + \frac{3}{4}g\omega\rho\beta_2^3 A_2^2\right)b_2\cos\tau$$

$$+ \frac{1}{\nu_{1,2}^2 - 1}\left(g\bar{\chi}_1\nu_{1,2}\beta_2 - 3g\omega\rho\beta_2\nu_{1,2}^2(v_0')^2 + \frac{3}{4}g\omega\rho\beta_2^3 A_2^2\right)a_2\sin\tau$$

$$+ \frac{3\nu_{1,2}}{2(\nu_{1,2}^2 - 1)}g\omega\rho\beta_2^2 v_0'(b_2^2 - a_2^2)\cos 2\tau \qquad (2.6.59)$$

$$+ \frac{3\nu_{1,2}}{2(\nu_{1,2}^2 - 1)}g\omega\rho\beta_2^2 v_0' a_2 b_2 \sin 2\tau$$

$$+ \frac{1}{4}(b_2^2 - 3a_2^2)\frac{g\omega\rho\beta_2^3}{\nu_{1,2}^2 - 9}b_2\cos 3\tau + \frac{1}{4}(a_2^2 - 3b_2^2)\frac{g\omega\rho\beta_2^3}{\nu_{1,2}^2 - 9}a_2\sin 3\tau$$

and

$$y_{1,0}^{(3)} = 0. \qquad (2.6.60)$$

The substitution of (2.6.54) into the fifth equation of system (2.6.53) gives

$$y_{0,1}^{(2)} + y_{0,1}^{(2)} = \left(-\alpha_{0,1}a_2 - \frac{1}{2}\gamma_2\varepsilon_2 a_2 + \bar{H}_2\beta_2 b_2 + \gamma_2 Q\right)\cos\tau$$

$$+ \left(-\alpha_{0,1}b_2 + \frac{1}{2}\gamma_2\varepsilon_2 b_2 - \bar{H}_2\beta_2 a_2 + \gamma_2 P\right)\sin\tau$$

$$+ \frac{1}{2}\gamma_2\varepsilon_2 a_2\cos 3\tau - \frac{1}{2}\gamma_2\varepsilon_2 b_2\sin 3\tau. \qquad (2.6.61)$$

The following expression is obtained from the condition of periodicity after transformations and after assuming that $P = Q$:

$$\alpha_{0,1}^4 + 2\alpha_{0,1}^2\left(\bar{H}_2^2\beta_2^2 - \frac{1}{4}\gamma_2^2\varepsilon_2^2 - \frac{P^2}{A_2^2}\gamma_2^2\right) + \left(\bar{H}_2^2\beta_2^2 - \frac{1}{4}\gamma_2^2\varepsilon_2^2\right)^2$$

$$+ \frac{1}{2}\gamma_2^4\varepsilon_2^2\frac{P^2}{A_2^2} - 2\gamma_2^2\beta_2^2\bar{H}_2\frac{P^2}{A_2^2}\left(\bar{H}_2\beta_2 - \gamma_2\varepsilon_2\right) = 0. \qquad (2.6.62)$$

Hence

$$\alpha_{0,1} = \pm\left(\frac{P^2}{A_2^2}\gamma_2^2 + \frac{1}{4}\gamma_2^2\varepsilon_2^2 - \bar{H}_2^2\beta_2^2\right.$$

$$\left.\pm\sqrt{\left(\frac{P}{A_2}\gamma_2\right)^4 + \gamma_2^4\varepsilon_2^2\frac{P^2}{A_2^2} - a\bar{H}_2\beta_2\frac{P^2}{A_2^2}\gamma_2^3\varepsilon_2}\right)^{\frac{1}{2}}. \qquad (2.6.63)$$

The particular integral of (2.6.61) is the following

$$y_{0,1}^{(2)} = \frac{\gamma_2 \varepsilon_2}{16}(a_2 \cos 3\tau + b_2 \sin 3\tau). \tag{2.6.64}$$

On the other hand, after substituting (2.6.54) into the second and seventh equation of (2.6.53), we find the particular solutions

$$\begin{aligned}
y_{0,1}^{(1)} &= \frac{1}{\nu_{1,2}^2 - 1}\left(-\frac{1}{2}\nu_{1,2}^2\gamma_1\varepsilon_2 a_2 + \nu_{1,2}\bar{H}_1\beta_2 b_2 + \gamma_1 Q\right)\cos\tau \\
&+ \frac{1}{2\nu_{1,2}^2 - 1}\left(\frac{1}{2}\nu_{1,2}^2\gamma_1\varepsilon_2 b_2 - \nu_{1,2}\bar{H}_1\beta_2 a_2 + \gamma_1 P\right)\sin\tau \\
&- \frac{\nu_{1,2}\gamma_1\varepsilon_2}{2(\nu_{1,2}^2 - 9)}a_2 \cos 3\tau - \frac{\nu_{1,2}\gamma_1\varepsilon_2}{2(\nu_{1,2}^2 - 9)}b_2 \sin 3\tau;
\end{aligned} \tag{2.6.65}$$

$$\begin{aligned}
y_{0,1}^{(3)} &= \bar{G} + \frac{1}{\nu_{3,2}^2 - 1}\left(\frac{1}{2}\nu_{3,2}^2\varepsilon_2 b_2 + P\right)\cos\tau \\
&+ \frac{1}{\nu_{3,1}^2 - 1}\left(\frac{1}{2}\nu_{3,2}^2\varepsilon_2 a_2 - Q\right)\sin\tau \\
&+ \frac{\nu_{3,2}^2\varepsilon_2}{2(\nu_{3,2}^2 - 9)}b_2 \cos 3\tau - \frac{\nu_{3,2}^2\varepsilon_2}{2(\nu_{3,2}^2 - 9)}a_2 \sin 3\tau.
\end{aligned} \tag{2.6.66}$$

Finally, let us consider the case of $\omega \cong p_3$. Periodic solutions are possible for the particular value of the parameter λ_3:

$$\lambda_3 \cong 1 + \varepsilon\frac{\alpha_{1,0}}{2} + \mu\frac{\alpha_{0,1}}{2} + \varepsilon^2\frac{\alpha_{2,0}}{2} + \mu^2\frac{\alpha_{0,2}}{2} + \varepsilon\mu\frac{\alpha_{1,1}}{2} + \ldots \tag{2.6.67}$$

Let us denote

$$\begin{aligned}
\lambda_1^2 &= \nu_{1,3}^2\lambda_3^2, \\
\lambda_2^2 &= \nu_{2,3}^2\lambda_3^2,
\end{aligned} \tag{2.6.68}$$

where $\nu_{1,3} = p_1/p_3$, $\nu_{2,3} = p_2/p_3$, and $\nu_{1,3}$ and $\nu_{2,3}$ are assumed not to be integers. Similarly to the previous considerations, assuming that

$$y_{0,0}^{(1)}(\tau) = y_{0,0}^{(2)}(\tau) = 0 \tag{2.6.69}$$

we obtain the following recurrent differential equation system from equation system (2.6.26)

$$y_{0,0}^{(1)\prime\prime} + \nu_{1,3}^2 y_{1,0}^{(1)} = (g + \Omega_1^2 y_{0,0}^{(3)})\left[\frac{\nu_{1,3}^2}{p_1^2} - \nu_{1,3}^2\bar{\chi}_1 v_0' + \omega\rho\nu_{1,3}^3(v_0')^3\right];$$

$$y_{0,1}^{(1)\prime\prime} + \nu_{1,3}^2 y_{0,1}^{(1)} = \gamma_1\nu_{1,3}^2 y_{0,0}^{(3)}\sin 2\tau + \gamma_1(P\sin\tau + Q\cos\tau);$$

$$\begin{aligned}
y_{2,0}^{(1)\prime\prime} + \nu_{1,3}^2 y_{2,0}^{(1)} &= -\nu_{1,3}^2\alpha_{1,0}y_{1,0}^{(1)} + (g + \Omega_1^2 y_{0,0}^{(3)})\left[\frac{\alpha_{1,0}}{p_3^2} - \nu_{1,3}^2\bar{\chi}_1\alpha_{1,0}v_0'\right. \\
&\left. -\nu_{1,3}\bar{\chi}_1(-\beta_1 y_{1,0}^{(1)\prime} + \beta_2 y_{1,0}^{(2)\prime} + \omega\rho\nu_{1,3}^2(v_0')^2\left(\frac{\alpha_{1,0}}{2}\nu_{1,3}v_0'\right.\right.
\end{aligned}$$

$$+ \beta_1 y_{1,0}^{(2)\prime}\Big)\Big] + \Omega_1^2 y_{1,0}^{(3)}\left[\frac{1}{p_3^2} - \nu_{1,3}^2 \bar{\chi}_1 v_0' + \omega \rho \nu_{1,3}^3 (v_0')^3\right];$$

$$y_{0,2}^{(1)\prime\prime} + \nu_{1,3}^2 y_{0,2}^{(1)} = -\nu_{1,3}^2 \alpha_{0,1} y_{0,1}^{(1)} + \nu_{1,3}^2 \gamma_1 \left(\varepsilon_1 y_{0,1}^{(1)} - \varepsilon_2 y_{0,1}^{(2)}\right) \cos 2\tau$$

$$+ \nu_{1,3} \bar{H}_1 \left(\beta_1 y_{0,1}^{(1)} - \beta_2 y_{0,1}^{(2)\prime}\right)$$

$$+ \nu_{1,3}^2 \gamma_1 \left(\alpha_{0,1} y_{0,0}^{(3)} + y_{0,1}^{(3)}\right) \sin 2\tau;$$

$$y_{1,1}^{(1)\prime\prime} + \nu_{1,3}^2 y_{1,1}^{(1)} = -\nu_{1,3}^2 \left(\alpha_{0,1} y_{0,0}^{(1)} + \alpha_{1,0} y_{0,1}^{(1)}\right) + \nu_{1,3}^2 \gamma_1 \left(\varepsilon_1 y_{1,0}^{(1)}\right.$$

$$\left. - \varepsilon_2 y_{1,0}^{(2)}\right) \cos 2\tau + \nu_{1,3} \bar{H}_1 \left(\beta_1 y_{1,0}^{(1)\prime} - \beta_2 y_{1,0}^{(2)\prime}\right)$$

$$+ \nu_{1,3}^2 \gamma_1 \left(\alpha_{1,0} y_{0,0}^{(3)} + y_{1,0}^{(3)}\right) \sin 2\tau + (g + \Omega_1^2 y_{0,0}^{(3)})\left[\frac{\alpha_{0,1}}{p_3^2}\right.$$

$$\left. - \nu_{1,2}^2 \bar{\chi}_1 \alpha_{0,1} v_0' - \nu_{1,3} \bar{\chi}_1 (-\beta_1 y_{0,1}^{(1)\prime} + \beta_2 y_{0,1}^{(2)})\right]$$

$$+ \omega \rho \nu_{1,3}^2 (v_0')^2 \left(\nu_{1,3}^2 v_0' \frac{\alpha_{0,1}}{2} - \beta_1 y_{0,1}^{(1)\prime} + \beta_2 y_{0,1}^{(2)\prime}\right) \quad (2.6.70)$$

$$+ \Omega_1^2 (y_{0,1}^{(3)} + y_{0,0}^{(3)} \cos 2\tau)\left(\frac{1}{p_3^2} - \nu_{1,3}^2 \bar{\chi}_1 v_0' + \omega \rho \nu_{1,3}^3 (v_0')^3\right);$$

$$y_{1,0}^{(2)\prime\prime} + \nu_{2,3}^2 y_{1,0}^{(2)} = (g + \Omega_1^2 y_{0,0}^{(3)})\left[\frac{\nu_{2,3}^2}{p_2^2} - \nu_{2,3} \bar{\chi}_2 v_0'' + \omega \rho \nu_{2,3}^2 (v_0'')^3\right];$$

$$y_{0,1}^{(2)\prime\prime} + \nu_{2,3}^2 y_{0,1}^{(2)} = \nu_{2,3}^2 \gamma_2 y_{0,0}^{(3)} \sin 2\tau + \gamma_2 (P \sin \tau + Q \cos \tau);$$

$$y_{2,0}^{(2)\prime\prime} + \nu_{2,3}^2 y_{2,0}^{(2)} = -\nu_{2,3}^2 \alpha_{1,0} y_{1,0}^{(2)} + (g + \Omega_1^2 y_{0,0}^{(3)})\left[\frac{\alpha_{1,0}}{p_3^2} - \nu_{2,3}^2 \bar{\chi}_2 v_0'' \alpha_{1,0}\right.$$

$$- \nu_{2,3}^2 \bar{\chi}_2 (-\beta_1 y_{1,0}^{(1)\prime} + \beta_2 y_{1,0}^{(2)\prime}) + \omega \rho \nu_{2,3}^2 (v_0'')^2$$

$$\left. \cdot \left(\frac{\alpha_{1,0}}{2} \nu_{2,3} v_0'' - \beta_1 y_{1,0}^{(1)\prime} + \beta_2 y_{1,0}^{(2)\prime}\right)\right]$$

$$+ \Omega_1^2 y_{1,0}^{(3)}\left[\frac{1}{p_3^2} - \nu_{2,3}^2 \bar{\chi}_2 v_0'' + \omega \rho \nu_{2,3}^2 (v_0'')^3\right];$$

$$y_{0,2}^{(2)\prime\prime} + \nu_{2,3}^2 y_{0,2}^{(2)} = -\nu_{2,3}^2 \alpha_{0,1} y_{0,1}^{(2)} + \nu_{2,3}^2 \gamma_2 (\varepsilon_1 y_{0,1}^{(1)} - \varepsilon_2 y_{0,1}^{(2)}) \cos 2\tau$$

$$+ \nu_{2,3} \bar{H}_2 (\beta_1 y_{0,1}^{(1)\prime} - \beta_2 y_{0,1}^{(2)\prime})$$

$$+ \nu_{2,3}^2 \gamma_2 (\alpha_{0,1} y_{0,0}^{(3)} + y_{0,1}^{(3)}) \sin 2\tau;$$

$$y_{1,1}^{(2)\prime\prime} + \nu_{2,3}^2 y_{1,1}^{(2)} = -\nu_{2,3}^2 (\alpha_{0,1} y_{1,0}^{(2)} + \alpha_{1,0} y_{0,1}^{(2)}) + \nu_{2,3}^2 \gamma_2 (\varepsilon_1 y_{1,0}^{(1)}$$

$$+ \varepsilon_2 y_{1,0}^{(2)}) \cos 2\tau - \nu_{2,3} \bar{H}_2 (\beta_1 y_{1,0}^{(1)\prime} - \beta_2 y_{1,0}^{(2)})$$

$$+ \nu_{2,3}^2 \gamma_2 (\alpha_{1,0} y_{0,0}^{(3)} + y_{1,0}^{(3)}) \sin 2\tau$$

$$+ (g + \Omega_1^2 y_{0,0}^{(3)})\left[\frac{\alpha_{0,1}}{p_3^2} - \nu_{2,3}^2 \bar{\chi}_2 \alpha_{0,1} v_0'' - \nu_{2,3} \bar{\chi}_2 (-\beta_1 y_{0,1}^{(2)\prime}\right.$$

$$+\beta_2 y_{0,1}^{(2)\prime}) + \omega\rho\nu_{2,3}^2 (v_0'')^2 \left(\nu_{2,3}v_0'' \frac{\alpha_{0,1}}{2}\right.$$
$$\left.- \beta_1 y_{0,1}^{(1)\prime} + \beta_2 y_{0,0}^{(2)\prime}\right)\Big] + \Omega_1^2 (y_{0,1}^{(3)} - y_{0,0}^{(3)}\cos 2\tau)$$
$$\cdot \left[\frac{1}{p_3^2} - \nu_{2,3}^2 \bar{\chi}_2 v_0'' + \omega\rho\nu_{2,3}^2 (v_0'')^3\right];$$

$$y_{0,0}^{(3)\prime\prime} + y_{0,0}^{(3)} = 0;$$
$$y_{1,0}^{(3)\prime\prime} + y_{1,0}^{(3)} = -\alpha_{1,0}y_{0,0}^{(3)};$$
$$y_{0,1}^{(3)\prime\prime} + y_{0,1}^{(3)} = -\alpha_{0,1}y_{0,0}^{(3)} + y_{0,0}^{(3)}\cos 2\tau + P\cos\tau - Q\sin\tau + \bar{G};$$
$$y_{2,0}^{(3)\prime\prime} + y_{2,0}^{(3)} = -\alpha_{1,0}y_{1,0}^{(3)} - \alpha_{2,0}y_{0,0}^{(3)};$$
$$y_{0,2}^{(3)\prime\prime} + y_{0,2}^{(3)} = -\alpha_{0,1}y_{0,1}^{(3)} - \alpha_{0,2}y_{0,0}^{(3)} - \varepsilon_1 y_{0,1}^{(1)}\sin 2\tau$$
$$+\varepsilon_2 y_{0,1}^{(2)}\sin 2\tau + \alpha_{0,1}y_{0,0}^{(3)}\cos 2\tau + y_{0,1}^{(3)}\cos 2\tau + \alpha_{0,1}\bar{G};$$
$$y_{1,1}^{(3)\prime\prime} + y_{1,1}^{(3)} = -\alpha_{1,1}y_{0,0}^{(3)} - \alpha_{1,0}y_{0,1}^{(3)} - \alpha_{0,1}y_{1,0}^{(3)} - \varepsilon_1 y_{1,0}^{(3)}$$
$$-\varepsilon_1 y_{1,0}^{(1)}\sin 2\tau + \alpha_{0,1}y_{0,0}^{(3)}\cos 2\tau + y_{0,1}^{(3)}\cos 2\tau + \alpha_{0,1}\bar{G};$$
$$y_{1,1}^{(3)\prime\prime} + y_{1,1}^{(3)} = -\alpha_{1,1}y_{0,0}^{(3)} - \alpha_{1,0}y_{0,1}^{(3)} - \alpha_{0,1}y_{1,0}^{(3)} - \varepsilon_1 y_{1,0}^{(1)}\sin 2\tau$$
$$+\varepsilon_2 y_{1,0}^{(2)}\sin 2\tau + \alpha_{1,0}y_{0,0}^{(3)}\cos 2\tau + y_{1,0}^{(3)}\cos 2\tau + \alpha_{1,0}\bar{G}.$$

After substituting

$$y_{0,0}^{(3)} = a_3\cos\tau + b_3\sin\tau \tag{2.6.71}$$

into the twelfth equation of system (2.6.70) we get

$$y_{1,0}^{(3)\prime\prime} + y_{1,0}^{(3)} = -\alpha_{1,0}(a_3\cos\tau + b_3\sin\tau). \tag{2.6.72}$$

For non-zero a_3 and b_3 from the condition of periodicity we get

$$\alpha_{1,0} = 0. \tag{2.6.73}$$

Hence,

$$y_{1,0}^{(3)} = 0. \tag{2.6.74}$$

Making use of (2.6.71) in the first and sixth equation of (2.6.70), we obtain their particular solutions

$$y_{1,0}^{(1)} = \left[\frac{g}{\nu_{1,3}^2} + \frac{\Omega_1^2}{\nu_{1,3}^2 - 1}(a_3\cos\tau + b_3\sin\tau)\right]$$
$$\cdot \left[\frac{\nu_{1,3}^2}{p_1^2} - \nu_{1,3}^2\bar{\chi}_1 v_0 + \omega\rho\nu_{1,3}^2 (v_0')^3\right], \tag{2.6.75}$$

$$y_{1,0}^{(2)} = \left[\frac{g}{\nu_{2,3}^2} + \frac{\Omega_1^2}{\nu_{2,3}^2 - 1} (a_3 \cos \tau + b_3 \sin \tau) \right]$$
$$\cdot \left[\frac{\nu_{2,3}^2}{p_2^2} - \nu_{2,3}^2 \bar{\chi}_2 v_0'' + \omega \rho \nu_{2,3}^2 (v_0'')^3 \right]. \tag{2.6.76}$$

After substituting (2.6.71) into the thirteenth equation in system (2.6.70), we have

$$y_{0,1}^{(3)\prime\prime} + y_{0,1}^{(3)} = \bar{G} + \left(-\alpha_{0,1} a_3 + \frac{1}{2} a_3 + P \right) \cos \tau \tag{2.6.77}$$

$$+ \left(-\alpha_{0,1} b_3 - \frac{1}{2} b_3 - Q \right) \sin \tau + \frac{1}{2} (a_3 \cos 3\tau + b_3 \sin 3\tau).$$

The condition of periodicity gives

$$\alpha_{0,1}^{(1)} = \frac{1}{2} + \frac{P}{a_3},$$
$$\alpha_{0,1}^{(2)} = -\frac{1}{2} - \frac{Q}{b_3}. \tag{2.6.78}$$

The particular solution of this equation is the following

$$y_{0,1}^{(3)} = \bar{G} - \frac{1}{16} (a_3 \cos 3\tau + b_3 \sin 3\tau). \tag{2.6.79}$$

When substituting (2.6.71) into the second and seventh equation of system (2.6.70), we obtain their particular integrals

$$y_{0,1}^{(1)} = \frac{\gamma_1}{\nu_{1,3}^2 - 1} \left(\frac{\nu_{1,3}^2}{2} b_3 + Q \right) \cos \tau + \frac{\gamma_1}{\nu_{1,3}^2 - 1} \left(\frac{\nu_{1,3}^2}{2} a_3 + P \right) \sin \tau$$

$$+ \frac{\nu_{1,3}^2 \gamma_1}{2(\nu_{1,3}^2 - 9)} (a_3 \sin 3\tau - b_3 \cos 3\tau), \tag{2.6.80}$$

$$y_{0,1}^{(2)} = \frac{\gamma_2}{\nu_{2,3}^2 - 1} \left(\frac{\nu_{2,3}^2}{2} b_3 + Q \right) \cos \tau + \frac{\gamma_2}{\nu_{2,3}^2 - 1} \left(\frac{\nu_{2,3}^2}{2} a_3 + P \right) \sin \tau$$

$$+ \frac{\nu_{2,3}^2 \gamma_2}{2(\nu_{2,3}^2 - 9)} (a_3 \sin 3\tau - b_3 \cos 3\tau). \tag{2.6.81}$$

After substituting (2.6.71) and (2.6.73) into the fourteenth equation of system (2.6.70), we obtain the following from the condition of the existence of periodic solutions:

$$\alpha_{2,0} = 0. \tag{2.6.82}$$

Analogously, taking (2.6.71), (2.6.79), (2.6.80), (2.6.81) into account in the fifteenth equation of system (2.6.70), we obtain equations which, after transformations, will assume the form

$$\alpha_{0,1}^{(1)} = \frac{\gamma_1 \varepsilon_1 \nu_{1,3}^2}{4} \left(\frac{\nu_{1,3}^2 - 1}{2} + \frac{1}{\nu_{1,3}^2 - 9} \right)$$

$$+ \frac{\gamma_2 \varepsilon_2 \nu_{2,3}^2}{4} \left(\frac{1}{\nu_{2,3}^2 - 1} + \frac{1}{\nu_{2,3}^2 - 9} \right)$$

$$+ \frac{1}{2} \alpha_{0,1}^{(1)} - \frac{1}{32} - \frac{P}{a_3} \left(\frac{\varepsilon_1 \gamma_1}{2(\nu_{1,3}^2 - 1)} - \frac{\varepsilon_2 \gamma_2}{2(\nu_{2,3}^2 - 1)} \right), \qquad (2.6.83)$$

$$\alpha_{0,2}^{(2)} = -\frac{\gamma_1 \varepsilon_1 \nu_{1,3}^2}{4} \left(\frac{1}{\nu_{1,3}^2 - 1} + \frac{1}{\nu_{1,3}^2 - 9} \right)$$

$$+ \frac{\gamma_2 \varepsilon_2 \nu_{2,3}^2}{4} \left(\frac{1}{\nu_{2,3}^2 - 1} + \frac{1}{\nu_{2,3}^2 - 1} \right)$$

$$- \frac{1}{2} \alpha_{0,1}^{(2)} - \frac{1}{32} - \frac{Q}{b_3} \left(\frac{\varepsilon_1 \gamma_1}{2(\nu_{1,3}^2 - 1)} + \frac{\varepsilon_2 \gamma_2}{2(\nu_{2,3}^2 - 1)} \right). \qquad (2.6.84)$$

The following algebraic equation system will be obtained from the condition of periodicity of the solutions of equation system (2.6.70) after substituting (2.6.71), (2.6.74), (2.6.75), (2.6.76) and (2.6.79) into the sixteenth equation of system (2.6.70):

$$-\alpha_{1,1} a_3 - \frac{\varepsilon_1 c_1 \Omega_1^2}{2(\nu_{1,3}^2 - 1)} b_3 + \frac{\varepsilon_2 c_2 \Omega_1^2}{2(\nu_{2,3}^2 - 1)} b_3 = 0,$$

$$-\alpha_{1,1} b_3 - \frac{\varepsilon_1 c_1 \Omega_1^2}{2(\nu_{1,3}^2 - 1)} a_3 + \frac{\varepsilon_2 c_2 \Omega_1^2}{2(\nu_{2,3}^2 - 1)} a_3 = 0, \qquad (2.6.85)$$

where

$$c_1 = \frac{\nu_{1,3}^2}{p_1^2} - \nu_{1,3}^2 \bar{\chi}_1 v_0' + \omega \rho \nu_{1,3}^2 (v_0')^3,$$

$$c_2 = \frac{\nu_{2,3}^2}{p_2^2} - \nu_{2,3}^2 \bar{\chi}_2 v_0' + \omega \rho \nu_{2,3}^2 (v_0')^3. \qquad (2.6.86)$$

From the condition of a non-zero solution of equation system (2.6.85) in relation to a_3 and b_3, we obtain

$$\alpha_{1,1}^{(1,2)} = \pm \left[\frac{\Omega_1^2}{2} \left(\frac{\varepsilon_2 c_2}{\nu_{2,3}^2 - 1} - \frac{\varepsilon_1 c_1}{\nu_{1,3}^2 - 1} \right) \right]. \qquad (2.6.87)$$

The coefficients of the series (2.6.67) are determined by expressions (2.6.73), (2.6.78), (2.6.83), (2.6.84) and (2.6.87).

2.6.4 Calculation Examples

The analytically obtained diagrams of parametric instability zones are presented below in order to illustrate the influence of particular parameters of the system on their magnitude and position. The physical parameters of the system are given in the form in which they occur in differential equation (2.6.12).

Figures 2.10 and 2.11 present the influence of the unbalance μP, the damping (μH_1), and the shape of the friction characteristic (α/β) on the magnitude of the parametric instability zones for p_1 and p_2, for the following data: $\Omega^2 = 900\,\mathrm{s}^{-2}$, $\Omega_1^2 = 480\,\mathrm{s}^{-2}$, $\omega_1^2 = 4800\,\mathrm{s}^{-2}$, $g = \mu G = 9.81\,\mathrm{m\,s}^{-2}$, $v_0 = 0.4\,\mathrm{ms}^{-1}$, $\varepsilon = 0.2$. On the basis of (2.6.14), $p_1 = 73.32\,\mathrm{s}^{-1}$, $p_2 = 28.35\,\mathrm{s}^{-1}$, and $p_3 = 69.28\,\mathrm{s}^{-1}$ have been obtained. The adequate coefficients assume the form $\gamma_1 = 0.833$, $\gamma_2 = 0.12$, $\beta_1 = 0.126$, $\beta_2 = -0.674$, $\varepsilon_1 = 1.176$, $\varepsilon_2 = 0.376$. The other quantities characterizing the system have been marked in the figures (when $\alpha/\beta = \alpha/\rho$).

The parametric instability zones presented in Figs 2.10 (for p_1) and 2.11 (for p_2) expand with the increase of the unbalance μP, while, depending on the value of the quotient α/β, this tendency can have different intensity. In the case of $\alpha/\beta = 0.5\,\mathrm{m^2s}^{-2}$, the doubling of the unbalance has caused the instability to be expanded twice for the zones corresponding to p_1 and p_2. For $\alpha/\beta = 1\,\mathrm{m^2s}^{-2}$, the unbalance has been increased three times, which has brought about a comparatively small expansion of the instability zones for p_1, while for p_2 the expansion is still almost doubled. In the case of large unbalance of the rotor, the changes of the quotient α/β do not influence the magnitude of the parametric instability zones. The influence of damping on the magnitude of the instability zones corresponding to the frequencies p_1 and p_2 is also very different. Small damping $(\mu H_1 = 0.05\,\mathrm{s}^{-1})$ causes considerable shift of the zone for p_2 in the direction of the growing value of the modulation depth μ $(\mu \geq 0.15)$. In the case of the double increase of damping the zone will not occur for $\mu \leq 0.3$.

The magnitude and position of the instability zones for p_1 are not sensitive to the changes of the damping coefficient. In the case of $\mu H_1 = 10\,\mathrm{s}^{-1}$ the proper zone of the frequency p_1 exists for $\mu \geq 0.034$. When $\mu H_1 = 20\,\mathrm{s}^{-1}$ damping is doubled, and the lower border of the occurrence of the zone is shifted to the value of $\mu = 0.07$.

The parametric instability zones for p_3 are presented in Fig. 2.12. The magnitude of the zones depends on the initial conditions of the system motion. The diagrams have been prepared on the assumption that $a_3 = b_3 = 0.01\,\mathrm{m}$, where $a_3 = y_3(0)$, $b_3 = \dot{y}_3(0)$. The calculations, in the case of the resonance coordinate y_3, have been performed with with a precision of up to second order, hence the inclination of the unstability zones in the direction of the growing values of the parameter λ_3^2 has appeared. For the first approximation, the zones remain symmetrical in relation to the straight line $\lambda_3^2 = 1$. As in the cases considered above, the increase of the unbalance considerably expands the instability zone. The changes of the value of the quotient α/β and damping have a negligible influence on the magnitude of the zone. Figure 2.13 presents the parametric instability zones for various

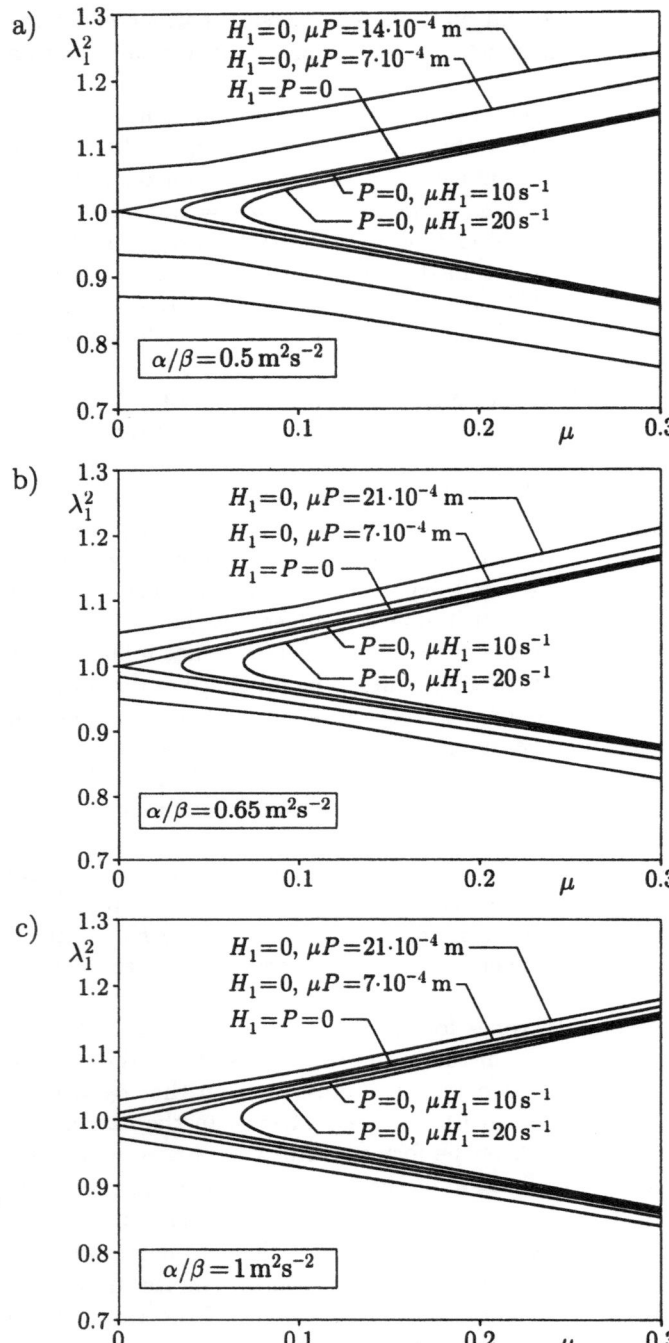

Fig. 2.10a-c. Instability zones for p_1

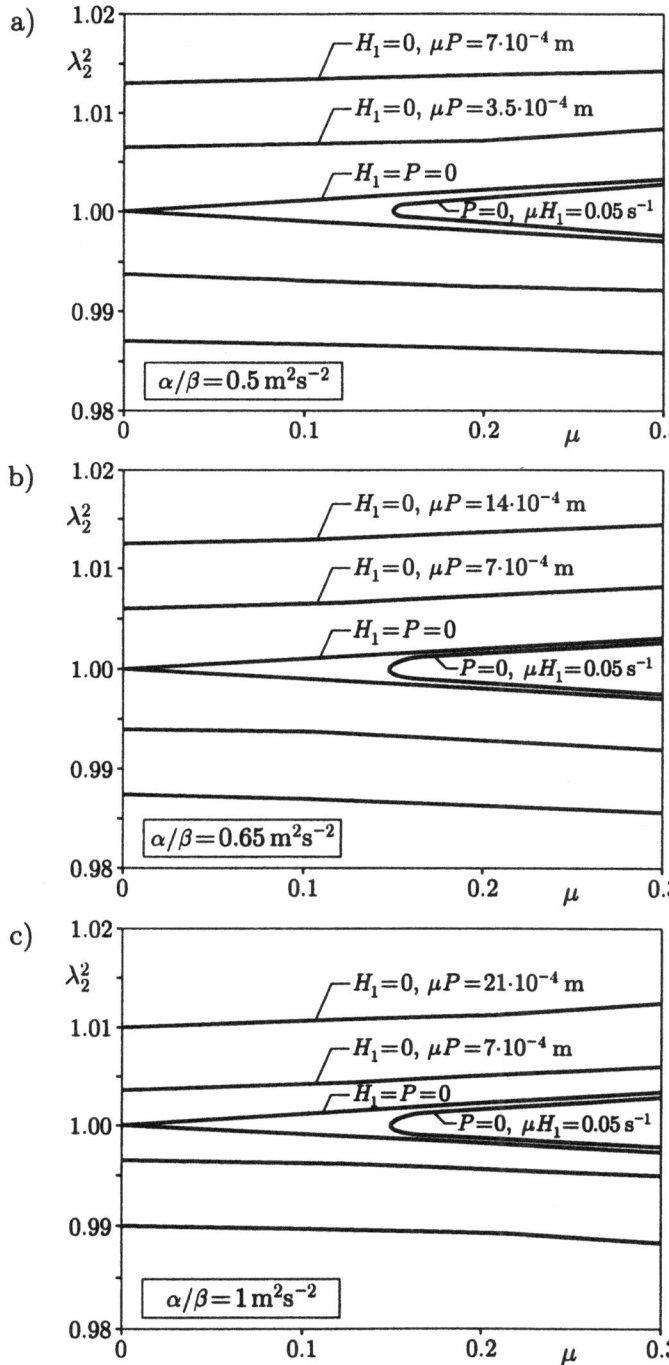

Fig. 2.11a-c. Instability zones for p_2

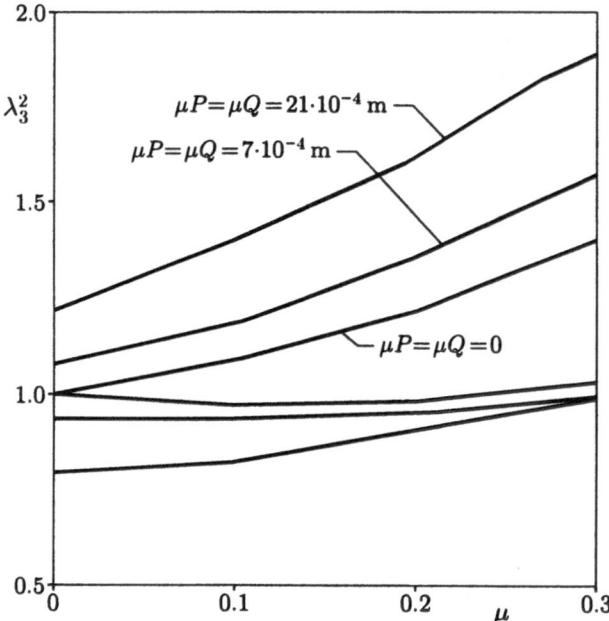

Fig. 2.12.
Instability zones for
p_3 ($\alpha/\beta = 0.5\,\mathrm{m^2 s^{-2}}$,
$\alpha/\beta = 0.65\,\mathrm{m^2 s^{-2}}$,
$\alpha/\beta = 1\,\mathrm{m^2 s^{-2}}$)

values of the parameters Ω^2, Ω_1^2 and ω_1^2. For the zones denoted by 1 we get
$\Omega^2 = 14400\,\mathrm{s^{-2}}$, $\Omega_1^2 = 1920\,\mathrm{s^{-2}}$, $\omega_1^2 = 19200\,\mathrm{s^{-2}}$; for the zones denoted by
2 we have $\Omega^2 = 3600\,\mathrm{s^{-2}}$, $\Omega_1^2 = 480\,\mathrm{s^{-2}}$, $\omega_1^2 = 4800\,\mathrm{s^{-2}}$, and for the zones
denoted by 3: $\Omega^2 = 900\,\mathrm{s^{-2}}$, $\Omega_1^2 = 120\,\mathrm{s^{-2}}$, $\omega_1^2 = 1200\,\mathrm{s^{-2}}$. In all the cases the
magnitudes of the other parameters are as follows: $\mu H_1 = 10\,\mathrm{s^{-1}}$, $\varepsilon = 0.2$,
$\alpha/\beta = 0.5\,\mathrm{m^2 s^{-2}}$, $v_0 = 0.5\,\mathrm{ms^{-1}}$, $\mu P = 0.0015\,\mathrm{m}$.

As shown in Fig. 2.13a,b the growth of the squares of frequencies Ω^2,
Ω_1^2, ω_1^2 (resulting from the increase in rigidity of the elastic elements in the
system, or from the decrease in the values of the masses) causes the instability
zones for p_1 and p_2 to expand. For example, when the parameters Ω^2, Ω_1^2,
and ω_1^2 increase by four times, it brings about an approximately doubled
expansion of the zones. The unstability zone for p_3 is not influenced by the
frequency changes in the system (Fig. 2.13c).

Figure 2.14a, b presents the influence of the velocity changes of the belt
v_0 on the magnitude of the instability zones for p_1 and p_2. Calculations have
been performed for the data denoted by 1, except for the velocity v_0, whose
value has been changed. In each case the increase in the belt velocity causes
the expansion of the instability zones. For $v_0 < 0.3\,\mathrm{ms^{-1}}$ these changes are less
evident. The influence of the velocity changes v_0 on the parametric instability
zone for p_3 is practically negligible.

We can summarize the obtained results as follows.

(1) The method of seeking a solution as a power series of the two perturbation parameters μ and ε used in the considerations makes it possible to

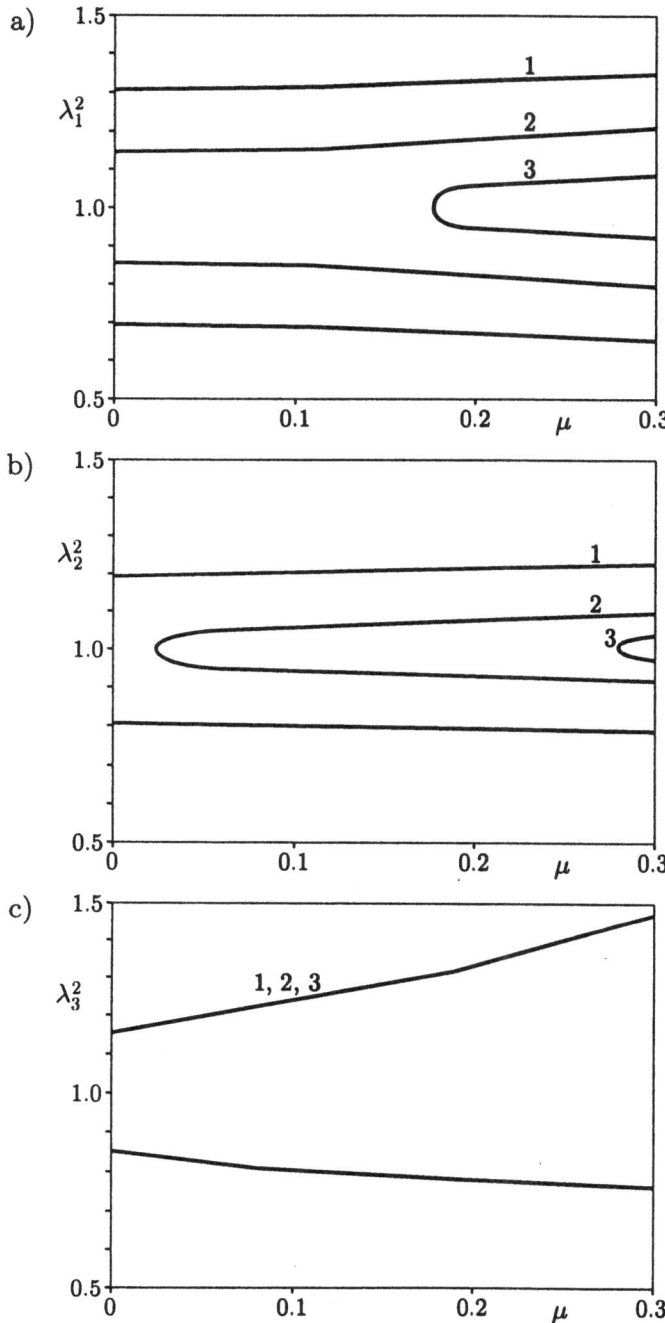

Fig. 2.13. Influence of parameter changes on the instability zones for: (a) p_1; (b) p_2; (c) p_3

investigate the single resonances of any order for the systems with weak nonlinearity and weakly modulated systems ($\mu \ll 1$).

When we perform calculations with a precision of up to second order, it turns out that the limits of instability zones incline in the direction of the growing values of the parameter λ_3^2 (Fig. 1.7). For the first approximation, the limits remain symmetrical in relation to the straight line $\lambda_3^2 = 1$.

(2) The parametric instability zones for p_1 and p_2 expand with the increase in the rotor unbalance. Depending on the value of the quotient α/p, this tendency has different intensity. In the case of $\alpha/\beta = 0.5\,\mathrm{m^2\,s^{-2}}$, the double increase in the unbalance has brought about a considerable expansion of the unstability zones, for p_1 as well as for p_2. For $\alpha/\beta = 1\,\mathrm{m^2\,s^{-2}}$ the unbalance, causes a rather small expansion of the instability zones for p_1, while for p_2 the expansion is still almost doubled. In the case of a lack

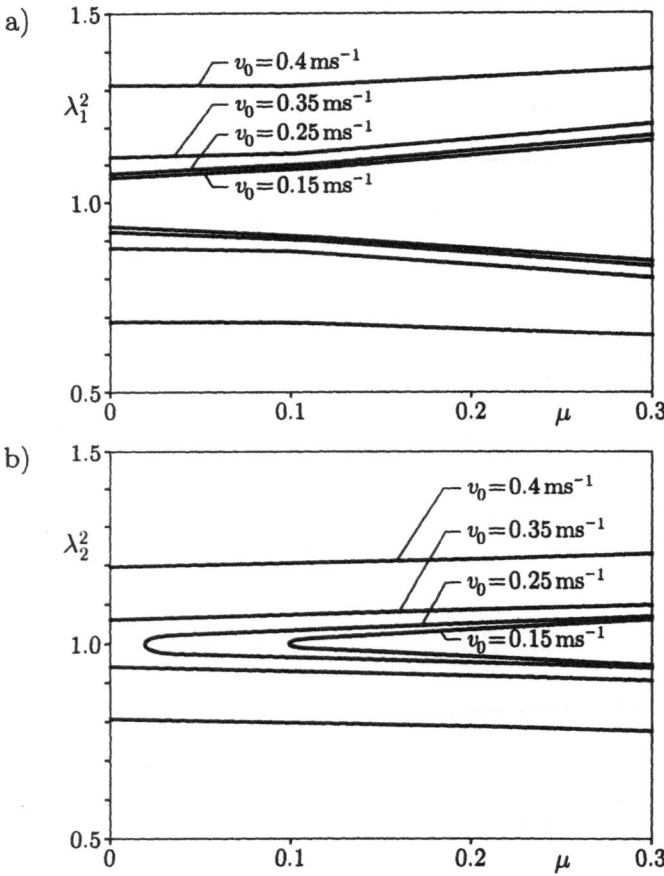

Fig. 2.14. Influence of the belt velocity changes v_0 on the instability zones for: (a) p_1; (b) p_2

of rotor unbalance the changes of the quotient α/β do not influence the magnitude of the parametric instability zones. The influence of damping on the magnitude of the instability zones corresponding to p_1 and p_2 is also very different. The minimum damping $(\mu H_1 = 0.05\,\mathrm{s}^{-1})$ causes a considerable shift of the zone for p_2 in the direction of the growing values of the modulation depth $(\mu H_1 = 0.15\,\mathrm{s}^{-1})$. The magnitude and position of the instability zones for p_1 are not so sensitive to the damping coefficient changes. The regularities indicated here are the more clear, the greater the difference between the values of the frequency p_1 and p_2 (i.e. for $p_1 \gg p_2$) is. The increase in the unbalance also produces a considerable expansion of the instability zone for p_3; however the changes of the parameter and of damping have no essential influence on the magnitude of the zone.

The growth of the frequency squares Ω^2, Ω_1^2 and ω_1^2 causes the expansion of the instability zones for p_1 and p_2. The parametric instability zone for the frequency p_3 is not sensitive to the frequency changes in the system. In the case of the belt velocity increase (v_0), the instability zones for p_1 and p_2 are expanded. This property is noticeable within the range of great velocities $(v_0 > 0.4\,\mathrm{m\,s}^{-1})$. The influence of the velocity changes v_0 on the parametric instability zone for p_3 is practically negligible.

(3) For the frequencies p_1 and p_2, the position of the instability zone limits does not depend in the first approximation on the initial conditions of the system motion. The magnitude of the instability zone limits for p_3 depends on these conditions.

2.7 Modified Poincaré Method

2.7.1 One-Degree-of-Freedom System

Consider a one-degree-of-freedom nonlinear system governed by the equation [33, 117]

$$\frac{d^2y}{dt^2} + \omega^2 y = \varepsilon Q\left(y, \frac{dy}{dt}, \varepsilon\right). \tag{2.7.1}$$

The above $\varepsilon > 0$ is a small perturbation parameter and the function Q is analytical with regard to its arguments $y, dy/dt$ and ε for $0 < \varepsilon \leq \varepsilon_0$.

We are going to find a periodic solution to (2.7.1) depending on the perturbation parameter ε. The system under consideration is autonomous, therefore we can arbitrarily take

$$\frac{dy}{dt}(0) = 0, \tag{2.7.2}$$

because starting with the initial conditions for t_0, (2.7.1) will not change with a shift of time.

For $\varepsilon = 0$ we have

$$\frac{d^2 y}{dt^2} + \omega^2 y_0 = 0, \qquad \dot{y}_0(0) = 0, \tag{2.7.3}$$

and a solution to (2.7.3) is given by

$$y_0(t) = A_0 \cos \omega t, \tag{2.7.4}$$

where the amplitude A_0 is not yet defined. However, contrary to the previous investigations and following Proskuriakov [65d], we are going to find an invariant orbit depending also on the second parameter $\delta = \delta(\varepsilon)$, where $\delta(0) = 0$.

Therefore, we have

$$y(0) = A_0 + \delta(\varepsilon). \tag{2.7.5}$$

On the basis of the assumption of analycity of F, this function will be analytical also with respect to $A_0 + \delta$.

We are focused on finding a periodic solution

$$y(t) = y(t + T), \tag{2.7.6}$$
$$\dot{y}(t) = y(t + T), \tag{2.7.7}$$

and the period T is defined as

$$T = T_0 + \alpha(\varepsilon), \tag{2.7.8}$$

where

$$T_0 = 2\pi \omega^{-1}, \tag{2.7.9}$$
$$\alpha(0) = 0. \tag{2.7.10}$$

From (2.7.6) and (2.7.7) we get

$$y(T_0 + \alpha, A_0 + \delta, \varepsilon) = y(0, A_0 + \delta, \varepsilon) = A_0 + \delta, \tag{2.7.11}$$
$$\dot{y}(T_0 + \alpha, A_0 + \delta, \varepsilon) = \dot{y}(0, A_0 + \delta, \varepsilon) = 0. \tag{2.7.12}$$

A general solution form of (2.7.1) is a function of the two parameters δ and ε:

$$y(t, A_0 + \delta, \varepsilon) = (A_0 + \delta) \cos \omega t + \sum_{k=1}^{K} y_k(t, A_0 + \delta) \varepsilon^k, \tag{2.7.13}$$

and

$$\dot{y}(t, A_0 + \delta, \varepsilon) = -\omega(A_0 + \delta) \sin \omega t + \sum_{k=1}^{K} \dot{y}_k(t, A_0 + \delta) \varepsilon^k. \tag{2.7.14}$$

From (2.7.13) for $t = 0$ and taking into account (2.7.2) and (2.7.5) we get

$$y_k(0, A_0 + \delta) = 0, \tag{2.7.15}$$
$$\dot{y}_k(0, A_0 + \delta) = 0. \tag{2.7.16}$$

We develop the right-hand side of (2.7.1) into a power series of ε in the neighbourhood of $\varepsilon = 0$ putting $\delta = 0$ into series (2.7.13) and (2.7.14). We obtain

$$
\varepsilon Q(y, \dot{y}, \varepsilon) = \varepsilon \left\{ Q(y_0, \dot{y}_0, 0) \Big|_{\varepsilon=\delta=0} + \varepsilon \left(\frac{\partial Q}{\partial y} \frac{dy}{d\varepsilon} \Big|_{\varepsilon=\delta=0} + \frac{\partial Q}{\partial \dot{y}} \frac{d\dot{y}}{d\varepsilon} \Big|_{\varepsilon=\delta=0} \right. \right.
$$
$$
+ \frac{\partial Q}{\partial \varepsilon} \Big|_{\varepsilon=\delta=0} \Big) + \varepsilon^2 \left(\frac{1}{2} \frac{\partial^2 Q}{\partial y^2} \left(\frac{dy}{d\varepsilon} \right)^2 \Big|_{\varepsilon=\delta=0} \right.
$$
$$
+ \frac{1}{2} \frac{\partial^2 Q}{\partial \dot{y}^2} \left(\frac{d\dot{y}}{d\varepsilon} \right) \Big|_{\varepsilon=\delta=0} + \frac{1}{2} \left(\frac{\partial^2 Q}{\partial \varepsilon^2} \right) \Big|_{\varepsilon=\delta=0}
$$
$$
+ \frac{\partial^2 Q}{\partial y \partial \dot{y}} \frac{dy}{d\varepsilon} \frac{d\dot{y}}{d\varepsilon} \Big|_{\varepsilon=\delta=0} + \frac{\partial^2 Q}{\partial y \partial \varepsilon} \frac{dy}{d\varepsilon} \Big|_{\varepsilon=\delta=0} + \frac{\partial^2 Q}{\partial \dot{y} \partial \varepsilon} \frac{d\dot{y}}{d\varepsilon} \Big|_{\varepsilon=\delta=0}
$$
$$
+ \frac{\partial Q}{\partial y} \frac{d^2 y}{d\varepsilon^2} \Big|_{\varepsilon=\delta=0} + \frac{\partial Q}{\partial \dot{y}} \frac{d^2 \dot{y}}{d\varepsilon^2} \Big|_{\varepsilon=\delta=0} \Big) + O(\varepsilon^3) \Big\}
$$
$$
= \varepsilon \left\{ Q(y_0, \dot{y}_0)_0 + \varepsilon \left(\left(\frac{\partial Q}{\partial y} \right)_0 y_1 + \left(\frac{\partial Q}{\partial \dot{y}} \right)_0 \dot{y}_1 + \left(\frac{\partial Q}{\partial \varepsilon} \right)_0 \right) \right.
$$
$$
+ \varepsilon^2 \left[\frac{1}{2} \left(\frac{\partial^2 Q}{\partial y^2} \right)_0 y_1^2 + \frac{1}{2} \left(\frac{\partial^2 Q}{\partial \dot{y}^2} \right)_0 \dot{y}_1^2 + \frac{1}{2} \left(\frac{\partial^2 Q}{\partial \varepsilon^2} \right)_0 \right.
$$
$$
+ \left(\frac{\partial^2 Q}{\partial y \partial \dot{y}} \right)_0 y_1 \dot{y}_1 + \left(\frac{\partial^2 Q}{\partial y \partial \varepsilon} \right)_0 y_1 + \left(\frac{\partial^2 Q}{\partial \dot{y} \partial \varepsilon} \right)_0 \dot{y}_1
$$
$$
+ \left(\frac{\partial Q}{\partial y} \right)_0 y_2 + \left(\frac{\partial Q}{\partial \dot{y}} \right)_0 \dot{y}_2 \right] + O(\varepsilon^3) \Big\}. \tag{2.7.17}
$$

Now we demonstrate that using the periodicity conditions we are able to solve the problem, i.e. to find A_0, $\alpha(\varepsilon)$, and then $\delta(\varepsilon)$.

To show this, let us begin with (2.7.12). From (2.7.14) and (2.7.2) for $T = T_0 + \alpha$ we obtain

$$
\dot{y}(T_0 + \alpha, A_0 + \delta, \varepsilon) = -\omega(A_0 + \delta) \sin \omega(T_0 + \alpha) \tag{2.7.18}
$$
$$
+ \sum_{k=1}^{K} \dot{y}_k(T_0 + \alpha, A_0 + \delta) \varepsilon^k = 0,
$$

which leads to the equation

$$
-\omega(A_0 + \delta) \sin \omega \alpha + \sum_{k=1}^{K} \dot{y}_k(T_0 + \alpha, A_0 + \delta) \varepsilon^k = 0. \tag{2.7.19}
$$

The above solution allows us to find $\alpha(\varepsilon)$. For this purpose, we first express α in an analytical form of ε

$$
\alpha(\varepsilon) = \varepsilon \alpha_1(A_0 + \delta) + \varepsilon^2 \alpha_2(A_0 + \delta) + \ldots + \varepsilon^k \alpha_k(A_0 + \delta). \tag{2.7.20}
$$

Each of α_k depends on A_0, T_0 and δ and can be developed into the Maclaurin series in the neighbourhood of $\delta = 0$:

$$\alpha_k(A_0 + \delta) = \alpha_k(A_0) + \frac{\partial \alpha_k}{\partial A_0}\delta + \frac{1}{2}\frac{\partial^2 \alpha_k}{\partial A_0^2}\delta^2 + \ldots, \quad k = 1,\ldots,K. \quad (2.7.21)$$

The corresponding α_k can be obtained by differentiating (2.7.19) with respect to ε and putting $\alpha = \delta = \varepsilon = 0$. We find

$$\varepsilon \ : \ -\omega^2 A_0\left(\frac{\partial \alpha}{\partial \varepsilon}\right)_0 + \dot{y}_1(T_0, A_0) = 0, \quad (2.7.22)$$

$$\varepsilon^2 \ : \ -\omega^2 A_0\left(\frac{\partial^2 \alpha}{\partial \varepsilon^2}\right)_0 + 2\left[\ddot{y}_1\left(\frac{\partial \alpha}{\partial \varepsilon}\right)_0 + \dot{y}_2\right] = 0, \quad (2.7.23)$$

$$\varepsilon^3 \ : \ -\omega^2 A_0\left(\frac{\partial^3 \alpha}{\partial \varepsilon^3}\right)_0 + 6\left\{\dot{y}_3 + \left(\frac{\partial^2 \alpha}{\partial \varepsilon^2}\right)_0 \ddot{y}_1 + \left(\frac{\partial \alpha}{\partial \varepsilon}\right)_0 \ddot{y}_2\right.$$

$$\left. +\frac{1}{2}\left(\frac{\partial \alpha}{\partial \varepsilon}\right)_0^2 \left[\dddot{y}_1 + \frac{1}{3}\omega^2\dot{y}_1\right]\right\}, \quad (2.7.24)$$

where $\partial^k \alpha / \partial \varepsilon^k = \alpha_k$.

If y_k are known then we can find α_k according to the subsequent equations (2.7.22), (2.7.23) and (2.7.24).

Now we illustrate how we obtain y_k. For this purpose we have to reconsider the left-hand side of (2.7.1). Taking into account (2.7.14), we obtain

$$\ddot{y}(t, A_0 + \delta, \varepsilon) = -\omega^2(A_0 + \delta)\cos\omega t + \sum_{k=1}^{K}\ddot{y}_k(t, A_0 + \delta)\varepsilon^k. \quad (2.7.25)$$

Finally, from (2.7.1), (2.7.17) and (2.7.25) we get the following recurrent set of linear differential equations (for $\delta = 0$)

$$\ddot{y}_1 + \omega^2 y_1 = Q(y_0, \dot{y}_0), \quad (2.7.26)$$

$$\ddot{y}_2 + \omega^2 y_2 = \left(\frac{\partial Q}{\partial y}\right)_0 y_1 + \left(\frac{\partial Q}{\partial \dot{y}}\right)_0 \dot{y}_1 + \left(\frac{\partial Q}{\partial \varepsilon}\right)_0, \quad (2.7.27)$$

$$\ddot{y}_3 + \omega^2 y_3 = \frac{1}{2}\left(\frac{\partial^2 Q}{\partial y^2}\right)_0 y_1^2 + \frac{1}{2}\left(\frac{\partial^2 Q}{\partial \dot{y}^2}\right)_0 \dot{y}_1^2 + \frac{1}{2}\left(\frac{\partial^2 Q}{\partial \varepsilon^2}\right)_0$$

$$+\left(\frac{\partial^2 Q}{\partial y \partial \dot{y}}\right)_0 y_1\dot{y}_1 + \left(\frac{\partial^2 Q}{\partial y \partial \varepsilon}\right)_0 y_1 + \left(\frac{\partial^2 Q}{\partial \dot{y} \partial \varepsilon}\right)_0 \dot{y}_1$$

$$+\left(\frac{\partial Q}{\partial y}\right)_0 y_2 + \left(\frac{\partial Q}{\partial \dot{y}}\right)_0 \dot{y}_2, \quad (2.7.28)$$

and so on, which allows us to find $y_k(A_0, T_0)$.

To conclude, using condition (2.7.12) we have found α_k, and therefore the unknown period $T = T_0 + \alpha(\varepsilon)$ is defined.

Using condition (2.7.11) we illustrate how to find A_0 and δ. For this purpose let us develop (2.7.11) into the Maclaurin series with respect to α

$$y(T_0, A_0 + \delta, \varepsilon) + \alpha\dot{y}(T_0, A_0 + \delta, \varepsilon) + \frac{1}{2}\alpha^2\ddot{y}(T_0, A_0 + \delta, \varepsilon)$$

$$+ \frac{1}{6}\alpha^3 \dddot{y}(T_0, A_0 + \delta, \varepsilon) + \ldots - A_0 - \delta = 0. \quad (2.7.29)$$

The left-hand side L of (2.7.29) can be described by the series

$$L = \varepsilon L_1(T_0, A_0 + \delta) + \varepsilon^2 L_2(T_0, A_0 + \delta) + \ldots + \varepsilon^k L_k(T_0, A_0 + \delta). \quad (2.7.30)$$

Each of the $L_k(T_0, A_0 + \delta)$ term can be developed into the Maclaurin series with respect of δ

$$L_k(T_0, A_0+\delta) = L_k(T_0, A_0) + \frac{\partial L_k}{\partial A_0}\delta + \frac{1}{2}\frac{\partial^2 L_k}{\partial A_0^2}\delta^2 + \ldots, \quad k = 1 \ldots, K. \quad (2.7.31)$$

Taking into account (2.7.31) in (2.7.30) we obtain the following evident form

$$L = \varepsilon \sum_{k=1}^{K} \left(L_k(T_0, A_0) + \frac{\partial L_k(T_0, A_0)}{\partial A_0}\delta + \frac{1}{2}\frac{\partial^2 L_k(T_0, A_0)}{\partial A_0^2}\delta^2 + \ldots \right.$$
$$\left. + \frac{1}{k!}\frac{\partial^2 L_k(T_0, A_0)}{\partial A_0^k}\delta^k \right) \varepsilon^{k-1} = 0. \quad (2.7.32)$$

From (2.7.29) for $\varepsilon = 0$ we get $y(T_0, A_0 + \delta) = A_0 + \delta$.

Each of the terms $y^{(m)} = d^m y/dt^m$ depends on A_0, T_0 and δ, and can be developed into the Maclaurin series in the neighbourhood of $\delta = 0$:

$$y^{(m)}(T_0, A_0 + \delta, \varepsilon) = y^{(m)}(T_0, A_0) + \frac{\partial y^{(m)}}{\partial A_0}\delta + \frac{1}{2}\frac{\partial^2 y^{(m)}}{\partial A_0^2}\delta^2 + \ldots$$
$$+ \frac{1}{k!}\frac{\partial^k y^{(m)}}{\partial A_0^k}\delta^k. \quad (2.7.33)$$

Now for $\delta = 0$ we compare the terms standing by the same powers of ε in (2.7.29), taking into account (2.7.20) and (2.7.30) and (2.7.32). First, we take $\delta = 0$ and from (2.7.29) we get

$$y(T_0, A_0, \varepsilon) + \alpha\dot{y}(T_0, A_0, \varepsilon) + \frac{1}{2}\alpha^2\ddot{y}(T_0, A_0, \varepsilon) \quad (2.7.34)$$
$$+ \frac{1}{6}\alpha^2\,\dddot{y}\,(T_0, A_0, \varepsilon) + \ldots - A_0 = 0.$$

Comparing the coefficients of the same powers of ε in (2.7.32) and (2.7.34), we obtain

$$\varepsilon^0 \; : \; L_1(T_0, A_0) = y_1(T_0, A_0) = 0, \quad (2.7.35)$$

$$\varepsilon^1 \; : \; L_2(T_0, A_0) = y_2(T_0, A_0) + \alpha_1\dot{y}_1(T_0, A_0) - \frac{1}{2}\alpha_1^2 A_0\omega^2, \quad (2.7.36)$$

$$\varepsilon^2 \; : \; L_3(T_0, A_0) = y_3(T_0, A_0) + \alpha_2\dot{y}_1(T_0, A_0) + \alpha_1\dot{y}_2(T_0, A_0)$$
$$- A_0\omega^2\alpha_1\alpha_2 - \frac{1}{2}\alpha_1^2\ddot{y}_1(T_0, A_0). \quad (2.7.37)$$

Equation (2.7.35), further called the amplitude equation, can possess two solutions: either $y_1(T_0, A_0) = 0$ or $y_1(T_0, A_0) \equiv 0$.

In the first case this equation allows us to find the amplitude A_0. If A_0 is real and not a multiplied root of the above equation, then it corresponds to the only real solution of (2.7.1).

In the latter case we divide (2.7.32) by ε and $L_2(T_0, A_0) = 0$ will serve as the amplitude equation, and so on.

We discuss further the first case. From (2.7.35) we calculate A_0, and then (2.7.32) defines the following implicit function of the two parameters δ and ε

$$
L(\varepsilon, \delta) = \varepsilon L_2 + \delta \frac{\partial L_1}{\partial A_0} + \varepsilon^2 L_3 + \varepsilon \delta \frac{\partial L_2}{\partial A_0} + \frac{1}{2} \delta^2 \frac{\partial^2 L_1}{\partial A_0^2}
$$

$$
+ \varepsilon^3 L_4 + \varepsilon^2 \delta \frac{\partial L_3}{\partial A_0} + \frac{1}{2} \varepsilon \delta^2 \frac{\partial^2 L_2}{\partial A_0^2} + \frac{1}{2} \delta^3 \frac{\partial^3 L_1}{\partial A_0^3} + \dots . \qquad (2.7.38)
$$

According to the implicit function theorem, the number of branches of the implicit function $\delta(\varepsilon)$ is defined by the smallest power of δ^k in the following series

$$
L(0, \delta) = \delta \frac{\partial L_1}{\partial A_0} + \frac{1}{2} \delta^2 \frac{\partial^2 L_1}{\partial A_0^2} + \frac{1}{6} \delta^3 \frac{\partial^3 L_1}{\partial A_0^3} + \dots + \frac{1}{k!} \delta^k \frac{\partial^k L_1}{\partial A_0^k} = 0. \quad (2.7.39)
$$

According to (2.7.35) $L_1(A_0) = y_1(A_0)$ and if all the derivatives up to the kth order $\partial L_1/\partial A_0$, $\partial^2 L_1/\partial A_0^2$, ..., $\partial^k L_1/\partial A_0^k$ vanish, then A_0 has k-multiplicity, i.e. A_0^k. In what follows k-multiplicity of A_0 defines the number of branches of $\delta(\varepsilon)$. Each of the k-branches can be described by the following series

$$
\delta(\varepsilon) = A_{1/m} \varepsilon^{1/m} + A_{2/m} \left(\varepsilon^{1/m} \right)^2 + A_{3/m} \left(\varepsilon^{1/m} \right)^3 + \dots, \qquad (2.7.40)
$$

where m equals one of the numbers $1, 2, \dots, k$.

In the case of fractional powers, the sum of the denominators cannot be greater than k.

For $m = 1$ the problem is reduced to the well-known situation of the classical series

$$
\delta(\varepsilon) = A_1 \varepsilon + A_2 \varepsilon^2 + \varepsilon^3 A_3 + \dots . \qquad (2.7.41)
$$

2.7.2 General Nonlinear Systems

2.7.2.1 Introduction.
Let us consider an autonomous nonlinear system in the general form

$$
\frac{\mathrm{d}X}{\mathrm{d}t} = F(X), \qquad (2.7.42)
$$

where:

1° F is continuous and $X \in \mathbb{R}^n$;
2° F fulfills the Lipschitz condition: $\|F(X) - F(Y)\| \leq C \|X - Y\|$, where C is constant;
3° a solution X is defined as: $X = X(t, X^{(0)})$, where $X^{(0)} = X(t_0)$ (further we take $t_0 = 0$).

Now we consider a set of first-order equations, which can govern not only mechanical, but also biological or chemical nonlinear dynamical systems. Of course, the set of second-order differential equations obtained using Newton's laws or the Lagrange equations can be easily reduced to (2.7.42).

We are going to find only the periodic solutions to (2.7.42) with a period $T > 0$ taking the initial conditions $X^{(0)}$ in such a way that

$$Z(T, X^{(0)}) = X(T, X^{(0)}). \tag{2.7.43}$$

$X(T, X^{(0)})$ is the solution to (2.7.42), and $X(0, X^{(0)}) = X^{(0)}$. If we find the real X and T which fulfil the equation

$$Z(T, X) = X, \tag{2.7.44}$$

the above map reduces the problem of finding the periodic solution of (2.7.42) to finding the fixed points of the map (2.7.44). Such an idea is often used during an application of numerical techniques to exhibit periodic solutions.

2.7.2.2 Autonomous System. Consider now the case where the right-hand side of (2.7.42) depends on a small parameter ε

$$\frac{\mathrm{d}x_s}{\mathrm{d}t} = \sum_{i=1}^{n} a_{si}x_i + \varepsilon F(x_1, \ldots, x_n), \qquad s = 1, \ldots, n, \tag{2.7.45}$$

where the same assumptions as for (2.7.42) are valid and $0 \le \varepsilon < \varepsilon_0$. Because (2.7.42) is autonomous, then for an arbitrary t_0 if we take $t + t_0$ instead of t in (2.7.45) this equation will not be changed. It implies that a solution to (2.7.45) does not depend on t_0 and we can take $t_0 = 0$.

Let us assume that for $\varepsilon = 0$ the linear system

$$\frac{\mathrm{d}x_s}{\mathrm{d}t} = \sum_{i=1}^{n} a_{si}x_i \tag{2.7.46}$$

has a family of periodic orbits with a period T_0. Let us assume that it corresponds to the existence of K pairs of ν_s eigenvalues $\pm r\mathrm{i}/T_0$, $r = 0, 1, 2, \ldots$, $\mathrm{i}^2 = -1$. Even if they are multiple values but with simple elementary divisors, system (2.7.45) can be reduced to the form

$$\frac{\mathrm{d}x_s}{\mathrm{d}t} = -\nu_s y_s + \varepsilon \varphi_s(x_1, \ldots, x_k, y_1, \ldots, y_k, z_1, \ldots, z_M, \varepsilon),$$

$$\frac{\mathrm{d}y_s}{\mathrm{d}t} = \nu_s x_s + \varepsilon \psi_s(x_1, \ldots, x_k, y_1, \ldots, y_k, z_1, \ldots, z_M, \varepsilon),$$

$$\frac{\mathrm{d}z_j}{\mathrm{d}t} = \sum_{i=1}^{M} p_{ij}z_i + \varepsilon h_j(x_1, \ldots, x_k, y_1, \ldots, y_k, z_1, \ldots, z_M, \varepsilon), \tag{2.7.47}$$

$$s = 1, 2, \ldots, K, \quad j = 1, 2, \ldots, M, \quad M + 2K = n.$$

For $\varepsilon = 0$ the linear system

$$\frac{\mathrm{d}x_s}{\mathrm{d}t} = -\nu_s y_s,$$

$$\frac{\mathrm{d}y_s}{\mathrm{d}t} = \nu_s x_s, \tag{2.7.48}$$

$$\frac{\mathrm{d}z}{\mathrm{d}t} = \sum_{i=1}^{M} p_{ij} z_i$$

has a family of T_0 periodic solutions depending on $2K$ arbitrary constants C_s and D_s of the form

$$x_s = C_s \cos \nu_s t - D_s \sin \nu_s t,$$

$$y_s = C_s \sin \nu_s t + D_s \cos \nu_s t, \tag{2.7.49}$$

$$z_j = 0,$$

$$s = 1, 2, \dots, K, \quad j = 1, \dots, n - 2K.$$

Because the system under consideration is autonomous, we can arbitrarily take one of $y_s = 0$ for $t_0 = 0$. For $s = 1$ we get

$$y_1 = D_1 = 0. \tag{2.7.50}$$

This means that instead of $2K$ variables we have to obtain $2K - 1$.

The solutions to system (2.7.47) have the following form

$$x_1 = (C_1 + C_1^*) \cos \nu_1 t + \varepsilon \int_0^t [\varphi_1 \cos \nu_1(t - \tau) - \psi_1 \sin \nu_1(t - \tau)] \, \mathrm{d}\tau,$$

$$y_1 = (C_1 + C_1^*) \sin \nu_1 t + \varepsilon \int_0^t [\varphi_1 \sin \nu_1(t - \tau) - \psi_1 \cos \nu_1(t - \tau)] \, \mathrm{d}\tau,$$

$$x_s = (C_s + C_s^*) \cos \nu_s t - (D_s + D_s^*) \sin \nu_s t$$

$$+ \varepsilon \int_0^t [\varphi_s \cos \nu_s(t - \tau) - \psi_s \sin \nu_s(t - \tau)] \, \mathrm{d}\tau,$$

$$y_s = (C_s + C_s^*) \sin \nu_s t + (D_s + D_s^*) \cos \nu_s t$$

$$+ \varepsilon \int_0^t [\varphi_s \sin \nu_s(t - \tau) + \psi_s \cos \nu_s(t - \tau)] \, \mathrm{d}\tau, \quad s = 2, \dots, K,$$

$$\{z\} = \mathrm{e}^{[P]t}\{z^*\} + \varepsilon \int_0^t \mathrm{e}^{[P](t - \tau)}\{h\} \mathrm{d}\tau. \tag{2.7.51}$$

The above solutions for $t = 0$ have the following initial values

$$x_1(0) = C_1 + C_1^*(\varepsilon),$$
$$y_1(0) = 0,$$
$$x_s(0) = C_s + C_s^*(\varepsilon),$$
$$y_s(0) = D_s + D_s^*(\varepsilon), \qquad s = 2, \ldots, K,$$
$$\{z\} = \{z^*(\varepsilon)\}, \tag{2.7.52}$$

where $\{z\} = \mathrm{col}(z_1, \ldots, z_{n-2K})$, $\{z^*\} = \mathrm{col}(z_1^*, \ldots, z_{n-2K}^*)$, $[P] = [p_{ij}]$, $\{h\} = \mathrm{col}(h_1, \ldots, h_{n-2K})$, and these solutions possess a new period

$$T = T_0 + \alpha(\varepsilon). \tag{2.7.53}$$

For $\varepsilon \to 0$ we have $\alpha(\varepsilon) \to 0$, $C_s^* \to 0$, $D_s^* \to 0$, $\{z^*\} \to 0$.

Now we show that solutions (2.7.51) fulfil (2.7.47). For this purpose we put (2.7.51) into (2.7.47) and we get

$$\frac{\mathrm{d}}{\mathrm{d}t} \left\{ \int_0^t [\varphi_s \cos \nu_s(t - \tau) - \psi_s \sin \nu_s(t - \tau)] \, \mathrm{d}\tau \right\}$$

$$= -\nu_s \int_0^t [\varphi_s \sin \nu_s(t - \tau) + \psi_s \cos \nu_s(t - \tau)] \, \mathrm{d}\tau + \varphi_s,$$

$$\frac{\mathrm{d}}{\mathrm{d}t} \left\{ \int_0^t [\varphi_s \sin \nu_s(t - \tau) + \psi_s \cos \nu_s(t - \tau)] \, \mathrm{d}\tau \right\}$$

$$= \nu_s \int_0^t [\varphi_s \cos \nu_s(t - \tau) - \psi_s \sin \nu_s(t - \tau)] \, \mathrm{d}\tau + \psi_s,$$

$$s = 1, \ldots, K,$$

$$\frac{\mathrm{d}}{\mathrm{d}t} \left[\int_0^t \mathrm{e}^{[P](t-\tau)} \{h\} \, \mathrm{d}\tau \right] = [P] \int_0^t \mathrm{e}^{[P](t-\tau)} \{h\} \, \mathrm{d}\tau + \{h\}. \tag{2.7.54}$$

To prove the identity of the above equations, we use the Laplace transformation, remembering that

$$x_1(\tau) * x_2(\tau) = \int_0^t x_1(\tau) x_2(t - \tau) \mathrm{d}\tau, \tag{2.7.55}$$

$$L[x_1(t) * x_2(t)] = X_1(s) X_2(s), \tag{2.7.56}$$

$$L[\cos \alpha t] = \frac{s}{s^2 + \alpha^2}, \tag{2.7.57}$$

$$L[\sin \alpha t] = \frac{\alpha}{s^2 + \alpha^2}. \tag{2.7.58}$$

Therefore, we find that in fact the following equations are fulfilled

$$\frac{s^2}{s^2 + \nu_s^2}L(\varphi_s) - \frac{s\nu_s}{s^2 + \nu_s^2}L(\psi_s)$$

$$= -\nu\left[\frac{\nu_s}{s^2 + \nu_s^2}L(\varphi_s) + \frac{s}{s^2 + \nu_s^2}L(\psi_s)\right] + L(\varphi_s),$$

$$\frac{s\nu_s}{s^2 + \nu_s^2}L(\varphi_s) + \frac{s^2}{s^2 + \nu_s^2}L(\psi_s)$$

$$= \nu_s\left[\frac{s}{s^2 + \nu_s^2}L(\varphi_s) - \frac{\nu_s}{s^2 + \nu_s^2}L(\psi_s)\right] + L(\psi_s),$$

$$s\frac{1}{s - [P]}L(h) = [P]\frac{1}{s - [P]}L(h) + L(h). \tag{2.7.59}$$

The temporarily unknown quantities C_s^*, D_s^* and z^*, which depend on ε, should be obtained in order to fulfil the periodicity conditions.

As system (2.7.47) is analytical, therefore also its solutions, as well as the period, are analytical because of the initial conditions and the parameter ε. The periodicity condition applied to solutions (2.7.51) will take the form

$$x_1(T) - x_1(0) = \varepsilon\int_0^T [\varphi_1\cos\nu_1(T - \tau) - \psi_1\sin\nu_1(T - \tau)]\,d\tau$$

$$+(C_1 + C_1^*)[\cos\nu_1 T - 1] = 0,$$

$$y_1(T) - y_1(0) = \varepsilon\int_0^T [\varphi_1\sin\nu_1(T - \tau) - \psi_1\cos\nu_1(T - \tau)]\,d\tau$$

$$+(C_1 + C_1^*)\sin\nu_1 T = 0,$$

$$x_s(T) - x_s(0) = \varepsilon\int_0^T [\varphi_s\cos\nu_s(T - \tau) - \psi_s\sin\nu_s(T - \tau)]\,d\tau$$

$$+(C_s + C_s^*)[\cos\nu_s T - 1] = 0,$$

$$y_s(T) - y_s(0) = \varepsilon\int_0^T [\varphi_s\sin\nu_s(T - \tau) + \psi_s\cos\nu_s(T - \tau)]\,d\tau$$

$$+(D_s + D_s^*)[\cos\nu_s T - 1] = 0,$$

$$\{z(T)\} - \{z(0)\} = \left(e^{[P]T} - [I]\right)\{z^*\} + \varepsilon\int_0^T e^{[P](T-\tau)}\{h\}\,d\tau = 0. \tag{2.7.60}$$

According to the above equations, we have n equations with the unknowns C_1^*, T, C_s^*, D_s^* ($s = 2, \ldots, K$). It is easy to show that

$$\sin\nu_s T = \sin\nu_s\alpha,$$

$$\cos\nu_s T = \cos\nu_s\alpha, \tag{2.7.61}$$

and from the second equations of (2.7.60) it follows that

$$\alpha(\varepsilon) = \varepsilon\alpha_1(\varepsilon). \tag{2.7.62}$$

For $C_1^* = C_s^* = D_s^* = \varepsilon = 0$ $(s = 2,\dots,K)$ we get from (2.7.60) n equations with the unknowns $2K - 1$, the initial conditions C_s, D_s $(D_1 = 0)$, and the period T_0.

In order to simplify further considerations we denote the left-hand side of (2.7.60) by Γ_i $(i = 1,\dots,n)$, and additionally we introduce the new variables

$$\begin{array}{lll}
C_1 = a_1, & C_1^* = a_1^*, & x_1 = \bar{x}_1, \\[4pt]
C_2 = a_2, & C_2^* = a_2^*, & \vdots \\[4pt]
\vdots & \vdots & x_K = \bar{x}_K, \\[4pt]
C_K = a_K, & C_K^* = a_K^*, & y_1 = \bar{x}_{K+1}, \\[4pt]
T_0 = a_{K+1}, & \alpha = a_{K+1}^*, & \vdots \\[4pt]
D_2 = a_{K+2}, & D_2^* = a_{K+2}^*, & y_K = \bar{x}_{2K}, \\[4pt]
\vdots & \vdots & z_1 = \bar{x}_{2K+1}, \\[4pt]
D_K = a_{2K}, & D_K^* = a_{2K}^*, & \vdots \\[4pt]
z_1^* = a_{2K+1}^*, & & z_{n-2K} = \bar{x}_n, \\[4pt]
\vdots & & \\[4pt]
z_{n-2K}^* = a_n^*. & &
\end{array} \tag{2.7.63}$$

The fundamental idea of the Poincaré method is to find n parameters a_i $(i = 1,\dots,n)$ and n functions $a_i^*(\varepsilon)$ $(i = 1,\dots,n)$ from the equations

$$\Gamma_i(a_1+a_1^*,\dots,a_{2K}+a_{2K}^*,a_{2K+1}^*,\dots,a_n^*,\varepsilon) = 0, \quad i = 1,2,\dots,n. \tag{2.7.64}$$

Because the general form of solutions (2.7.51) can not always be expressed by elementary functions (after necessary integrations), therefore to omit the potential difficulties, we look for the solution as a power series of the perturbation parameter ε of the form

$$\bar{x}_i = R_{i,0} + \sum_{l=1}^{\infty} R_{i,l}(a_1 + a_1^*,\dots,a_{2K} + a_{2K}^*,a_{2K+1}^*,\dots,a_n^*)\varepsilon^l, \tag{2.7.65}$$

where

$$R_{1,0} = (a_1 + a_1^*)\cos\nu_1 t,$$
$$R_{2,0} = (a_2 + a_2^*)\cos\nu_2 t - (a_{K+2} + a_{K+2}^*)\sin\nu_2 t,$$
$$\vdots$$
$$R_{K,0} = (a_K + a_K^*)\cos\nu_K t - (a_{2K} + a_{2K}^*)\sin\nu_K t,$$
$$R_{K+1,0} = (a_1 + a_1^*)\sin\nu_1 t,$$
$$R_{K+2,0} = (a_2 + a_2^*)\sin\nu_2 t + (a_{K+2} + a_{K+2}^*)\cos\nu_2 t,$$

$$\vdots$$

$$R_{2K,0} = (a_K + a_K^*) \sin \nu_K t + (a_{2K} + a_{2K}^*) \cos \nu_K t,$$

$$\left\{ \begin{array}{c} R_{2K+1,0} \\ \vdots \\ R_{n,0} \end{array} \right\} = e^{[P]t} \left\{ \begin{array}{c} a_{2K+1}^* \\ \vdots \\ a_n^* \end{array} \right\}. \tag{2.7.66}$$

For $t = 0$ from (2.7.65) and (2.7.66) we obtain

$$R_{1,0}(t = 0) = a_1 + a_1^*(\varepsilon),$$
$$R_{2,0}(t = 0) = a_2 + a_2^*(\varepsilon),$$

$$\vdots$$

$$R_{K,0}(t = 0) = a_K + a_K^*(\varepsilon),$$
$$R_{K+1,0}(t = 0) = 0,$$
$$R_{K+2,0}(t = 0) = a_{K+2} + a_{K+2}^*(\varepsilon),$$

$$\vdots$$

$$R_{2K,0}(t = 0) = a_{2K} + a_{2K}^*(\varepsilon),$$
$$R_{2K+1,0}(t = 0) = a_{2K+1}^*(\varepsilon),$$

$$\vdots$$

$$R_{n,0}(t = 0) = a_n^*(\varepsilon), \tag{2.7.67}$$

and $R_{i,l}(t = 0) = 0$ for $i = 1, \ldots, n$.

In general it can happen that some of the functions Γ_i have the following structure $\Gamma_i = \varepsilon \bar{\Gamma}_i$. The necessary periodicity condition is then equivalent to either $\varepsilon = 0$ or $\bar{\Gamma}_i = 0$. The first assumption implies that the function $R_{i,0}(t)$ is periodic for all parameters a_i. In the second case we get

$$\bar{\Gamma}_i = \sum_{l=1}^{\infty} R_{i,l}(a_1 + a_1^*, \ldots, T, \ldots, a_{2K} + a_{2K}^*, a_{2K+1}^*, a_n^*)\varepsilon^{l-1} = 0. \tag{2.7.68}$$

For $\varepsilon = 0$ we have $a_i^*(0) = 0$, and from (2.7.68) we obtain

$$\bar{\Gamma}_i(a_1, \ldots, a_{2K}, 0, \ldots, 0, 0) = 0, \quad i = 1, \ldots, 2K. \tag{2.7.69}$$

Let us suppose that the algebraic nonlinear set of equations possesses one or a few solutions which we denote by

$$a_1 = \bar{a}_1,$$

$$\vdots \tag{2.7.70}$$

$$a_{2K} = \bar{a}_{2K},$$

and which can be multiplied. Then, we constrain such a solution to the periodicity condition

$$\Gamma_i(\bar{a}_1 + a_1^*, \ldots, \bar{a}_{2K} + a_{2K}^*, a_{2K+1}^*, \ldots, a_n^*, \varepsilon) = 0 \quad i = 1, \ldots, n. \quad (2.7.71)$$

When we introduce $a_1^* = \ldots = a_n^* = \varepsilon = 0$, then we have the left-hand side of (2.7.71) equal to zero. We are going to find the functions $a_i^*(\varepsilon)$ $(i = 1, \ldots, n)$ on the basis of the theory of the implicit functions with the condition $a_i^*(0) = 0$.

The functional determinant of (2.7.64) for (2.7.70) and for $a_1^* = \ldots = a_n^* = \varepsilon = 0$ has the form

$$\Delta_0 = \begin{vmatrix} \frac{\partial \Gamma_1}{\partial a_1} & \cdots & \frac{\partial \Gamma_1}{\partial a_n} \\ \vdots & & \\ \frac{\partial \Gamma_n}{\partial a_1} & \cdots & \frac{\partial \Gamma_n}{\partial a_n} \end{vmatrix} = 0. \quad (2.7.72)$$

If for $\bar{a}_1, \ldots, \bar{a}_n$ we get $\Delta_0 \neq 0$, then we have a simple solution, and in this case the unknown functions can be expressed by

$$a_i^*(\varepsilon) = \varepsilon A_{i1} + \varepsilon^2 A_{i2} + \varepsilon^3 A_{i3} + \ldots, \quad (2.7.73)$$

which fulfils the condition $a_i^*(0) = 0$.

If we get $\Delta_0 = 0$, then the solution is multiple and if the rank of the matrix built on the determinant is equal to $n-1$, then from (2.7.71) we exclude $n-1$ quantities. We finally obtain a new nonlinear algebraic equation of the form

$$F(\bar{a}_i + a_i^*, \varepsilon) = 0, \qquad i = 1, \ldots, n, \quad (2.7.74)$$

where for $2K+1 < i \leq n$ we have $\bar{a}_i = 0$, and i is equal to one of the numbers from 1 to n.

For the unknown $\bar{a}_i$ the problem reduces to finding $a_i^*(\varepsilon)$ as the implicit function of ε in the form of the power series of either the total or fractional powers of the parameter ε. The number of the solution branches as well as their series representations depend on the multiplicity solutions of (2.7.69). The latter is defined by the first nonvanishing term of $\partial^k F / \partial a_i^*$.

In the case when the rank of the matrix is equal to $n - 1$, n is large, or when the rank of that matrix is less than $n - 1$, there is still a method proposed by McMillan [64d].

2.8 Hopf Bifurcation

Bifurcation from equilibrium to dynamics is associated with the so-called Hopf bifurcation [116]. Up to now there have been many papers addressing this phenomenon, which is encountered in many branches of science [26, 27, 89, 91, 98, 15d, 16d, 18d–20d, 45d, 46d]. Here, we would like to present some examples of the analytical approach based on both the perturbation and harmonic balance methods similar to those obtained by Huseyin and his co-workers.

We begin with an analytical method of determining the post-critical family of the periodic solutions of some autonomous ordinary nonlinear differential equations dependent on one bifurcation parameter. It is known [116] that when the bifurcation parameter reaches the critical point, and when the system of linearized differential equations originating from the initial system of equations has complex conjugate eigenvalues of the Jacobian (the other values have non-zero real terms), and after fulfilling additionally the Hopf conditions, equilibrium path stability is lost and a family of periodic solutions appears. Denoting the bifurcation parameter as "η", the following case has been considered: for $\eta < \eta_c$ (where η_c is a critical point) the system has a stable equilibrium path. For $\eta = \eta_c$ the complex conjugate eigenvalues of the Jacobian of the characteristic equation occur (the others have negative non-zero terms), and together with an increase in η ($\eta > \eta_c$) the real terms of the eigenvalues become positive. The Hopf conditions ensure that the intersection of the imaginary axis with the complex conjugate eigenvalues occurs with non-zero velocity.

A method for searching for bifurcation solutions was presented in [91, 98]. Nevertheless, it is troublesome to use because of the time-consuming calculations for equations of dimension bigger than two. As the Hopf bifurcation occurs in nonlinear systems, it would be desirable to use for their analysis the analytical methods widely known in the field of nonlinear vibrations. The most popular and effective are the methods of harmonic balancing and the perturbation method. The main defect of the harmonic balancing method, which makes it useless for Hopf bifurcation analysis, is the necessity of *a priori* knowledge about the solution. An advantage of the perturbation method lies in constructing the solution by the subsequent solving of the perturbation equations of the linear differential equations, when it is only necessary to know the solutions of the undisturbed differential equation system. The combination of the two methods makes it possible to solve the Hopf problem (the method of harmonic balancing is used to solve each of the perturbation equations of the linear differential equations). We consider the system of differential equations, whose characteristic equation is of the form

$$(\sigma - \sigma_1)(\sigma - \sigma_2)P(\sigma) = 0, \qquad (2.8.1)$$

where

$$\sigma_{1,2} = \xi(\eta) \pm i\omega(\eta), \quad \omega(\eta_c) \neq 0, \quad \xi(\eta_c) = 0, \quad \frac{\partial \xi(\eta_c)}{\partial \eta} \neq 0,$$

and the roots of the polynomial $P(\sigma) = 0$ have negative real parts.

It results from the centre manifold theorem [88] that the critical subsystem is mainly responsible for the bifurcation and bifurcated solution, and for the qualitative assesment of the bifurcated solution it is possible to limit oneself only to the solution of the two-dimensional critical differential equation. Here this solution serves as the initial approximate solution of the full nonlinear differential equation system, and the "detailed" solution is determined by the method of successive approximations. The latter also makes it possible to solve the problem where there are nonanalytical nonlinearities.

Let us consider the differential equation system having the form

$$\frac{d}{dt}(x) = F(\eta, x), \quad x \in \mathbb{R}^n, \tag{2.8.2}$$

where η is the parameter vector and $F(\eta, x)$ is a nonlinear function, analytical in the state variables η and x. For the purpose of the further analysis it has been assumed that η is a one-dimensional bifurcation parameter. Let x_0 fulfil the equation

$$F(\eta, x_0) = 0. \tag{2.8.3}$$

Examination of the stability of the equilibrium path x_0 is known to be limited to the determination of the eigenvalues of the Jacobian $F_x(\eta, x_0)$, where

$$F_x(\eta, x_0) = \left(\frac{\partial F_i}{\partial x_j}\right)_{x=x_0}, \quad (i, j = 1, \ldots, n). \tag{2.8.4}$$

Let the equilibrium path x_0 for $\eta < \eta_c$ (the critical value of the parameter) be the stable solution of system (2.8.2). On the other hand, when $\eta = \eta_c$ the two complex conjugate eigenvalues cross the imaginary axis with non zero velocity, i.e. let

$$\sigma_1 = \xi(\eta) + i\omega(\eta), \quad \sigma_2 = \xi(\eta) - i\omega(\eta) \quad \text{and} \quad \xi(\eta_c) = 0,$$

$$\omega(\eta_c) = \omega_c \neq 0, \quad \text{and} \quad \left.\frac{\partial \xi}{\partial \eta}\right| \eta = \eta_c = \xi_\eta(\eta_c) \neq 0. \tag{2.8.5}$$

For $\eta > \eta_c$, the real parts σ_1, σ_2 become positive. A family of periodic solutions is created at the critical point.

Let us assume that equation system (2.8.2) can be presented in the form

$$\dot{u} = K_u(\eta)u + K_v(\eta)v + \tilde{K}(\eta, u, v), \quad u \in \mathbb{R}^2,$$

$$\dot{v} = S_u(\eta)u + S_v(\eta)v + \tilde{S}(\eta, u, v), \quad v \in \mathbb{R}^{n-2}, \tag{2.8.6}$$

where $x = \text{colon}(u, v)$, and the matrices $K_{(*)}(\eta)$, and $S_{(*)}(\eta)$ are the linear parts of the expansion of $F(\eta, x)$ into the Taylor series in the equilibrium path x_0.

Let the characteristic equation (2.8.6) have the form of (2.8.1), while $P(\sigma) = 0$ has roots with negative real parts, and σ_1 and σ_2 are the eigenvalues of the two-dimensional matrix $K_u(\eta)$. The matrix K_u, known also as the critical matrix, decides about the Hopf bifurcation.

From the centre manifold theorem it follows that in the neighbourhood of the equilibrium path $x_0 = 0$ there exists a function $v = f(u)$ which in a sufficiently close neighbourhood $x_0 = 0$ has the property $\partial f/\partial u = 0$. This allows us to assume to a first approximation that

$$v = f(u) = 0. \tag{2.8.7}$$

Taking into account (2.8.7) in (2.8.6), we obtain the following

$$\dot{u} = K_u(\eta)u + \tilde{K}(\eta, u, 0), \quad u \in \mathbb{R}^2. \tag{2.8.8}$$

Later we shall assume that $\tilde{K}(\eta, u, 0) = \tilde{K}(\eta, u)$.

Let us develop the matrix $K(\eta)$ into a Taylor series in the neighbourhood of the critical point

$$K_u(\eta) = K_u(\eta_c) + K_{u\eta}(\eta - \eta_c) + \frac{1}{2}K_{u\eta\eta}(\eta - \eta_c)^2 + \ldots, \tag{2.8.9}$$

where

$$K_{u\eta} = \left.\frac{\partial K_u(\eta)}{\partial \eta}\right|_{\eta=\eta_c} \qquad \text{etc.}$$

Let $u(\eta) = 0$ be the solution of (2.8.8) for $\eta < \eta_c$, and for $\eta = \eta_c$ the periodic solution $u(t; \varepsilon) = u(t + T; \varepsilon)$ of the period T bifurcates, which is dependent on one formally assumed small parameter ε connected with the amplitude. After the transformations this parameter can be arbitrarily assumed to be $\varepsilon = 1$.

In order to obtain the period $T = 2\pi$, we shall introduce the dimensionless time $\tau = \omega t$ and, as a result, we will obtain the following expression from (2.8.8):

$$\omega(\varepsilon)\frac{d}{dt}u(\tau; \varepsilon) = K(\eta)u(\tau; \varepsilon) + \tilde{K}(\eta, u(\tau; \varepsilon)). \tag{2.8.10}$$

The periodic solution $u(\tau; \varepsilon)$ will be sought in the form of a certain Fourier series, where the amplitudes and the frequencies depend on the parameter ε:

$$u_i(\tau; \varepsilon) = \sum_{k=0}^{K}(p_{ik}(\varepsilon)\cos k\tau + r_{ik}(\varepsilon)\sin k\tau). \tag{2.8.11}$$

Because the system (2.8.10) is autonomous, then $r_{11}(\varepsilon) \equiv 0$. Moreover, $p_{ik}(\varepsilon)$, $r_{ik}(\varepsilon)$, $\eta(\varepsilon)$ and $\omega(\varepsilon)$ are developed into a power series of the parameter ε of the following form

$$p_{ik}(\varepsilon) = p_{iko} + p'_{ik}\varepsilon + \frac{1}{2}p''_{ik}\varepsilon^2 + \ldots,$$

$$r_{ik}(\varepsilon) = r_{iko} + r'_{ik}\varepsilon + \frac{1}{2}r''_{ik}\varepsilon^2 + \ldots, \tag{2.8.12}$$

$$\eta(\varepsilon) = \eta_c + \eta'\varepsilon + \frac{1}{2}\eta''\varepsilon^2 + \ldots,$$

$$\omega(\varepsilon) = \omega_c + \omega'\varepsilon + \frac{1}{2}\omega''\varepsilon^2 + \ldots,$$

where $p_{iko} = r_{iko} = 0$, because $u_i(\tau; 0) = 0$ at the critical point.

The solution of $u_i(\tau; \varepsilon)$ is also sought in the form of a power series

$$u_i(\tau; \varepsilon) = u'_i(\tau)\varepsilon + \frac{1}{2}u''_i(\tau)\varepsilon^2 + \ldots, \tag{2.8.13}$$

where

$$u_i^{(\cdot)}(\tau) = \sum_{k=0}^{K} (p_{ik}^{(\cdot)} \cos k\tau + r_{ik}^{(\cdot)} \sin k\tau).$$ (2.8.14)

We can now proceed in two ways. We can either introduce relations (2.8.11)–(2.8.14) into (2.8.10) and by comparing terms in the same power ε obtain the perturbation equations of the linear differential equations, or obtain these equations by means of successive differentiation of (2.8.10) with respect to ε.

As an example, let us consider the mechanical system with $1\frac{1}{2}$ degrees of freedom, presented in Fig. 2.15. The vibration equations of the system have the form

$$m_1 u_{1tt} = -k_1 u_1 - k(u_1 - u_3)^3 + (\alpha u_1^2 - \eta)u_{1t},$$
$$c u_{3t} = -k_3 u_3 - k(u_3 - u_1)^3.$$ (2.8.15)

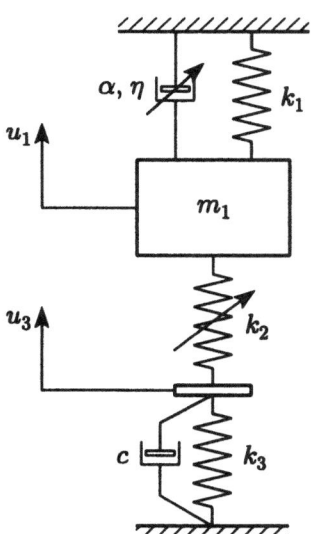

Fig. 2.15. Mechanical system with $1\frac{1}{2}$ degrees of freedom

The bifurcation parameter η is related to damping, and the other coefficients in (2.8.15) are positive. After applying the first step of the perturbations, we obtain

$$\omega u_{1\tau} = \frac{1}{m_1} u_2 - \frac{1}{3m_1} \alpha u_1^3 + \frac{1}{m_1} \eta u_1,$$
$$\omega u_{2\tau} = -k_1 u_1 - k(u_1 - u_3)^3,$$ (2.8.16)
$$\omega u_{3\tau} = -\frac{k_3}{c} u_3 - \frac{k}{c}(u_3 - u_1)^3.$$

The roots of the characteristic equations (2.8.16) are

$$\sigma_{1,2} = \frac{1}{2}\left(\frac{\eta}{m_1} \pm \sqrt{\left(\frac{\eta}{m_1}\right)^2 - \frac{4k_1}{m_1}}\right),$$

$$\sigma_3 = -\frac{k_3}{c}. \tag{2.8.17}$$

For $\eta = \eta_c = 0$ we have

$$\sigma_{1,2} = \pm i\omega_c = \pm i\sqrt{\frac{k_1}{m_1}}. \tag{2.8.18}$$

The first two equations of (2.8.16), after the assumption of $u_3 = 0$, have the form

$$\omega u_{1r} = \frac{1}{m_1}\eta u_1 + \frac{1}{m_1}u_2 - \frac{\alpha}{3m_1}u_1^3,$$

$$\omega u_{2r} = -k_1 u_1 - k u_1^3, \tag{2.8.19}$$

and we get

$$\sqrt{\frac{k_1}{m_1}}u_{1r}' = \frac{1}{m_1}u_2',$$

$$\sqrt{\frac{k_1}{m_1}}u_{2r}' = -k_1 u_1'. \tag{2.8.20}$$

Having taken into account $r_{11}' = 0$, we will obtain the following solution of Eqs. (2.8.20)

$$u_1' = p_{11}' \cos\tau,$$

$$u_2 = -\sqrt{k_1 m_1}p_{11}' \sin\tau. \tag{2.8.21}$$

From the second system of perturbation equations, we obtain

$$p_{21}'' = p_{22}'' = r_{21}'' = r_{22}'' = \eta' = \omega' = 0. \tag{2.8.22}$$

Finally, the third system of perturbation equations, after taking into account (2.8.22), has the form

$$\frac{1}{6}\omega_c u_{1r}''' + \frac{1}{2}\omega'' u_{1r}' = \frac{1}{6m_1}u_2''' + \frac{1}{2m_1}\eta'' u_1' - \frac{\alpha}{3m_1}(u_1')^3,$$

$$\frac{1}{6}\omega_c u_{2r}''' + \frac{1}{2}\omega'' u_{2r}' = -\frac{1}{6}k_1 u_1''' - k(u_1')^3. \tag{2.8.23}$$

Comparing the terms in $\sin\tau$ and $\cos\tau$ in (2.8.23), we obtain

$$-\frac{1}{6}\omega_c p_{11}''' = \frac{1}{2}\omega'' p_{11}' + \frac{1}{6m_1}r_{21}''',$$

$$\frac{1}{6}\omega_c r_{11}''' = \frac{1}{6m_1}p_{21}''' + \frac{1}{2m_1}\eta'' p_{11}' - \frac{9\alpha}{4m_1}(p_{11}')^3,$$

$$\omega_c p_{21}''' = k_1 r_{11}''',$$

$$\frac{1}{6}\omega_c r_{21}''' = -\frac{k_1}{6}p_{11}''' - \frac{3}{4}k(p_{11}') + \frac{1}{2}\omega''\sqrt{k_1 m_1}p_{11}',$$

and therefore we have

$$r_{11}''' = p_{21}''' = r_{21}''' = 0,$$

$$p_{11}''' = -\frac{9}{4}\frac{k}{k_1}(p_{11}')^3,$$

$$\omega'' = \frac{3}{4}\frac{k}{\sqrt{k_1 m_1}}(p_{11}')^2,$$

$$\eta'' = \frac{9}{2}\alpha(p_{11}')^2.$$

(2.8.25)

Comparing the terms in $\sin 3\tau$ and $\cos 3\tau$, we obtain

$$-\frac{1}{2}\omega_c p_{13}''' = \frac{1}{6m_1}r_{23}''',$$

$$\frac{1}{2}\omega_c r_{13}''' = \frac{1}{6m_1}p_{23}''' - \frac{3\alpha}{4m_1}(p_{11}')^3,$$

$$\frac{1}{2}\omega_c p_{23}''' = k_1 r_{13}''',$$

$$\frac{1}{2}\omega_c r_{23}''' = -\frac{k_1}{6}p_{13}''' + \frac{1}{4}k(p_{11}')^3.$$

(2.8.26)

and then we get

$$p_{13}''' = -\frac{3}{16}\frac{k}{k_1}(p_{11}')^3,$$

$$r_{13}''' = \frac{27}{16}\frac{\alpha}{\sqrt{m_1 k_1}}(p_{11}')^3,$$

$$p_{23}''' = \frac{9}{16}\alpha(p_{11}')^3,$$

$$r_{23}''' = \frac{9}{16k}\sqrt{\frac{m_1}{k_1}}(p_{11}')^3.$$

(2.8.27)

From the third equation of (2.8.16), we obtain

$$\omega u_{3\tau} = -\frac{k_3}{c}u_3 + \frac{k}{c}u_1^3.$$

(2.8.28)

After equating the terms in $\sin \tau$ and $\cos \tau$ of (2.8.28), we obtain

$$r_{31}''' = \frac{3k(p_{11}')^3}{4c\left(\sqrt{\frac{k_1}{m_1}} + \left(\frac{k_3}{c}\right)^2\sqrt{\frac{m_1}{k_1}}\right)},$$

$$p_{31}''' = \frac{3k_3 k(p_{11}')^3}{4c^2\left(\frac{k_1}{m_1} + \left(\frac{k_3}{c}\right)^2\right)}.$$

(2.8.29)

On the other hand, after equating the terms in $\sin 3\tau$ and $\cos 3\tau$ of (2.8.28), we obtain

$$r_{33}''' = \frac{k(p_{11}')^3}{4c\left(3\sqrt{\frac{k_1}{m_1}} + \frac{1}{3}\left(\frac{k_3}{c}\right)^2\sqrt{\frac{m_1}{k_1}}\right)},$$

$$p_{33}''' = \frac{k_3 k(p_{11}')^3}{12c^2\left(3\frac{k_1}{m_1} + \frac{1}{3}\left(\frac{k_3}{c}\right)^2\right)}. \tag{2.8.30}$$

We shall limit ourselves to terms in $(p_{11}')^3$ in calculations. The periodic bifurcation solution has the form

$$u_1 = p_{11}'\cos\tau - \frac{3}{8}\frac{k}{k_1}(p_{11}')^3\cos\tau - \frac{1}{32}(p_{11}')^3\cos 3\tau$$

$$+ \frac{9}{32}\frac{\alpha}{\sqrt{m_1 k_1}}(p_{11}')^3\sin 3\tau, \tag{2.8.31}$$

$$u_2 = -\sqrt{m_1 k_1}p_{11}'\sin\tau + \frac{3}{16}\alpha(p_{11}')^3\cos 3\tau + \frac{3}{16}k\sqrt{\frac{m_1}{k_1}}(p_{11}')^3\sin 3\tau,$$

$$u_3 = \frac{3k_3 k(p_{11}')^3}{4c^2\left(\frac{k_1}{m_1} + \left(\frac{k_3}{c}\right)^2\right)}\cos\tau + \frac{3k(p_{11}')^3}{4c\left(\sqrt{\frac{k_1}{m_1}} + \left(\frac{k_3}{c}\right)^2\sqrt{\frac{m_1}{k_1}}\right)}\sin\tau$$

$$+ \frac{k(p_{11}')^3}{4c\left(3\sqrt{\frac{k_1}{m_1}} + \frac{1}{3}\left(\frac{k_3}{c}\right)^2\sqrt{\frac{m_1}{k_1}}\right)}\cos 3\tau$$

$$+ \frac{kk_3(p_{11}')^3}{12c^2\left(3\frac{k_1}{m_1} + \frac{1}{3}\left(\frac{k_3}{c}\right)^2\right)}\sin 3\tau,$$

Finally, we obtain the amplitude parameter–frequency relations

$$\omega = \sqrt{\frac{k_1}{m_1} + \frac{3}{8}\frac{k}{\sqrt{k_1 m_1}}(p_{11}')^2} \tag{2.8.32}$$

and the parameter–amplitude relations

$$\eta = \frac{9}{4}\alpha(p_{11}')^2. \tag{2.8.33}$$

2.9 Stability Control of Vibro-Impact Periodic Orbit

2.9.1 Introduction

It is well known that mechanical vibro-impact systems have been widely employed in both theoretical and applied mechanics for a long time. Vibro-impact dynamics can be observed in many real engineering systems, such as hammer-like devices, ball-and-race dynamics in a ball bearing assembly, wheel-rail impact dynamics, etc. [41d, 63d].

Nowadays again this field of research has attracted strong interest, but in the framework of theories of modern dynamical systems. Recent industrial examples (tube fretting wear through vibro-impact behaviour in nuclear reactors or impacts between old and high buildings excited by earthquakes) belong to additional but not satisfactorily solved questions of discontinuous dynamical systems.

There are two parallel branches of investigations in the framework of vibro-impact dynamics. The first one is based on a better approximation of laws for impact motion and restitution coefficients, and it is more involved in the physics of materials. The second branch includes control of steady-state vibro-impact motion with the possibility of stability changes (either to destabilize or to stabilize the vibro-impact attractor).

Recently many papers have appeared, which are devoted to control of nonlinear oscillators, including also control of chaotic orbits [66d, 145].

In general, these methods could be devided for feedback control with a time delay [71d], sliding mode control [146], repetitive control [43d], iterative learning control [25], adaptive control [146], and so on. The main purpose of these methods is to control complicated systems, even with imprecise knowledge of their mathematical models. However, the control of the attractor or repeller is based on the numerical observations of the results by the introduction of "helping" control coefficients. Theoretical prediction are rather not given. Here we address one, not yet satisfactorily solved problem of vibro-impact dynamics control with delay feedback and we give an analytical prediction of the proper choice of control parameters.

2.9.2 Control of Vibro-Impact Periodic Orbits

We analyse the following one-degree-of-freedom vibro-impact system with one clearance presented in Fig. 2.16.

The equation of dynamics is as follows:

$$\ddot{x} + c\dot{x} + \alpha^2 x = P_0 \cos \omega t + A\left[x(t) - x(t-T)\right] +$$
$$+ B\left[\dot{x}(t) - \dot{x}(t-T)\right] \quad \text{for} \quad x < s, \qquad (2.9.1)$$
$$\text{and} \quad x_+ = x_-, \quad \dot{x}_+ = -R_r\dot{x}_- \quad \text{for} \quad x \geq s$$

where: $P_0 = y_0 k_2/m$, $c = c_1/m$, $\alpha^2 = (k_1 + k_2)/m$, $A = k_2 a_1/m$, $B = k_2 b_1/m$, and $T = 2\pi/\omega$ is the period of the periodic orbit being stabilized.

A key point of this control is that a periodic solution possesses the same period as the excitation, i.e. $x_0 = x_0(t - T)$ and x_0 is a particular solution of both the controlled and uncontrolled system [52d]. The delay loop is switched off where perturbations are not present. In the case of perturbations the controller causes the perturbation to vanish more quickly than in

a)

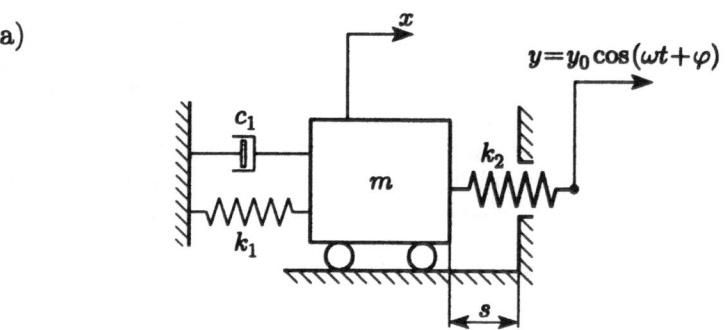

b)

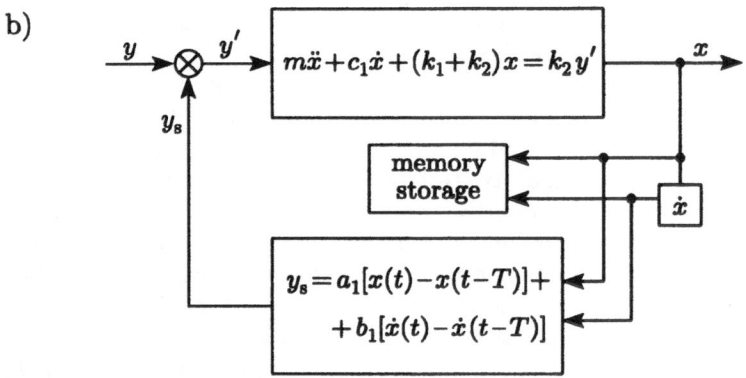

Fig. 2.16. One-degree-of-freedom kinematically excited vibro-impact system with one clearance (**a**) and its control diagram (**b**) (s denotes the clearance)

the case without control. The problem of analytical estimation of the influence of control coefficients for periodic orbit stability cannot be solved in a standard way. Here we propose the following approach. Because in practise the differences $x(t)-x(t-T)$ and $\dot{x}(t)-\dot{x}(t-T)$ are small we express them by introducing the small parameter ε, which allows us then to apply the KBM method formally and next to take $\varepsilon = 1$ [27d].

We assume damping of the same order as ε and from (2.9.1), we obtain

$$\ddot{x} + \alpha^2 x = P_0 \cos \omega t + \varepsilon A \left[x(t) - x(t - T) \right]$$
$$+ \varepsilon B \left[\left(1 - \frac{c}{B} \right) \dot{x}(t) - \dot{x}(t - T) \right]. \qquad (2.9.2)$$

Introducing

$$x = z + \frac{P_0}{\alpha^2 - \omega^2} \cos \omega t, \qquad (2.9.3)$$

we get

$$\ddot{z} + \alpha^2 z = \varepsilon f(a, \eta, \psi), \tag{2.9.4}$$

where

$$\varepsilon f(a, \eta, \psi) = \varepsilon A \left[z + \frac{P_0}{\alpha^2 - \omega^2} \cos \omega t - z(t - T) \right.$$
$$\left. - \frac{P_0}{\alpha^2 - \omega^2} \cos \omega(t - T) \right] + \varepsilon B \left[\left(1 - \frac{c}{B} \right) \right.$$
$$\left. \cdot \left(\dot{z} - \frac{P_0 \omega}{\alpha^2 - \omega^2} \sin \omega t \right) - \dot{z}(t - T) + \frac{P_0 \omega}{\alpha^2 - \omega^2} \sin \omega(t - T) \right],$$

$$\eta = \omega t, \qquad \psi = \alpha t.$$

Using the KBM method we have truncated the ε series up to order $O(\varepsilon)$ and we have obtained

$$\frac{da}{dt} = \frac{1}{2}(B - c)a + \frac{Aa}{2\alpha} \sin \alpha T - \frac{Ba}{2} \cos \alpha T,$$
$$\frac{d\psi}{dt} = \alpha - \frac{A}{2\alpha} + \frac{A}{2\alpha} \cos \alpha T + \frac{1}{2} B \sin \alpha T. \tag{2.9.5}$$

For $A = B = 0$ we get the uncontrolled solution, which supperts the validity of our approach.

Therefore, we analyse the following equivalent solution

$$x = \frac{P_0}{\alpha^2 - \omega^2} \cos(\omega t + \varphi) + e^{Rt}(C \cos \alpha_0 t + D \sin \alpha_0 t), \tag{2.9.6}$$

where: $C = s - P_0 \cos \varphi / \sqrt{(\alpha_0^2 - \omega^2)^2 + c^2 \omega^2}$, $D = (C/\sin 2\beta\lambda)(e^{\beta c} - \cos 2\beta\lambda)$, $\lambda^2 = \alpha^2 - \omega^2$, $\beta = \pi k/\omega$, $\cos \varphi = (\sqrt{(\alpha^2 - \omega^2)^2 + c^2 \omega^2}/P_0)[s - (R_\tau + 1)\dot{x}_- \sin 2\beta\lambda / \lambda(2 \cos 2\beta\lambda - e^{-\beta c} - e^{\beta c})]$.

After integration of (2.9.5), we get

$$a(t) = C_0 e^{Rt}, \quad \psi(t) = \alpha_0 t + \theta_0,$$
$$R = \frac{1}{2}(B - c) + \frac{A}{2\alpha} \sin \alpha T - \frac{B}{2} \cos \alpha T, \tag{2.9.7}$$
$$\alpha_0 = \alpha - \frac{A}{2\alpha} + \frac{A}{2\alpha} \cos \alpha T + \frac{1}{2} B \sin \alpha T,$$

and according to (2.9.7) and (2.9.6) one obtains

$$C = C_0 \cos \theta_0, \quad D = -C_0 \sin \theta_0, \quad C_0 = \sqrt{C^2 + D^2}.$$

2.9.3 Stability Control

From (2.9.6) it is seen that when $R < 0$ the assumed solution is stabilised more quickly in comparison to the case of $R = 0$. However, the problem of the stability investigation of the vibro-impact state is much more subtle.

Before impact number l, the mass possesses the velocity x_{l-}. This causes the following perturbation solution to occur

$$x + \delta x_l = e^{R\tau_l}\left[(C + \delta C_l)\cos\alpha_0\tau_l + (D + \delta D_l)\sin\alpha_0\tau_l\right]$$

$$+ F\cos(\omega\tau_l + \varphi + \delta\varphi_l), \quad F = \frac{P_0}{\alpha^2 - \omega^2}. \tag{2.9.8}$$

A new time τ is measured from the l-th impact $\tau_l = \tau + \delta\tau_l$. For example, the next impact occurs for $\tau_{l+1} = 2\pi/\omega + \delta T_l$, where δT_l denotes the perturbation period $T = 2\pi/\omega$.

After some calculations we get

$$\delta x_l = e^{R\tau}\left(-C\alpha_0\delta\tau_l\sin\alpha_0\tau + \delta C_l\cos\alpha_0\tau + D\alpha_0\delta\tau_l\cos\alpha_0\tau\right.$$

$$+\delta D_l\sin\alpha_0\tau + R\delta\tau_l C\cos\alpha_0\tau + R\delta\tau_l D\sin\alpha_0\tau)$$

$$-F\delta\varphi_l\sin(\omega\tau + \varphi) - F\omega\delta\tau_l\sin(\omega\tau + \varphi),$$

$$\delta\dot x_l = e^{R\tau}\left[2R\alpha_0\delta\tau_l\left(D\cos\alpha_0\tau - C\sin\alpha_0\tau\right) + R\delta C_l\cos\alpha_0\tau\right.$$

$$+R\delta D_l\sin\alpha_0\tau + R^2\delta\tau_l C\cos\alpha_0\tau + R^2\delta\tau_l D\sin\alpha_0\tau$$

$$-\alpha_0^2\delta\tau_l\left(C\cos\alpha_0\tau + D\sin\alpha_0\tau\right) - \delta C_l\alpha_0\sin\alpha_0\tau$$

$$+ \delta D_l\alpha_0\cos\alpha_0\tau] - F\omega\delta\varphi_l\cos(\omega\tau + \varphi) - F\omega^2\delta\tau_l\cos(\omega\tau + \varphi).$$

The following boundary conditions are introduced

$$l: \quad \tau = 0, \quad \delta\tau_l = 0, \quad \delta x_l = 0, \quad \delta\dot x_l = \delta\dot x_{l+}, \tag{2.9.9}$$

$$l+1: \quad \tau = \frac{2\pi k}{\omega} + \delta\tau_l, \quad \delta\tau_l = \delta T_l, \quad \delta x_l = 0, \quad \delta\dot x_l = \delta\dot x_{(l+1)-}.$$

After some calculations we obtain the equations

$$\delta C_l - F\delta\varphi_l\sin\varphi = 0, \tag{2.9.10}$$

$$e^{2\beta R}\left\{\delta C_l\cos 2\beta\alpha_0 + \delta D_l\sin 2\beta\alpha_0 + \frac{1}{\omega}\left(\delta\varphi_{l+1} - \delta\varphi_l\right)\right.$$

$$\left.\cdot\left[(\alpha_0 D + RC)\cos 2\beta\alpha_0 + (RD - \alpha_0 C)\sin 2\beta\alpha_0\right]\right\} - \delta C_{l+1} = 0,$$

$$R_r e^{2\beta R}\left\{\delta C_l\left(R\cos 2\beta\alpha_0 - \alpha_0\sin 2\beta\alpha_0\right) + \delta D_l\left(R\sin 2\beta\alpha_0\right.\right.$$

$$+ \alpha_0\cos 2\beta\alpha_0) + \frac{1}{\omega}\left(\delta\varphi_{l+1} - \delta\varphi_l\right)\left[\left(R^2 C - \alpha_0^2 C\right.\right.$$

$$+ 2R\alpha_0 D)\cos 2\beta\alpha_0 + \left(R^2 D - \alpha_0^2 D - 2R\alpha_0 C\right)\sin 2\beta\alpha_0\right]\}$$

$$+R\delta C_{l+1} + \alpha_0\delta D_{l+1} - \left(R_r + 1\right)F\omega\delta\varphi_{l+1}\cos\varphi = 0,$$

where $\beta = \pi k/\omega$ and $R_r \leq 1$ as usual denotes the restitution coefficient. Assuming that

$$\delta\varphi_l = \delta\varphi_0 + \sum_{i=1}^{l}\omega\delta T_i, \tag{2.9.11}$$

and introducing

$$\delta C_l = a_1 \gamma^l, \quad \delta D_l = a_2 \gamma^l, \quad \delta \varphi_l = a_3 \gamma^l \tag{2.9.12}$$

we get the following characteristic equation

$$b_2 \gamma^2 + b_1 \gamma + b_0 = 0, \tag{2.9.13}$$

where

$$b_2 = \alpha_0 \left\{ F \sin \varphi - \frac{1}{\omega} e^{2\beta R} \left[(\alpha_0 D + RC) \cos 2\beta \alpha_0 \right. \right.$$
$$\left. \left. + (RD - \alpha_0 C) \sin 2\beta \alpha_0 \right] \right\},$$

$$b_1 = e^{2\beta R} \left\{ \frac{1}{\omega} \alpha_0 \left[(RC + \alpha_0 D) \left(\cos 2\beta \alpha_0 - R_r e^{2\beta R} \right) \right. \right.$$
$$\left. + (RD - \alpha_0 C) \sin 2\beta \alpha_0 \right] - (R_r + 1) F\omega \cos \varphi \sin 2\beta \alpha_0$$
$$\left. + F \sin \varphi \left[(R_r - 1) \alpha_0 \cos 2\beta \alpha_0 + (R_r + 1) R \sin 2\beta \alpha_0 \right] \right\}, \tag{2.9.14}$$

$$b_0 = R_r \alpha_0 e^{4\beta R} \left[\frac{1}{\omega} (RC + \alpha_0 D) - F \sin \varphi \right].$$

Note that $C(s)$ could be obtained using a similar approach but without the perturbations.

Therefore, the problem of stability is reduced to consideration of the second-order characteristic equation (2.9.12). If the roots of (2.9.13) are $|\gamma_{1,2}| < 1$, then according to the assumed solutions (2.9.12) δC_l, δD_l and $\delta \varphi_l$ approach zero for $l \to \infty$, and the solutions will be asymptotically stable. We can easily estimate the stability regions, which are defined by the following inequalities

$$\left| \frac{b_2}{b_0} \right| < 1 \quad \text{and} \quad \left| \frac{b_1}{b_0 + b_2} \right| < 1. \tag{2.9.15}$$

Taking into account (2.9.14) it is now easy to find parameters of the system (or a delay loop) which fulfil inequalities (2.9.15). Additionally, for mechanical reasons, we have $x(t) \le s$.

2.9.4 Simulation Results

During numerical simulations we used the MATLAB-Simulink package and the MATLAB-model for (2.9.1) with the boundary conditions (Fig. 2.17).

We have taken the following parameters: $m = 1[\text{kg}]$, $k_1 = 3[\text{N/m}]$, $k_2 = 1[\text{N/m}]$, $c_1 = 0$ [Ns/m], $x_0 = 1[\text{m}]$, $T = 0.4\pi[\text{s}]$, $R_r = 0.65$, $s = 0.01[\text{m}]$, $a_l = -0.01[\text{N/m}]$, $b_l = -0.045[\text{Ns/m}]$.

For these parameters according to (2.9.7) we get $R = -0.04$, which shows that the delay loop control coefficients A and B allow us to obtain quicker

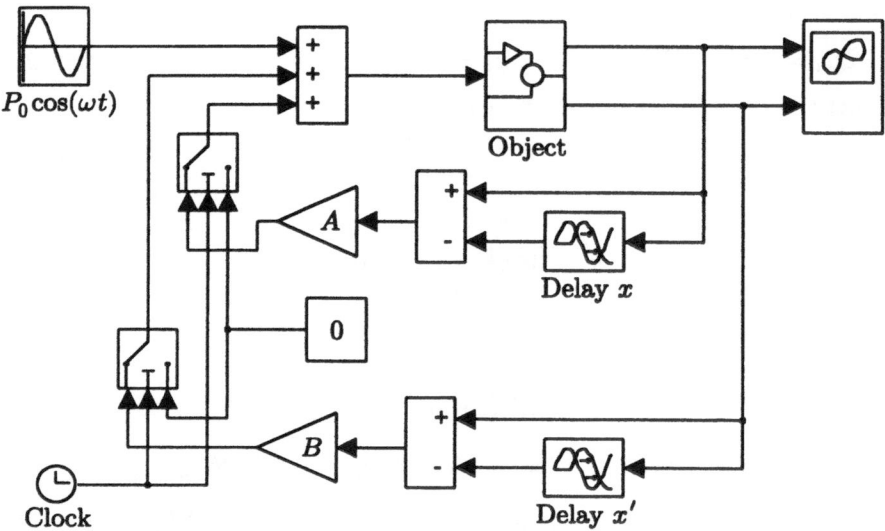

$P_0 \cos(\omega t)$

Object

Delay x

0

A

B

Clock

Delay x'

Fig. 2.17. MATLAB-Simulink model of the vibro-impact system presented in Fig. 2.16

damping of free oscillations in the solution (2.9.6) than without control (Fig. 2.18). Additionally, for the given parameters we have found from Eq. (2.9.13) that $|\gamma_{1,2}|$ lie closer to the origin for the system with the control coefficients than without control (with the delay loop $|\gamma_{1,2}| = 0.62$, whereas without the loop $|\gamma_{1,2}| = 0.65$).

For the given parameters numerical simulations confirm the analytical predictions.

It can be seen in Fig. 2.18 that with control the transients vanish more quickly than in the case without control. In the case presented above the periodic orbit is achieved after about 24.1 seconds for the system analysed without the delay loop and after 21.8 seconds for the system analysed with the delay loop ($|\mu| \leq 10^{-4}$), respectively.

2.10 Normal Modes of Nonlinear Systems with n Degrees of Freedom

2.10.1 Definition

The normal mode of a finite-dimensional system behaves like a conservative system having a single degree of freedom. In this case all position coordinates can be well defined by

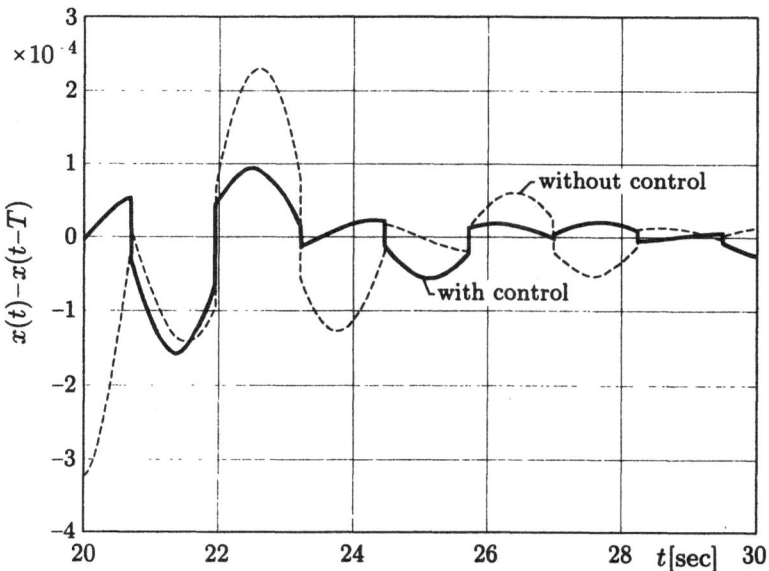

Fig. 2.18. Difference between two transients $x(t) - x(t-T)$ approaching a periodic orbit for the system $(a_1 = -0.01, b_1 = -0.045)$

$$x_i = p_i(x), \quad x \equiv x_1, \quad i = 2, 3, \ldots, n, \tag{2.10.1}$$

$p_i(x)$ being analytical functions.

Rosenberg [135] gets credit for being the first to introduce broad classes of essentially nonlinear conservative systems allowing normal vibrations with rectilinear trajectories in a configurational space

$$x_i = k_i x_1, \quad i = 2, 3, \ldots, n. \tag{2.10.2}$$

"A phenomenon like 'vibration in normal modes' exists in nonlinear systems; in a purely verbal manner, this motion may be described as vibration in which all masses move periodically "in unison", among all the properties of normal vibrations we consider asymptotic the basic one that the normal coordinates decouple the equation of motion" [135].

For instance, homogenous systems whose potential is an even homogenous function of the variables refer to such a class. It is interesting to note that the number of modes of normal vibrations in the nonlinear case can exceed the number of degrees of freedom. This remarkable property has no analogy in the linear (non-degenerate) case.

Naturally this definition may now be called "naive", and the problem of a more accurate definition was studied by many authors [55d]. But it is very clear from the intuitive point of view and we accept Rosenberg's definition form the asymptotic standpoint in our further investigations.

In systems of a more general type, trajectories of normal vibrations are curvilinear. Lyapunov [54d] showed that solutions of this kind exist in non-linear finite-dimensional systems with an analytical first integral, which are close to generating linear systems.

2.10.2 Free Oscillations and Close Natural Frequencies[1]

For systems with two degrees of freedom which have quadratic nonlinearities, the internal resonance at the frequency ratio 1:2 has been studied, along with the internal resonance for systems with cubic nonlinearities and the frequency ratio 1:3 [119]. In recent years attention has turned to mode interactions (of an internal resonanse type) for close natural frequencies. Experimantal observations and solutions of particular problems show that this effect is relevant to the description of oscillatory processes in suspension bridges [1], cylindrical shells and other constructions [41–52, 54–77, 28d–35d]. However, the literature does not contain any general analysis of mode interactions of free oscillations in nonlinear systems with close natural frequencies. In particular, we do not know what types of oscillation modulation are possible, what determines the degree and period of energy exchange in a system, what is the number of steady-state regimes (without modulation), which of them are stable, etc.

Consider a nonlinear oscillatory system (initially, in general, with damping), described by the equations

$$\ddot{u}_k + 2\mu_* \dot{u}_k + \omega_*^2 u_k = b_{kk} u_k^3 + b_{12} u_1^k u_2^{3-k}, \qquad k = 1, 2. \tag{2.10.3}$$

The frequencies ω_1 and ω_2 are assumed to be close, and the damping factors for the two modes are taken to be the same.

Equations (2.10.3) represent the general case of a system with symmetric potentials (when $\mu_* = 0$) that include terms of the second and fourth degree. They are similar to the Duffing equation for systems with one degree of freedom and describe a broad class of mechanical systems (in general, no restrictions are imposed on the coefficients b_{ij}).

In accordance with the method of multiple scales we introduce the "fast" and "slow" times $T_0 = t$, $T_1 = \varepsilon T_0$, $T_2 = \varepsilon^2 T_0$ (the time T_1 will not be necessary below). We will seek a solution of system (2.10.3) in the form of an expansion

$$u_k = \varepsilon u_{k1}(T_0, T_1, T_2, \ldots) + \varepsilon^3 u_{k3}(T_0, T_1, T_2, \ldots) + \ldots \tag{2.10.4}$$

(terms of order ε^2 vanish for a system with cubic nonlinearities).

The smallness of μ_* and the difference in frequency are introduced through the conditions

$$\mu_* = \varepsilon^2 \mu, \quad \omega_2^2 = \omega^2 + \varepsilon^2 \sigma, \quad \omega_1 \equiv \omega. \tag{2.10.5}$$

---

[1] By courtesy Ye.V. Ladygina, A.I. Manevitch [106]

Taking into account that

$$\frac{d}{dt} = D_0 + \varepsilon D_1 + \varepsilon^2 D_2 + \ldots; \qquad D_0 = \frac{\partial}{\partial T_0}, \; D_1 = \frac{\partial}{\partial T_1}, \; D_2 = \frac{\partial}{\partial T_2},$$

$$\frac{d^2}{dt^2} = D_0^2 + 2\varepsilon D_0 D_1 + \varepsilon^2(2D_0 D_2 + D_1^2) + \ldots,$$

we obtain the following systems of equations for the two approximations

$$D_0^2 u_{k1} + \omega^2 u_{k1} = 0, \tag{2.10.6}$$

$$D_0^2 u_{k3} + \omega^2 u_{k3} = -2D_0(D_2 u_{k1} + \mu u_{k1}) + b_{kk} u_{k1}^3$$
$$+ b_{12} u_{11}^k u_{21}^{3-k} - \delta_{2k} \sigma u_{k1}, \tag{2.10.7}$$

where δ_{ij} is the Kronecker delta.

The solution of system (2.10.6) is written in the form

$$u_{1k} = A_k(T_2) \exp(i\omega T_0) + \bar{A}_k(T_2) \exp(-i\omega T_0), \tag{2.10.8}$$

where the bars denote complex conjugats.

Substituting (2.10.8) into system (2.10.7), from the condition that there are no secular terms in the resulting equations, we have

$$-2i\omega(A_k' + \mu A_k) + 3b_{kk} A_k^2 \bar{A}_k + 2b_{12} A_k A_{3-k} \bar{A}_{3-k}$$
$$+ b_{12} A_k A_{3-k}^2 - \delta_{2k} \sigma A_k = 0, \tag{2.10.9}$$

where the prime denotes differentiation with respect to T_2.

Putting the complex amplitude in the exponential form $A_k = 1/2 a_k \exp(i\theta_k)$ ($k = 1, 2$), we separate (2.10.9) into real and imaginary parts and obtain a system of equations governing the amplitude and phase modulation of both modes:

$$(a_k^2)' + 2\mu a_k^2 = (-1)^k b_{12}(4\omega)^{-1} a_1^2 a_2^2 \sin 2\gamma, \tag{2.10.10}$$

$$8\omega\theta_k' = -3b_{kk} a_k^2 - b_{12} a_{3-k}^2 (2 + \cos 2\gamma) + 4\delta_{2k}\sigma, \quad \gamma = \theta_2 - \theta_1. \tag{2.10.11}$$

Having eliminated $\sin 2\gamma$, we obtain from (2.10.10) the integral

$$a_1^2 + a_2^2 = E \exp(-2\mu T_2) = E \exp(-2\varepsilon^2 \mu t), \tag{2.10.12}$$

where the arbitrary constant E is proportional to the energy of the system (in the first approximation). In particular, for a conservative system ($\mu = 0$)

$$a_1^2 + a_2^2 = E. \tag{2.10.13}$$

From (2.10.11) we obtain the equation for the phase difference γ:

$$8\omega\gamma' = (3b_{11} - 2b_{12})a_1^2 + (2b_{12} - 3b_{22})a_2^2 + b_{12}(a_2^2 - a_1^2)\cos 2\gamma + 4\sigma. \tag{2.10.14}$$

For further analysis it is convenient [1] to introduce the new variable $\xi = a_1^2/E$ ($0 \leq \xi \leq 1$). Then, from (2.10.11) with $k = 1$ and (2.10.14) we obtain a system of equations governing the amplitude–frequency modulation in ξ, γ variables:

$$\xi' = -2\mu\xi + \Gamma_0\xi(1-\xi)\sin 2\gamma,$$

$$\gamma' = \xi\Gamma_1 + \Gamma_2 + \Gamma_0(\frac{1}{2} - \xi)\cos 2\gamma. \tag{2.10.15}$$

Here

$$\Gamma_0 = \frac{b_{12}E}{4\omega}, \qquad \Gamma_1 = \frac{(3b_{11} - 4b_{12} + 3b_{22})E}{8\omega},$$

$$\Gamma_2 = \frac{(2b_{12} - 3b_{22})E + 4\sigma}{8\omega}. \tag{2.10.16}$$

Without loss of generality we assume $\Gamma_0 \neq 0$, because otherwise $b_{12} = 0$ and system (2.10.3) decomposes into two decoupled equations.

We will perform the further analysis for the case of a conservative system ($\mu = 0$). Dividing the second equation of (2.10.15) by the first one, we obtain

$$\frac{d\gamma}{d\xi} = \frac{\xi(\Gamma_1 - \Gamma_0 \cos 2\gamma) + \Gamma_2 + 1/2\Gamma_0 \cos 2\gamma}{\Gamma_0\xi(1-\xi)\sin 2\gamma}. \tag{2.10.17}$$

The solution of this ordinary differential equation is

$$\Gamma_0\xi(1-\xi)\cos 2\gamma + \Gamma_1\xi^2 + 2\Gamma_2\xi = C, \tag{2.10.18}$$

where C is a constant of integration and determines the trajectory in the (ξ, γ) plane of the "amplitude-phase portrait" of the system (the AP-portrait). Eliminating γ from the first equation of (2.10.15) and (2.10.18), we obtain

$$\frac{1}{\Gamma_0^2}\left(\frac{d\xi}{dT_2}\right)^2 = F_1^2(\xi) - F_2^2(\xi), \tag{2.10.19}$$

$$F_1(\xi) = \xi(1-\xi), \qquad F_2(\xi) = \frac{\Gamma_1\xi^2 + 2\Gamma_2\xi - C}{\Gamma_0}.$$

The form of this equation is identical with that derived in [1] for the case when $\omega_2 = 3\omega_1$, but the functions $F_1(\xi)$ and $F_2(\xi)$ are of a different form.

The condition

$$|F_1| \geq |F_2| \tag{2.10.20}$$

for the solution of (2.10.19) to exist means that solutions correspond to the parabolic segments $F_2(\xi)$ inside the domain bounded by the parabolic arcs $\pm F_1(\xi)$ over the interval $[0,1]$ (Fig. 2.19). The points of intersection of the parabolas $F_2(\xi)$ with the axis ξ, as can be seen from (2.10.18), correspond to the condition $\cos 2\gamma = 0$, i.e. $\gamma = \pm(2n+1)\pi/4$ ($n = 0, 1, 2, \ldots$), or the values $\xi = 0$ and $\xi = 1$. The points of intersection of the parabolas $F_2(\xi)$ and $F_1(\xi)$ correspond to extremal values of the functions $\xi(T_2)$ and $a_k(T_2)$, respectively ($k = 1, 2$). It follows from (2.10.15) (for $\mu = 0$) that at these points $\sin 2\gamma = 0$ (if $\xi \neq 0$ and $\xi \neq 1$), i.e. $\gamma = \pm n\pi/2$ ($n = 0, 1, 2, \ldots$), and that the points on the lower curve ($F_1 < 0$) correspond to the even values of n and those on the upper curve to the odd values. Consequently, the minimum and maximum values of ξ determining the amplitude modulations and the degree of energy exchange between the modes are equal to the roots of the equations $F_1(\xi) = \pm F_2(\xi)$, i.e. the equations

$$\xi^2(\mp\Gamma_1 - \Gamma_0) + \xi(\Gamma_0 \mp 2\Gamma_2) \pm C = 0, \tag{2.10.21}$$

where the upper and lower signs correspond to the points of intersection of the parabola $F_2(\xi)$ with the upper and lower curves $\pm F_1(\xi)$, respectively, while the value of C is governed by the initial values ξ_0 and γ_0.

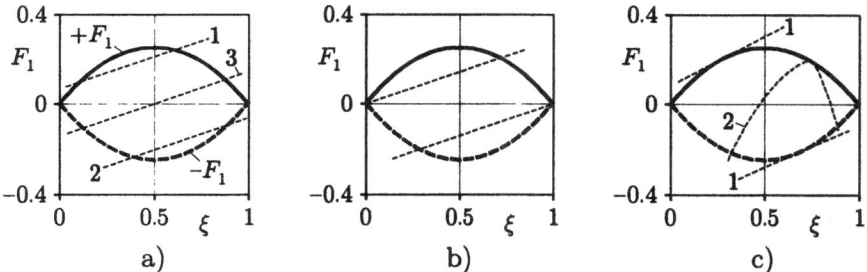

Fig. 2.19a-c. Graphical representation of the oscillatory regime

The solutions of (2.10.21) and the construction of a "characteristic graph" (Fig. 2.19) give a graphical representation of the oscillatory regime. There are two basic ways in which the curves $F_1(\xi)$ and $\pm F_2(\xi)$ can intersect in a "coarse" system, corresponding to the two basic oscillatory regimes:

1. both the points of intersection lie on the same parabola $+F_1(\xi)$ or $-F_1(\xi)$ (curves 1 and 2 in Fig. 2.19a);
2. the parabola $F_2(\xi)$ intersects both the parabolas $+F_1(\xi)$ and $-F_1(\xi)$ in the interval $[0, 1]$ (curve 3).

In the first case, the phase difference γ oscillates about the value $\gamma = \pm n\pi/2$. Synchronization of the oscillations proceeds "on average" over the modulation period: at the times when the extrema a_1 and a_2 are achieved the oscillations on both degrees of freedom proceed either in phase or in antiphase, if both the points of intersection lie on the lower branch, or the phase difference in these instants is equal to $\pi/2$, $(3\pi/2)$ if both the points lie on the $F_1 > 0$ branch.

In the second case the phase difference increases monotonically, running through the sequential values $n\pi/2$ ($n = 0, 1, \dots$) at the extremal times. These two types of oscillatory regime with oscillating and monotonically increasing phase differences will respectively be called modulations of the first and second type.

The solution of (2.10.19) has the form

$$\pm\frac{1}{|\Gamma_0|}\int_{\xi_0}^{\xi} \left(F_1^2(\xi) - F_2^2(\xi)\right)^{-1/2} d\xi = T_2 - T_{20}, \quad \xi_0 = \xi(T_{20}). \tag{2.10.22}$$

Suppose $\xi_1, \dots, \xi_4$ are the roots of the fourth-degree polynomial $F_1^2(\xi) - F_2^2(\xi)$, arranged in increasing order, with ξ_2 and ξ_3 lying inside the domain bounded by the parabolas $\pm F_1(\xi)$. The modulation semiperiod (for the oscillating phase case) corresponds to ξ varying over the interval (ξ_2, ξ_3), and so the modulation period is equal to

$$T^* = \frac{2}{|\Gamma_0^2 - \Gamma_1^2|^{1/2}} \int\limits_{\xi_2}^{\xi_3} ((\xi - \xi_1)(\xi - \xi_2)(\xi - \xi_3)(\xi - \xi_4))^{-1/2} \, d\xi. \quad (2.10.23)$$

For "non-coarse" systems one must consider singular cases for the position of the $F_2(\xi)$ curve (Figs 2.19b and c): the passage of $F_2(\xi)$ through the points $\xi = 0$ or $\xi = 1$ (Fig. 2.19b), and "external" touching of the parabolas $F_1(\xi)$ and $F_2(\xi)$ (corresponding to lines 1 and 2 in Fig. 2.19c). In these cases two of the roots ξ_j coincide: in the first case $\xi_1 = \xi_2 = 0$ or $\xi_3 = \xi_4 = 1$, and in cases 2 and 3 $\xi_2 = \xi_3$. Because the improper integral in (2.10.23) diverges when two of the roots ξ_j coincide, the modulation period tends to infinity as these regimes are approached. These are "boundary" regimes separating the modulations of the two types distinguished above (lines 1 and 3) and associated with separatrices in the plane (ξ, γ), or regimes of stationary oscillations without modulation (curve 2). We remark that the "aperiodic" oscillations described in [1d] correspond to these boundary regimes.

Consider the possible AP-portraits in the plane (ξ, γ) which are given by integral (2.10.18) and which graphically describe the oscillatory modes of the system.

Stationary points corresponding to oscillations with no modulation are found using (2.10.15) from the system of equations

$$\xi(1 - \xi) \sin 2\gamma = 0,$$
$$\xi\Gamma_1 + \Gamma_2 + \Gamma_2(1/2 - \xi) \cos 2\gamma = 0, \quad (2.10.24)$$

which can have the solutions

$$\xi = 0, \quad \cos 2\gamma = -\frac{2\Gamma_2}{\Gamma_0}, \quad (2.10.25)$$

$$\xi = 1, \quad \cos 2\gamma = \frac{2(\Gamma_1 + \Gamma_2)}{\Gamma_0}, \quad (2.10.26)$$

$$\gamma = \pm\frac{n\pi}{2} \quad (n = 0, 1, 2, \ldots), \quad \xi = \xi_*^\pm = \frac{\pm\Gamma_0/2 - \Gamma_2}{\Gamma_1 \pm \Gamma_0}. \quad (2.10.27)$$

These solutions exist when the following conditions are satisfied:

(1) $|2\Gamma_2| \leq |\Gamma_0|$,   (2) $2|\Gamma_1 + \Gamma_2| \leq |\Gamma_0|$,
(3) $0 \leq \xi_*^+ \leq 1$,   (4) $0 \leq \xi_*^- \leq 1$.   (2.10.28)

Using the periodicity with respect to γ, we confine ourselves to the plane rectangle $(0 \leq \xi \leq 1; 0 \leq \gamma \leq \pi)$. The stationary points (2.10.25)–(2.10.27) can be positioned on the boundary lines of this rectangle and on the mid-line $\gamma = \pi/2$ (with not more than one point on a line). It is easiest to investigate the nature of a stationary point with the help of (2.10.18), considering the form of the integral curves in the neighbourhood of the stationary point. The stationary points at $\xi = 0$ and $\xi = 1$ are saddle points and therefore unstable. From this it follows that the presence of a second degree of freedom makes oscillations along the first generalized coordinate unstable if the stationary points (2.10.25) or (2.10.26) exist.

In the neighbourhoods of the stationary points on the lines $\gamma = \pm n\pi/2$ the trajectories can be of either elliptic or hyperbolic type, and consequently these stationary points can be stable or unstable. The stability conditions for odd and even n, respectively, have the forms

$$(5)\ \Gamma_0(\Gamma_0 + \Gamma_1) > 0, \qquad (6)\ \Gamma_0(\Gamma_0 - \Gamma_1) < 0. \tag{2.10.29}$$

On the characteristic graph (Fig. 2.19c) the "externally" touching hyperbolas (curve 1) correspond to the stable stationary points and the "internally" touching ones (curve 2) to the unstable points.

The stationary points on the lines $\gamma = \pm n\pi/2$ correspond to synchronous single-frequency modes, i.e. normal oscillations of the nonlinear system. It follows from (2.10.3) and (2.10.8) that the points on the lines $\gamma = 0$ and $\gamma = \pi$ correspond in the (u_1, u_2) configuration space to the two straight lines $u_2 = \pm h u_1$, where

$$h = \frac{a_2}{a_1} = \left(\frac{1 - \xi_*^-}{\xi_*^-}\right)^{1/2} = \left(\frac{\Gamma_0/2 - \Gamma_1 - \Gamma_2}{\Gamma_0/2 + \Gamma_2}\right)^{1/2}.$$

The stationary points on the lines $\gamma = \pi/2$ and $3\pi/2$ correspond to the ellipses

$$\frac{u_1^2}{\xi_*^+} + \frac{u_2^2}{1 - \xi_*^+} = E\varepsilon^2,$$

and the points on the lines $\xi = 0$ and $\xi = 1$ correspond to straight lines along the axes Ou_2 and Ou_1.

The separatrices pass through the possible unstable stationary points. For separatrices passing through the "left" points (2.10.25) one should put $C = 0$ in (2.10.18). We obtain equations for two branches:

$$(1)\ \xi = 0, \qquad (2)\ \cos 2\gamma = -\frac{\xi \Gamma_1 + 2\Gamma_2}{\Gamma_0(1 - \xi)}, \tag{2.10.30}$$

which exist when condition 1 of (2.10.28) is satisfied. The "right" separatrix, passing through the stationary points (2.10.26), exists when condition 2 is satisfied. The equations of the branches of this separatrix are obtained from (2.10.18) with $C = \Gamma_1 + 2\Gamma_2$

$$(1)\ \xi = 1, \qquad (2)\ \cos 2\gamma = \frac{(\xi + 1)\Gamma_1 + 2\Gamma_2}{\Gamma_0 \xi}. \tag{2.10.31}$$

The central separatrix (CS) passing through the stationary points (2.10.27) for odd (or even) n exists when condition 3 (condition 4) is satisfied and condition 5 (condition 6) is violated. Substituting the coordinates of points

(2.10.27) into (2.10.18), we obtain $C = (-\Gamma_2 \pm \Gamma_0/2)^2/(\mp\Gamma_0 - \Gamma_1)$ and the equations for the branches of the central separatrix

$$\xi = B \pm \sqrt{B^2 - D},$$
$$B = \frac{\Gamma_0 \cos 2\gamma + \Gamma_2}{\Gamma_0 \cos 2\gamma - \Gamma_1}, \qquad D = \frac{(\Gamma_2 \mp \Gamma_0/2)^2}{(\Gamma_1 \pm \Gamma_0)(\Gamma_1 - \Gamma_0 \cos 2\gamma)}. \qquad (2.10.32)$$

The stationary points and separatrices possess the following properties.

1. If the "left" stationary points (2.10.25) exist (i.e. condition 1 is satisfied), then in the rectangle $(0 \le \xi \le 1, 0 \le \gamma \le \pi)$ there is at least one "intermediate" stationary point (2.10.27) on the line $\gamma = \pi/2$ or $\gamma = 0$, and this point is stable.

Indeed, when condition 1 is satisfied the sign of the numerators in condition 3, 4 is given by the sign of their first term, and for their moduli we have $|\pm\Gamma_0/2 - \Gamma_2| \le |\Gamma_0|$. If the signs of Γ_1 and Γ_0 are the same, then the sign of the denominator in condition 3 is the same as the sign of the numerator, and because we then have $|\Gamma_1 + \Gamma_0| > |\Gamma_0|$, condition 3 is satisfied and, clearly, condition 5. In the case of opposite signs for Γ_1 and Γ_0, the signs of the numerator and denominator in condition 4 are the same and $|\Gamma_1 - \Gamma_0| > |\Gamma_0|$, so that conditions 4 and 6 are satisfied.

A similar assertion holds for the "right" stationary points (2.10.26).

2. If one stationary unstable point (2.10.27) exists on the line $\gamma = \pi/2$ (or $\gamma = 0$), then a stable stationary point exists on the line $\gamma = 0$ (or $\gamma = \pi/2$); and there are no separatrices (2.10.30) and (2.10.31).

Suppose condition 3 is satisfied and condition 5 is not satisfied (i.e. the stationary point at $\gamma = \pi/2$ is unstable). Then Γ_0 and Γ_1 have opposite signs and $|\Gamma_1| > |\Gamma_0|$. It follows from condition 3 (because the sign of the denominator is governed by the sign of Γ_1 and is opposite to the sign of Γ_0) that the signs of Γ_0 and Γ_2 are the same and $|\Gamma_2| > |\Gamma_1|/2$. Condition 1 is therefore violated. Considering the case $\Gamma_1 > 0$ and $\Gamma_1 < 0$ separately, and taking into account that the sign of Γ_1 is opposite to the signs of Γ_0 and Γ_2 and that $|\Gamma_1| > |\Gamma_0|$, we find that in both cases condition 2 is violated, and the (right) inequality in condition 4 is also violated, which proves the assertion.

These properties enable us to describe the various possible AP-portraits in the plane (ξ, γ). Each side separatrix (SS) joins two unstable stationary points at $\xi = 0$ or $\xi = 1$. The branches of these separatrices surround a single stable stationary point at $\gamma = \pi/2$ or $\gamma = 0$ $(0 < \xi < 1)$. One can verify that if, for example, between the "left" separatrices there is a point on the line $\gamma = 0$, then the abscissa of the point of intersection of the separatrix with the line $\gamma = 0$ is twice the abscissa of the stationary point ξ; it is obvious that $\xi < 1/2$. A similar property is satisfied by the right separatrix: here it is necessary for the stationary point surrounded by its branches to be in the right half of the rectangle. The separatrix originating from $\xi = 0$ cannot intersect the line $\xi = 1$, and conversely.

The branches of the CS join the two unstable stationary points (2.10.27), corresponding to even or odd values of n, and surrounding the stable stationary points. The CS cannot intersect the lines $\xi = 0$ or $\xi = 1$. Inside the domains surrounded by the SS or CS a modulation regime of the first type exists, and outside these domains, there is a regime of the second type.

Thus, four qualitatively different types of the AP-portrait, governed by conditions 1–6, are possible and they are shown in Fig. 2.20

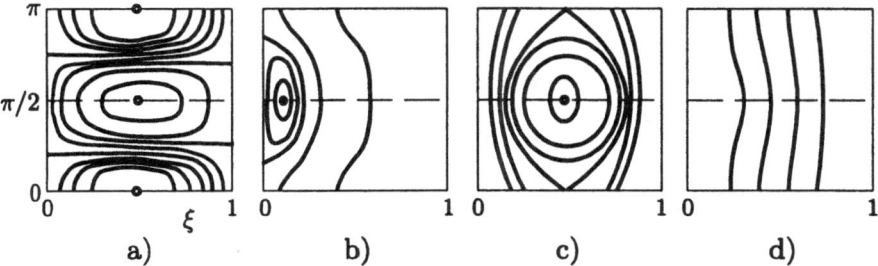

Fig. 2.20a–d. Four qualitatively different types of the AP-portrait

1. Conditions 1 and 2 are satisfied. There are stable stationary points at $\gamma = n\pi/2$ (2.10.27) for even and odd n, in the left section ($\xi < 1/2$) and right section ($\xi > 1/2$) of the rectangle, i.e. three stable normal modes exist (and two trivial unstable ones $u_k = 0$, $k = 1, 2$). Each of the stationary points is surrounded by the corresponding SS; there are no CS (Fig. 2.20a).
2. Only one of conditions 1 and 2 is satisfied. There is a stable stationary point (2.10.27) only for an odd or even n, and only one SS (on the left if condition 1 is satisfied, and on the right if condition 2 is satisfied); there are no CS (Fig. 2.20b). All the normal modes, apart from the single $u_k = 0$, $k = 1$ or $k = 2$ mode, exist and stable modes also exist that are either rectilinear (if condition 4 is satisfied), or elliptic (when condition 3 is satisfied).
3. Neither condition 1 nor 2 is satisfied, but condition 3 is satisfied. The stable and unstable stationary points (2.10.27) alternate (with the point for odd n being stable if condition 5 is satisfied). There is a CS, but no SS (Fig. 2.20c). Three normal modes exist, where either the rectilinear one is stable (when condition 6 is satisfied), or the elliptic one is stable (when condition 5 holds).
4. Conditions 1–3 are not satisfied. There are no stationary points (normal modes) or separatrices. All oscillatory modes are of modulation type 2, with the modulation being relatively small if compared with cases 1–3 (Fig. 2.20d).

In cases 1–3 one can distinguish subcases. In case 1 there are two subcases distinguished by the position of the left stationary point: on the line $\gamma = 0$ or on $\gamma = \pi/2$. Similarly, in case 3 the stationary point can be stable at $\gamma = 0$ or at $\gamma = \pi/2$. Four subcases are possible for case 2: a left or right separatrix, and a stationary point at $\gamma = 0$ or $\gamma = \pi/2$. The corresponding AP-portraits can be obtained from those shown in Fig. 2.20.

We introduce the parameters

$$\alpha_1 = \frac{b_{11}}{b_{12}}, \qquad \alpha_2 = \frac{b_{22}}{b_{12}}, \qquad \sigma^0 = \frac{4\sigma}{b_{12}E}. \tag{2.10.33}$$

Then, conditions 1–6 can be represented in the form

$$
\begin{array}{lll}
(1) & 3\alpha_2 - 3 \le \sigma^0 \le 3\alpha_2 - 1, & \\
(2) & -3\alpha_1 + 1 \le \sigma^0 \le -3\alpha_1 + 3, & \\
(3) & -3\alpha_1 + 1 \le \sigma^0 \le 3\alpha_2 - 1 & \text{for } \alpha_1 + \alpha_2 > \tfrac{2}{3}, \\
& 3\alpha_2 - 1 \le \sigma^0 \le -3\alpha_1 + 1 & \text{for } \alpha_1 + \alpha_2 < \tfrac{2}{3}, \\
(4) & 3\alpha_2 - 3 \le \sigma^0 \le -3\alpha_1 + 3 & \text{for } \alpha_1 + \alpha_2 < 2, \\
& -3\alpha_1 + 3 \le \sigma^0 \le 3\alpha_2 - 3 & \text{for } \alpha_1 + \alpha_2 > 2, \\
(5) & \alpha_1 + \alpha_2 > \tfrac{2}{3}, & \\
(6) & \alpha_1 + \alpha_2 < 2. &
\end{array}
\tag{2.10.34}
$$

Unlike σ and E, the dimensionless frequency detuning parameter σ^0 does not depend on the choice of ε and can be written in the following form

$$\sigma^0 = \frac{4\sigma_*}{b_{12}(u_1^2(0) + u_2^2(0))}, \qquad (\sigma_* = \varepsilon^2\sigma = \omega_2^2 - \omega_1^2). \tag{2.10.35}$$

As can be seen from (2.10.34), the type of AP-portrait is determined by the relative positions of the points

$$c_1 = 3\alpha_2 - 3, \quad c_2 = 3\alpha_2 - 1, \quad d_1 = -3\alpha_1 + 1, \quad d_2 = -3\alpha_1 + 3 \tag{2.10.36}$$

and the quantity σ^0. Four possible positions of the intervals (c_1, c_2) and (d_1, d_2) are shown in Fig. 2.21 $(c_2 < d_1, c_1 < d_1 < c_2, c_1 < d_2 < c_2, d_2 < c_2)$. The type of the AP-portrait (easily determined from (2.10.34)) is shown above the intervals. In case (a) the interval (c_2, d_1) contains the stable stationary point at $\gamma = 0$ (π), i.e. the rectilinear normal mode is stable, and the unstable one is at $\gamma = \pi/2$ $(3\pi/2)$ (i.e. elliptic). In case (d) these points (and normal oscillations) "exchange" stability.   •

Figure 2.21 graphically demonstrates the influence of the parameter σ^0 on the system behaviour. If σ^0 lies in the interval

$$\delta_1 < \sigma^0 < \delta_2, \qquad \delta_1 = \min(c_1, d_1), \qquad \delta_2 = \max(c_2, d_2), \tag{2.10.37}$$

then we have the AP-portraits of types 1–3 with stationary points and pronounced modulation of the amplitude and phase (energy exchange). If σ^0 lies outside this interval, the AP-portait of type 4 is indicated with relatively small modulation. Thus, condition (2.10.37) allows one to specify the smallness of the frequency detuning parameter. The minimum width of interval (2.10.37) is 2. The centre of the interval is the point

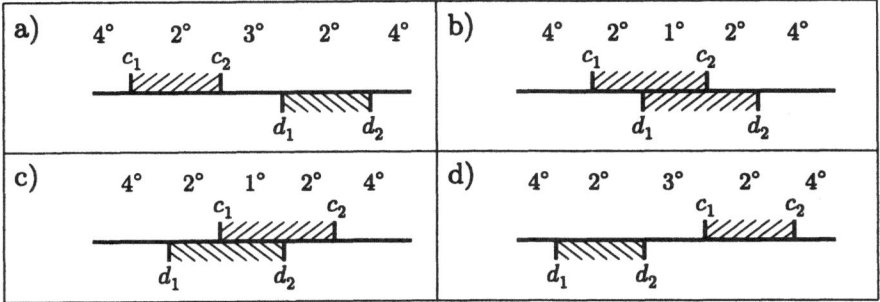

Fig. 2.21a-d. Four possible positions of the intervals (c_1, c_2) and (d_1, d_2)

$$\alpha_* = \frac{3}{2}(\alpha_2 - \alpha_1) = \frac{3}{2}\frac{b_{22} - b_{11}}{b_{12}}.$$

In the case when $b_{11} = b_{22}$ we have $\alpha_1 = \alpha_2$: $\alpha_* = 0$, i.e. interval (2.10.37) is symmetric in relation to the origin.

If $\alpha_1 \neq \alpha_2$ the interval is displaced relative to the origin and for sufficiently large $|\alpha_2 - \alpha_1|$ (or $|b_{22} - b_{11}|$) the point $\sigma^0 = 0$ can turn out to lie outside the interval. One must also take into account that the sign of σ^0 is governed by the sign of b_{12} (one can always put $\sigma_* > 0$, i.e. $\omega_2 > \omega_1$). The sign-constancy condition on σ^0 indicates either the positive or negative part of interval (2.10.37) (if it exists). Two conclusions follow from this:

1. it is not necessary for the large modulation to correspond to the smaller value of σ^0: combinations of the coefficients b_{ij} are possible with the AP-portraits of types 1–3 for intervals with σ^0 far from the point 0;
2. for certain combinations of b_{ij} only the type 4 AP-portraits are possible, irrespective of the energy and frequency separation.

In the above analysis there is a natural separation of the influence of the oscillation energy and the ratio of the initial amplitudes of the two modes on the energy exchange. The quantity E acts on σ^0 according to (2.10.35) (increasing E is equivalent to decreasing σ), and together with σ_* it therefore determines the type of AP-portrait. The initial amplitude ratio ξ_0 determines the phase trajectory in a given AP-portrait.

Consider the special case when $b_{11} = b_{22} = 0$, $b_{12} \neq 0$. Then $\alpha_1 = \alpha_2 = 0$, $c_1 = -3$, $c_2 = -1$, $d_1 = 1$, $d_2 = 3$, i.e. we have Fig. 2.21, case (a). Condition 5 is not satisfied, while condition 6 is satisfied. For $-3 < \sigma^0 \leq -1$ we have the type 2 AP-portrait with a left separatrix and stable stationary point at $\gamma = 0$ (π), i.e. with rectilinear normal oscillations. For $-1 < \sigma^0 < 1$ the AP-portrait is of type 3 and has a stable stationary point at $\gamma = 0$ (π) and an unstable one at $\gamma = \pi/2$ $(3\pi/2)$, i.e. with stable rectilinear normal modes and an unstable elliptic mode. When $1 < \sigma^0 < 3$ the AP-portrait is of type 2 with a right separatrix and stable rectilinear normal modes. Finally, for $\sigma^0 < -3$ and $\sigma^0 > 3$ the AP-portrait is of type 4.

In conclusion we note that the numerical integrations of (2.10.3) performed for the purpose of estimating the accurancy of the solution obtained by the multiple scales method demonstrated almost complete agreement between the analytic and numerical solutions in all the cases considered with an arbitrary choice of $\varepsilon \leq 0.1$ and amplitudes of up to 0.5 (the error in determining the amplitude was of the order of 0.1%). But when ε was increased beyond 0.1, the error increased rapidly. For example, when $\varepsilon = 0.15$ the error in the amplitude computation reached 30%.

2.11 Nontraditional Asymptotic Approaches

2.11.1 Choice of Asymptotic Expansion Parameters

Introducing a small parameter into the nonlinear problems is a very delicate and nontrivial matter.

We consider in this section an elementary illustrative problem: finding the roots of a fifth-degree polynomial [49, 10d].

We are concerned here with finding the real root x_0 of the polynomial equation

$$x^5 + x = 1. \qquad (2.11.1)$$

We have chosen the degree of this polynomial to be 5 because it is high enough to be sure that there is no quadrature formula for the roots. However, one can be sure that there is a *unique* real root x_0 and that this root is positive because the function $x^5 + x$ is monotone increasing. Using Newton's method we compute

$$x_0 = 0.75487767\ldots \qquad (2.11.2)$$

There are several conventional perturbative approaches that we could use to find x_0. One such approach, which we will call the *weak-coupling* perturbation theory, requires that we introduce a perturbative parameter ε in front of the x^5 term [49]:

$$\varepsilon x^5 + x = 1. \qquad (2.11.3)$$

Now, x depends on ε and we assume that $x(\varepsilon)$ has a formal power series expansion in ε:

$$x(\varepsilon) = a_0 + a_1\varepsilon + a_2\varepsilon^2 + a_3\varepsilon^3 + \cdots. \qquad (2.11.4)$$

To find the coefficients we substitute (2.11.4) into (2.11.3) and expand the result in an asymptotic series in powers of ε. We find that the coefficients a_n are integers which oscillate in sign and grow rapidly as n increases:

$$a_0 = 1, \quad a_1 = -1, \quad a_2 = 5, \quad a_3 = -35,$$
$$a_4 = 285, \quad a_5 = -2530, \quad a_6 = 23751, \qquad (2.11.5)$$

etc. In fact we can find a closed-form expression of a_n valid for all n,

$$a_n = [(-1)^n(5n)!]\,[n!(4n+1)!]\,,\tag{2.11.6}$$

from which we can determine the radius of convergence R for the series in (2.11.4):

$$R = \frac{4^4}{5^5} = 0.08192.\tag{2.11.7}$$

Evidently, if we try to use the weak-coupling series in (2.11.4) directly to calculate $x(1)$, we will fail miserably. Indeed, using the seven coefficients in (2.11.5) for $\varepsilon = 1$ gives

$$x(1) = \sum_{n=0}^{6} a_n = 21476,$$

which is a poor approximation to the true value of $x(1)$ in (2.11.2)!

Of course, we can improve the prediction enormously first by coupling the [3/3] Padé approximants and then by evaluating the result at $\varepsilon = 1$. Now we obtain the result

$$x(1) = 0.76369,\tag{2.11.8}$$

which differs from the correct answer in (2.11.2) by 1.2%.

A second conventional perturbative approach is to use a strong-coupling expansion. Here, we introduce a perturbative parameter ε in front of the term x in (2.11.1) [49]:

$$x^5 + \varepsilon x = 1.\tag{2.11.9}$$

As before, x depends on ε and we assume that $x(\varepsilon)$ has a formal series expansion in powers of ε:

$$x(\varepsilon) = b_0 + b_1\varepsilon + b_2\varepsilon^2 + b_3\varepsilon^3 + \cdots.\tag{2.11.10}$$

Determining the coefficients of this series is routine and we find that

$$b_0 = 1,\quad b_1 = -\frac{1}{5},\quad b_2 = -\frac{1}{25},\quad b_3 = -\frac{1}{125},$$

$$b_4 = 0,\quad b_5 = \frac{21}{15625},\quad b_6 = \frac{78}{78125},\tag{2.11.11}$$

etc. Again, we can find a closed-form expression for b_n valid for all n,

$$:b_n = -\frac{\Gamma\left(\frac{4n-1}{5}\right)}{5\Gamma\left(\frac{4-n}{5}\right)n!}\tag{2.11.12}$$

from which we can determine the radius of convergence R of the series in (2.11.10)

$$R = \frac{5}{4^{4/5}} = 1.64938\ldots.\tag{2.11.13}$$

Now, $\varepsilon = 1$ lies inside the circle of convergence so it is easy to compute $x(1)$ by summing the series (2.11.10) directly. Using the coefficients listed in (2.11.11), we have

$$x(1) = \sum_{n=0}^{6} b_n = 0.75434, \qquad\qquad (2.11.14)$$

which differs from the true result in (2.11.2) by 0.07%, a vast improvement over the weak-coupling approach.

Now we use the δ-expansion method to find the root x_0. We introduce a small parameter δ in the exponent of the nonlinear term in (2.11.1) [49],

$$x^{1+\delta} + x = 1, \qquad\qquad (2.11.15)$$

and seek an expansion for $x(\delta)$ as a series of powers of δ:

$$x(\delta) = c_0 + c_1\delta + c_2\delta^2 + c_3\delta^3 + \cdots . \qquad\qquad (2.11.16)$$

The coefficients of this series may be computed easily. The first few are

$$c_0 = \frac{1}{2}, \quad c_1 = \frac{1}{4}\ln 2, \quad c_2 = -\frac{1}{8}\ln 2,$$

$$c_3 = -\frac{1}{48}\ln^3 2 + \frac{1}{32}\ln^2 2 + \frac{1}{16}\ln 2,$$

$$c_4 = \frac{1}{32}\ln^3 2 - \frac{3}{64}\ln^2 2 - \frac{1}{32}\ln 2,$$

$$c_5 = \frac{1}{480}\ln^5 2 - \frac{7}{768}\ln^4 2 - \frac{3}{128}\ln^3 2 + \frac{3}{64}\ln^2 2 + \frac{1}{64}\ln 2,$$

$$c_6 = -\frac{1}{192}\ln^5 2 + \frac{35}{1536}\ln^4 2 + \frac{5}{768}\ln^3 2 - \frac{5}{128}\ln^2 2 - \frac{1}{128}\ln 2,$$

etc.

The radius of convergence of the δ series in (2.11.16) is 1. A heuristic argument for this conclusion is as follows. The radius of convergence is determined by the location of the nearest singularity of $x(\delta)$ in the complex -δ plane. To find this singularity we differentiate (2.11.15) with respect to δ and solve the resulting equation for $x'(\delta)$:

$$x'(\delta) = -\frac{x^{1+\delta}\ln x}{1 + x^\delta(1+\delta)}.$$

Since $x(\delta)$ is singular where its derivative ceases to exist we look for zeroes of the denominator

$$1 + x^\delta(1+\delta) = 0.$$

We solve this equation simultaneously with (2.11.15) to eliminate δ and obtain a single equation satisfied by x:

$$0 = x\ln x + (1-x)\ln(1-x).$$

The solution to this equation corresponding to the smallest value of $|\delta|$ is $x = 0$. From (2.11.15) we therefore see that $\delta = -1$ is the location of the nearest singularity in the complex-δ plane. In fact, as δ decreases below -1, (2.11.15) abruptly ceases to have a real root. This abrupt transition accounts for the singularity in the function $x(\delta)$.

Clearly, to compute x_0 it is necessary to evaluate series (2.11.16) at $\delta = 4$. For this large value of δ we use the coefficients in (2.11.1) and convert the Taylor series to [3/3] Padé approximants. Evaluating the Padé approximant at $\delta = 4$ gives

$$x(\delta = 4) = 0.75448, \tag{2.11.17}$$

which differs from the exact answer in (2.11.2) by 0.05%.

The δ series continues to provide excellent numerical results as we increase the order of the perturbation theory. If we compute all the coefficients up to c_{12} and then convert (2.11.16) to a [6/6] Padé approximant, we obtain

$$x(\delta = 4) = 0.75487654, \tag{2.11.18}$$

which differs from x_0 in (2.11.2) by 0.00015%.

Last but not least, we may introduce a small parameter in our equation in the following way [10d]:

$$x^{\varepsilon^{-1}} + x = 1, \quad \varepsilon \ll 1.$$

After substituting $x = y^{\varepsilon}$ one obtains

$$y + y^{\varepsilon} = 1.$$

Taking into account the relation

$$\varepsilon^{\varepsilon} = 1 + \varepsilon \ln \varepsilon + \dots$$

we may represent y in the form

$$y = \varepsilon + o(\varepsilon \ln \varepsilon).$$

Then, we have

$$x \approx \varepsilon^{\varepsilon},$$

and for $\varepsilon = 1/5$, $x \approx 0.724780$ (the error of the first approximation is only 3.9%).

2.11.2 δ-Expansions in Nonlinear Mechanics [49]

Let us start with the solution of a simple nonlinear differential equation.

Consider the nonlinear ordinary differential equation problem

$$y'(x) = [y(x)]^n, y(0) = 1. \tag{2.11.19}$$

The exact solution of this problem is

$$y(x) = [1 - (n-1)x]^{-1/(n-1)}. \tag{2.11.20}$$

To solve (2.11.19) approximately using the δ expansion, we let $n = 1 + \delta$ and solve

$$y'(x) = [y(x)]^{1+\delta}. \tag{2.11.21}$$

To solve (2.11.21) perturbatively, we can seek a solution $y(x)$ in the form of a series in the powers of δ:

$$y(x) = y_0(x) + \delta y_1(x) + \delta^2 y_2(x) + \ldots. \tag{2.11.22}$$

For example, $y_0(x)$ satisfies the linear differential equation problem

$$y_0' = y_0(x), \quad y_0(0) = 1,$$

whose solution is

$$y_0 = e^x.$$

Indeed, all functions $y_n(x)$ satisfy *linear* differential equations which are easy to solve. We find that

$$y_1(x) = \frac{1}{2}e^x x^2, \quad y_2(x) = e^x \left[\frac{1}{3}x^3 + \frac{1}{8}x^4 \right],$$

etc. The reason for using a perturbative approach if that, in general, even when one cannot solve the nonlinear differential equation, the differential equation for the perturbation coefficients $y_0(x)$, $y_1(x)$, $y_2(x)$, $\ldots$ are always linear and therefore can be solved in quadrature form.

For the particular problem (2.11.21) a closed-form solution exists. Therefore, we can determine the radius of convergence R of series (2.11.22):

$$R = \frac{1}{|x|}.$$

We have computed the series in (2.11.22) up to the δ^{10} term. Let us examine the numerical accuracy of the δ series.

The exact value of $y(x)$ at $x = 1/4$ for the case $n = 4$ $(\delta = 3)$ is

$$y\left(\frac{1}{4}\right) = 1.587401. \tag{2.11.23}$$

Directly summing the δ series $\sum_0^n \delta^n y_n(1/4)$ gives 1.284 when $n = 0$ (19% error), 1.404 when $n = 1$ (11.5% error), 1.470 when $n = 2$ (7.4% error), 1.5099 when $n = 3$ (4.9% error), 1.5626 when $n = 6$ (1.6% error), and 1.58128 when $n = 10$ (0.39% error). We can also compute a Padé approximant from the δ series and then set $\delta = 3$. The [3/3] Padé approximant gives 1.58692 (0.03% error) and the [5/5] Padé approximant gives 1.587395 (3.7×10^{-4}% error). It is numerical results such as these that encourage us to use the δ expansion to solve difficult nonlinear differential equations.

Now we turn to a more complicated problem.

The classical anharmonic oscillator is defined by the nonlinear ordinary differential equation

$$\frac{d^2 y}{dt^2} + y + \varepsilon y^3 = 0. \tag{2.11.24}$$

We impose the conventional initial conditons

$$y(0) = 1, \quad y'(0) = 0. \tag{2.11.25}$$

Our objective here will be to find the period of the anharmonic oscillator. It is well known that the initial-value problem (2.11.24) can be solved exactly in terms of elliptic functions and the period T can be expressed exactly as an elliptic integral

$$T = 4 \int_0^{\frac{\pi}{2}} d\theta \left[1 + \frac{\varepsilon}{2}(1 + \sin^2 \theta) \right]^{-\frac{1}{2}}. \tag{2.11.26}$$

The integral in (2.11.26) can be expanded as a series in the powers of ε

$$T = 2\pi \left[1 + \frac{3}{8}\varepsilon + \frac{21}{256}\varepsilon^2 + \cdots \right]^{-1}. \tag{2.11.27}$$

One cannot use the conventional perturbation theory to find the period T for small $|\varepsilon|$. It is true that when $|\varepsilon|$ is small the exact solution $y(t)$ approximates the motion of a harmonic oscillator of period 2π. However, solving the Duffing equation perturbatively requires some subtlety. If we seek a conventional perturbative solution for $y(t)$ as a series in powers of ε, we find that there is a resonant coupling between successive orders in the perturbation theory. As a result, the coefficient of ε in the pertubation series for $y(t)$ grows linearly with t, the coefficient of ε^2 grows quadratically with t, the coefficient of ε^3 grows like t^3, etc. Thus the perturbative solution is only valid for times t which are small compared with $1/\varepsilon$. At such short times we cannot use the perturbation expansion for $y(t)$ to obtain the series expansion in (2.11.27).

More sophisticated perturbative methods have been devised which enable us to calculate $y(t)$ perturbatively for times $t \sim 1/\varepsilon$ and thus to obtain the series in (2.11.27). One such method is called multiple-scale perturbation theory.

We will attack (2.11.24) using the δ expansion and will find that here, too, the methods of multiple scale perturbation theory must be used. To use the δ expansion we replace y^3 by $y^{1+2\delta}$ and consider the differential equation

$$\frac{d^2}{dt^2} + y + (\omega^2 - 1)y^{1+2\delta} = 0, \quad y(0) = 1, \quad y'(0) = 0. \tag{2.11.28}$$

In (2.11.28) we have found that it is convenient to set

$$\varepsilon = \omega^2 - 1, \tag{2.11.29}$$

so that when $\delta = 0$, (2.11.28) describes a classical harmonic oscillator whose frequency is ω. Note, also, that $y^{2\delta}$ is to be interpreted as the positive quantity $(y^2)^\delta$. Thus when we expand $y^{2\delta}$ as a series in powers of δ we obtain

$$y^{2\delta} = 1 + \delta \ln(y^2) + \frac{\delta^2}{2}[\ln(y^2)]^2 + \frac{\delta^3}{6}[\ln(y^3)]^3 + \cdots, \tag{2.11.30}$$

in which the argument of the logarithm is always positive and no complex numbers appear.

Let us begin by trying to find a conventional perturbative solution to (2.11.28) as a series in powers of δ

$$y(t) = \sum_{n=0}^{\infty} \delta^n y_n(t). \tag{2.11.31}$$

Substituting (2.11.31) into (2.11.28) and using (2.11.30) we obtain a sequence of linear equations and associated initial conditions which must be solved. The first few read as

$$\frac{d^2 y_0}{dt^2} + \omega^2 y_0 = 0, \quad y_0(0) = 1, \quad y_0'(0) = 0, \tag{2.11.32}$$

$$\frac{d^2 y_1}{dt^2} + \omega^2 y_1 = -(\omega^2 - 1)y_0 \ln(y_0^2), \quad y_1(0) = y_1'(0) = 0, \tag{2.11.33}$$

$$\frac{d^2 y_2}{dt^2} + \omega^2 y_2 = -(\omega^2 - 1)\left\{ y_1 \ln(y_0^2) + 2y_1 + \frac{1}{2}y_0[\ln(y_0^2)]^2 \right\}, \tag{2.11.34}$$

$$y_2(0) = y_2'(0) = 0,$$

$$\frac{d^2 y_3}{dt^2} + \omega^2 y_2 = -(\omega^2 - 1)\left\{ y_2 \ln(y_0^2) + 2y_2 + \frac{y_1^2}{y_0} + \frac{1}{2}y_1[\ln(y_0^2)]^2 \right.$$

$$\left. + 2y_1 \ln(y_0^2) + \frac{1}{6}y_0[\ln(y_0^2)]^3 \right\}, \tag{2.11.35}$$

$$y_3(0) = y_3'(0) = 0.$$

The solution to (2.11.32) is

$$y_0 = \cos(\omega t). \tag{2.11.36}$$

We can solve (2.11.33) using the method of order. We let

$$y_1(t) = \cos(\omega t)u - 1(t). \tag{2.11.37}$$

The equation satisfied by $u_1(t)$ is

$$\cos(\omega t)\frac{d^2 u_1}{dt^2} - 2\omega \sin(\omega t)\frac{du_1}{dt} = -(\omega^2 - 1)\cos(\omega t)\ln[\cos^2(\omega t)], \tag{2.11.38}$$

which has $\cos(\omega t)$ as its integrating factor

$$\frac{d}{dt}\left[\cos^2(\omega t)\frac{du_1}{dt}\right] = -(\omega^2 - 1)\cos^2(\omega t)\ln[\cos^2(\omega t)]. \tag{2.11.39}$$

Two integrations of (2.11.39) give from (2.11.37)

$$y_1(t) = -\cos(\omega t)(\omega^2 - 1)\int_0^t \frac{ds}{\cos^2(\omega s)}\int_0^s dr\, \cos^2(\omega r)\ln[\cos^2(\omega r)]. \tag{2.11.40}$$

The integral with respect to s in (2.11.40) can be performed by interchanging the orders of integration:

$$y_1 = \frac{-\cos(\omega t)(\omega^2 - 1)}{\omega} \int_0^t dr \ \cos^2(\omega r) \ln[\cos^2(\omega r)][\tan(\omega t) - \tan(\omega t)]$$

$$= \cos(\omega t)\frac{\omega^2 - 1}{2\omega^2}\{\cos^2(\omega t) - 1 - \ln[\cos^2(\omega t)]\cos^2(\omega t)\}$$

$$- \sin(\omega t)\frac{\omega^2 - 1}{\omega^2} \int_0^{\omega t} dx \ \cos^2 x \ln(cos^2 x). \tag{2.11.41}$$

Note that the integral in (2.11.41) grows linearly with t for large t because the integrand is a positive periodic function. However, we know that the exact solution to (2.11.28) is a *bounded* function. Hence, (2.11.41) can only be valid for times that are short compared with $1/\delta$. This problem appears because each successive order in the perturbation theory is resonantly coupled to the previous orders. To see this, note that (2.11.33) is the differential equation for a driven harmonic oscillator of the natural frequency ω. The driving term (the inhomogeneous part of the differential equation) *also* has the frequency ω because it is a functional of y_0. Thus the oscillator described by (2.11.33) is driven on resonance and the solution exhibits secular behaviour (it grows with t).

We can still try to use the expression in (2.11.41) to infer a value for the period of the oscillator accurate to order δ. We assume that the period T of (2.11.28) is itself a series in powers of δ:

$$T = \frac{2\pi}{\omega} + \frac{a}{\omega}\delta + \cdots. \tag{2.11.42}$$

To determine the coefficient a in (2.11.42) we require that after a quarter-period the amplitude of the oscillator

$$y(t) = y_0(t) + \delta y_1(t)$$

will fall from $y = 1$ at $t = 0$ to $y = 0$ at $t = T/4$. Evaluating the expressions for y_0 in (2.11.36) and y_1 in (2.11.41) we obtain, to order δ,

$$0 = \cos\left(\frac{\pi}{2} - \frac{a\delta}{4}\right) - \delta\frac{\omega^2 - 1}{\omega^2} \int_0^{\frac{\pi}{2}} dx \ \cos^2 x \ln(\cos^2 x),$$

or, neglecting terms which are higher order in δ,

$$a = -4\frac{\omega^2 - 1}{\omega^2} \int_0^{\frac{\pi}{2}} dx \ \cos^2 x \ln(\cos^2 x) = \pi\frac{1 - \omega^2}{\omega^2}(1 - 2\ln 2). \tag{2.11.43}$$

Thus, to the leading order in δ, the period of the oscillator is

$$T = \frac{2\pi}{\omega}\left[1 + \delta\frac{\omega^2 - 1}{2\omega^2}(2\ln 2 - 1)\right]. \tag{2.11.44}$$

Let us reexamine the problem in (2.11.28) using the methods of multiple-scale perturbation theory. We assume that there are two time scales in the problem: a short-time scale described by the variable t and a long-time scale described by the variable

$$\tau \equiv \delta t.$$

We then seek a solution to (2.11.24) of the form

$$y(t) = Y_0(t, \tau) + \delta Y_1(t, \tau) + \ldots, \tag{2.11.45}$$

where the initial conditions in (2.11.25) become

$$Y_0(0, 0) = 1, \quad \frac{\partial Y_0}{\partial t}(0, 0) = 0,$$

$$Y_1(0, 0) = 0, \quad \frac{\partial Y_0}{\partial \tau}(0, 0) + \frac{\partial Y_1}{\partial t}(0, 0) = 0, \tag{2.11.46}$$

etc.

If we substitute (2.11.45) into (2.11.28), we obtain

$$\frac{\partial^2}{\partial t^2}Y_0(t, \tau) + \omega^2 Y_0(t, \tau) = 0 \tag{2.11.47}$$

to the zeroth order in δ and

$$\frac{\partial^2}{\partial t^2}Y_1(t, \tau) + \omega^2 Y_1(t, \tau) = -2\frac{\partial^2 Y_0}{\partial t \partial \tau} + (1 - \omega^2)Y_0 \ln(y_0^2) \tag{2.11.48}$$

to the first order in δ. The most general real solution to (2.11.47) is

$$Y_0(t, \tau) = A(\tau)e^{i\omega t} + A^*(\tau)e^{-i\omega t}, \tag{2.11.49}$$

where $A(\tau)$ is to be determined.

Using (2.11.49) we can evaluate the right-hand side of (2.11.48):

$$(1 - \omega^2)[A(\tau)e^{i\omega t} + A^*(\tau)e^{-i\omega t}]\ln\left[(2|A|^2)\left(1 + \frac{A^2 e^{2i\omega t}}{2|A|^2} + \frac{A^{*2}e^{-2i\omega t}}{2|A|^2}\right)\right]$$

$$-2[i\omega e^{i\omega t}A'(\tau) - i\omega e^{-i\omega t}A^{*\prime}(\tau). \tag{2.11.50}$$

Now, we expand the logarithm in (2.11.50) to identify all terms proportional to $e^{i\omega t}$ and $e^{2i\omega t}$. Such terms oscillate at the frequency ω and thus give rise to the seculiar behavior in Y_1. The coefficient of $e^{i\omega t}$ is

$$-2i\omega A'(\tau) + (1 - \omega^2)A(\tau)\ln(2|A|^2) - \frac{1 - \omega^2}{2}A(\tau)\sum_{k=1}^{\infty}\frac{1}{k}4^{-1}\begin{bmatrix} 2k \\ k \end{bmatrix}$$

$$+\frac{1 - \omega^2}{2}A(\tau)\sum_{k=0}^{\infty}\frac{1}{2k + 1}4^{-k}\begin{bmatrix} 2k + 1 \\ k + 1 \end{bmatrix}. \tag{2.11.51}$$

Evaluating the sums in (2.11.51) gives

$$-2i\omega A'(\tau) + (1 - \omega^2)A(\tau)\ln(2|A|^2)$$
$$- (1 - \omega^2)A(\tau)\ln 2 + (1 - \omega^2)A(\tau). \tag{2.11.52}$$

Thus, the condition that there is no secular behaviour in $Y_1(t, \tau)$ is that the expression in (2.11.52) (as well as its complex conjugate) vanishes:

$$- 2i\omega A'(\tau) + (1 - \omega^2)A(\tau)[1 - \ln(|A^2|)] = 0. \tag{2.11.53}$$

To solve (2.11.53) we let

$$A(\tau) = R(\tau)e^{i\theta(\tau)}, \tag{2.11.54}$$

substitute (2.11.54) into (2.11.53), and decompose the result into its real and imaginary parts:

$$R'(\tau) = 0,$$
$$\theta'(\tau) = \frac{\omega^2 - 1}{2\omega}(1 + 2\ln R). \tag{2.11.55}$$

Hence, $R(\tau)$ is a constant

$$R(\tau) = R_0, \tag{2.11.56}$$

and $\theta(\tau)$ is a linear function of τ,

$$\theta(\tau) = \frac{\omega^2 - 1}{2\omega}(1 + 2\ln R_0)\tau + \theta_0. \tag{2.11.57}$$

The initial conditions in (2.11.46) imply that $R_0 = 1/2$ and $\theta_0 = 0$, thus our final result for $T_0(t, \tau)$ is

$$Y_0(t, \tau) = \cos\left[\omega t + \tau\frac{\omega^2 - 1}{2\omega}(1 - 2\ln 2)\right]. \tag{2.11.58}$$

Finally, we eliminate τ in favour of δt to obtain the MSA result

$$T_{MSA} = \frac{2\pi}{\omega - \delta\frac{\omega^2 - 1}{2\omega}(2\ln 2 - 1)}, \tag{2.11.59}$$

which we expand to order δ:

$$T_{MSA} = \frac{2\pi}{\omega}\left[1 + \delta\frac{\omega^2 - 1}{\omega^2}(2\ln 2 - 1)\right] + O(\delta^2). \tag{2.11.60}$$

To our surprise, (2.11.60) agrees exactly with the order-δ result we obtained in (2.11.44) using the δ-perturbation method at a quarter-period.

It is a long but routine calculation to carry out the δ-perturbation series to order δ^2. Using the quarter-period method we find that at $\delta = 1$,

$$T = \frac{1}{\omega}\left[2\pi + 0.5238\frac{\omega^2 - 1}{\omega^2} + 0.6041\left(\frac{\omega^2 - 1}{\omega^2}\right)^2\right]. \tag{2.11.61}$$

Table 2.1. Comparison of the exact value of the period of the anharmonic oscillator with the period calculated from the order-δ quarter-period method (same as MSA) and the order-δ^2 quarter-period method

ε	ω	T(exact)	T(order δ)	T(order δ^2)
1	$\sqrt{2}$	4.76802	4.87195	4.73488
3	2	3.52114	3.59669	3.50794
8	3	2.41289	2.45397	2.40871

In Table 2.1 we compare three results: the exact numerical calculation of the period T; the order-δ quarter-period calculation, which is the same as the order-δ MSA result in (2.11.60); and the order-δ^2 quarter-period calculation in (2.11.61). We set $\delta = 1$ and look at three values of $\varepsilon = \omega^2 - 1$. As expected, the MSA and order-δ results are excellent, having an accuracy of about 2%. The order-δ^2 results are even better, having a relative error of less than 0.5%.

2.11.3 Asymptotic Solutions for Nonlinear Systems with High Degrees of Nonlinearity

As an example, one can consider the equation

$$x^{\cdot\cdot} + x^n = 0, \qquad n = 2k + 1, \qquad k = 1, 2, \ldots,$$

for which we will seek a single-parameter family of periodic solutions which are skew-symmetric with respect to the origin of coordinates in the limit as $n \to \infty$.

Let us introduce the function $\xi = x/A$ (A is the amplitude) for which the inequality $0 \le |\xi| \le 1$ holds. Note that the function ξ is continuous and periodic.

The initial equation can then be represented as follows:

$$\xi^{\cdot\cdot} + A^{n-1}\xi^n = 0. \tag{2.11.62}$$

We will expand the function ξ^n in a series in $1/n$ as $n \to \infty$. In order to do this, we first transform the function

$$\varphi = \begin{cases} \xi^n, & 0 \le \xi \le 1 \\ 0, & \xi > 1 \end{cases}$$

using the Laplace transformation [152] $\varphi(\xi) \to p^{-n-1}\gamma(n + 1, p)$.

On expanding the incomplete gamma function $\gamma(n + 1, p)$ in a series in $1/n$ and on carrying out the inverse transformation in a term-by-term manner (this procedure is justified in [164]), we obtain

$$\varphi = \delta(\xi - 1)(n + 1)^{-1} - \delta(\xi - 1)(n + 1)^{-1}(n + 2)^{-1} + \ldots,$$

where $\delta(\cdot)$ is the delta function.

We will now make the change of variable $t = \tau/\omega$ in (2.11.62).

On retaining just the principal term in the sum and putting

$$\omega^2 = A^{n-1}/(n+1) \qquad (2.11.63)$$

(since $0 \leq |\xi| \leq 1$), we have the equation

$$\frac{d^2\xi_0}{d\tau^2} = -\delta(\xi_0 - 1) \qquad (2.11.64)$$

for determining the periodic function ξ_0.

Integration of (2.11.64) taking into account the skew symmetry with respect to the origin of the coordinates yields in the initial variables

$$x_0 = A\omega t. \qquad (2.11.65)$$

Expression (2.11.63), which can be treated as an amplitude–frequency dependence, and the solutions over a quarter of the period argee with those obtained by another method in [55d].

Solution (2.11.65) does not satisfy the boundary conditions

$$\frac{d\xi_0}{d\tau} = 0 \quad \text{for} \quad \tau = 1.$$

The additional solution is the boundary layer y, and we represent ξ_0 as

$$\xi = \xi_0 + y. \qquad (2.11.66)$$

Here $y \ll \xi_0$.

One has from the boundary condition for $\tau = 1$

$$\frac{dy}{d\tau} = -\frac{d\xi_0}{d\tau} = -1.$$

This condition leads to the following asymptotical estimation

$$\eta \sim \frac{\xi_0}{n}; \qquad \frac{dy}{d\tau} \sim ny.$$

Substituting (2.11.66) into the governing equation and taking into account the asymptotic estimations, in the first approximation one has

$$\omega^2\frac{d^2y}{d\tau^2} + \xi_0^n = 0.$$

A solution of this equation in the initial variables is

$$y = \frac{-(A\omega)^n t^{n+2}}{(n+1)(n+2)}$$

and coincides with the solution obtained in [55d] by another approach.

The approaches proposed in 2.10.2 and 2.10.3 give the possibility of obtaining a solution of the nonlinear differential equation containing the term $\chi^{1+\delta}$ for $\delta \to 0$ and $\delta \to \infty$. Matching these limiting asymptotics by, for example, two-point Padé approximants leads to the solution for any value of the parameter δ.

It is worth noting that the asymptotic approach based on the distributionals now has many applications [57, 76, 161, 34d, 35d].

2.11.4 Square-Well Problem of Quantum Theory

The problems of strong interactions and the nonperturbative approach are the main ones in field theory. In particular, solving the Schrödinger equation for $N \to \infty$ is very interesting from this point of view. In [54] such a solution was obtained on the basis of δ-expansions in connection with the matching asymptotic procedure. Unfortunately, a substantial number of expansion terms has to be engaged in time-consuming calculations. In this work we propose new techniques which give us the possibility constructing expansions in the degree of $\varepsilon = 1/N$ and to obtain good results using only lower terms of the asymptotic expansions.

We consider the equation

$$\psi_{xx} + x^{2N}\psi - E\psi = 0, \tag{2.11.67}$$

accompanied by the boundary conditions

$$\psi(\pm\infty) = 0. \tag{2.11.68}$$

We seek an expansion of the eigenvalue $E(N)$ as the series in the power $1/N$ for $N \to \infty$.

If we suppose $N = \infty$, we have from (2.11.67), (2.11.68)

$$\psi_{xx} + E\psi = 0, \tag{2.11.69}$$
$$\psi(\pm 1) = 0. \tag{2.11.70}$$

The eigenvalues of the problem (2.11.69), (2.11.70) are

$$E_n = 0.25\pi^2(n+1)^2, \qquad n = 0, 1, 2, \ldots. \tag{2.11.71}$$

where n labels the energy level. A comparsion of these results with the exact values of $E_0(N)$ obtained numerically for $n = 0$ (see table 2.2) reveals that an acceptable accuracy can be reached only for high N, therefore a construction of a more exact solution is needed.

Real minimal solutions of this equation for various N are listed in the Table 2.2.

Table 2.2. Comparison of exact and approximate values of eigenvalues $E(N)$

N	Exact value E	E_0	Error %	N	Exact value E	E_0	Error %
1	1.0000	0.9100	9.0	200	2.3379	2.3383	0.02
2	1.0604	1.0422	1.72	500	2.4058	2.4032	0.01
4	1.2258	1.2385	1.04	1500	2.4431	2.4428	0.01
10	1.5605	1.5731	0.81	3500	2.4558	2.4555	0.01
50	2.1052	2.1074	0.10				

Let us consider the function φ

$$\varphi = \begin{cases} x^n, & 0 \le x \le 1, \\ 0, & x > 1. \end{cases}$$

Now we split the function φ in $1/N$ for $N \to \infty$. For this purpose we transform the function φ by the Laplace transformation [152]

$$\varphi(x) \to p^{-N-1}\gamma(N+1,p),$$

where $\gamma(\ldots)$ is the incomplete γ-function, and p are parameter of the Laplace transform.

After splitting $\gamma(N+1,p)$ on $1/N$ and inverting the Laplace transform term by term one obtains

$$\varphi = \delta(x-1)(2N+1)^{-1} + \delta^{(1)}(x-1)(2N+1)^{-1}(2N+2)^{-1} + \ldots$$
$$+\delta^{(i-1)}(x-1)(2N+1)^{-1}(2N+2)^{-1}\ldots(2N+i)^{-1} + \ldots. \quad (2.11.72)$$

Thus, in the interval $0 \le x \le 1$, (2.11.67) may be represented by

$$\psi_{1xx} + \varphi\psi_1 - E\psi_1 = 0. \quad (2.11.73)$$

Seeking a solution of equation (2.11.73) in the form of the expansion

$$\psi_1 = \sum_{k=0}^{\infty} \psi_{1k}(2N+1)^{-1}; \qquad E = \sum_{k=0}^{\infty} E^{(k)}(2N+1)^{-1},$$

after splitting into $(2N+1)^{-1}$, one obtains a recurrent system of equations

$$-\psi_{10xx} - E^{(0)}\psi_{10} = 0, \quad (2.11.74)$$
$$-\psi_{11xx} - E^{(0)}\psi_{11} - E^{(1)}\psi_{10} + \delta(x-1)\psi_{10} = 0, \quad (2.11.75)$$

$$\cdots\cdots$$

Solving equation (2.11.74) for the case of symmetry with respect to the line $x = 0$ (the antisymmetric case may be considered similarly), one obtains

$$\psi_1 = C\cos\lambda x, \qquad \lambda = (E^{(0)})^{1/2}. \quad (2.11.76)$$

Now let us consider the zone $x > 1$. For the zero-order approximation one can neglect the term $E\psi_2$

$$\psi_{2xx}^{(0)} - x^{2N}\psi_2^{(0)} = 0. \quad (2.11.77)$$

A boundary condition for $x \to \infty$ can be formulated as follows:

$$\psi_2^{(0)} \to 0 \quad \text{for} \quad x \to \infty. \quad (2.11.78)$$

The solution of the boundary value problem (2.11.77), (2.11.78) is the following

$$\psi_2^{(0)} = C_1 x^{1/2} K_\nu(\nu x^{N+1}), \quad (2.11.79)$$

where K_ν is the Bessel function, $\nu = 0.5/(N+1)$.

For $x = 1$ the solutions ψ_1 and ψ_2 must be matched. Then one obtains

$$\text{for} \quad x = 1 \quad \psi_1^{(i)} = \psi_2^{(i)}, \tag{2.11.80}$$

$$\psi_{1x}^{(i)} = \psi_{2x}^{(i)}, \quad i = 0, 1, 2, \ldots$$

From (2.11.80), assuming $i = 1$, we obtain a transcendental equation for λ:

$$-\text{ctg}\,\lambda = \frac{4\lambda K_\nu(\nu)}{2K_\nu(\nu) - K_{1-\nu}(\nu) - K_{1+\nu}(\nu)}.$$

Now, we consider the equation of a higher-order approximation for $x > 1$. If we suppose ψ_2 in the form $\psi_2 = \bar{\psi}(\bar{x})$, where $\bar{x} = x^{N+1}/(N+1)$, we have the following equation for the function $\bar{\psi}(\bar{x})$:

$$\bar{\psi}_{\bar{x}\bar{x}} + Nx^{-N-1}\bar{\psi}_{\bar{x}} + Ex^{-2N}\bar{\psi} - \bar{\psi} = 0. \tag{2.11.81}$$

After expanding the function x^{-2N} and x^{-N-1} in a series of the power $1/(2N+1)$ and $1/(N+2)$ as described above, one can obtain

$$x^{-2N} = \sum_{i=0}^{\infty} (-1)^i \delta^{(i)} (1 - 1/x)(2N+1)^{-1} \ldots (2N+1+i)^{-1}, \tag{2.11.82}$$

$$x^{-N-1} = \sum_{i=0}^{\infty} (-1)^i \delta^{(i)} (1 - 1/x)(N+2)^{-1} \ldots (N+1+i)^{-1}. \tag{2.11.83}$$

Substituting expressions (2.11.82), (2.11.83) into (2.11.81) and splitting it into $1/N$, we have a recurrent sequence of equations, whose solution gives us the possibility formulating the boundary conditions for $\psi_1^{(i)}$.

In conclusion, we may say that the approach proposed above is the natural asymptotic method for solving the differential equations which contain the term $x^{1+\delta}$ for $\delta \to \infty$. A similar asymptotical approach for the case of small δ was proposed in [30d]. Matching solutions for $\delta \to 0$ and $\delta \to \infty$ by means of a two-point Padé approximant one can obtain a solution for any value of δ.

2.12 Padé Approximants

2.12.1 One-Point Padé Approximants: General Definitions and Properties

The principal shortcoming of the perturbation methods is the local nature of solutions based on them. As the technique of asymptotic integration is well developed and widely used, such problems as elimination of the locality of expansion, evaluation of the convergence domain, construction of uniformly suitable solutions, are very urgent.

There exist a lot of approaches to these problems [94, 154]. The method of analytic continuation (for example, the Euler transformation $\bar{\varepsilon} = \varepsilon(1+\varepsilon)^{-1}$) requires a knowledge of the positions of the singularities' of the sought function of the parameter ε [154, 70d]. It is useful to apply those methods in cases when a great number of expansion components is known. It is then possible, using, for example, the Domb–Sikes diagram [40, 70d], to determine the positions of the singularities' and to perform analytic continuation. A significant number of expansion components is also necessary to apply the methods of generalized summation. Not diminishing the merits of these techniques, let us, however, note that in practice only a few of the first components of the perturbation theory are usually known. Lately, the situation has indeed changed a little due to the application of computers. However, up till now, there are usually 3–5 components available of the perturbation series, and exactly from this segment of the series we have to extract all available information. For this purpose the method of Padé approximants (PA) may be very useful [4, 5, 123, 144, 64d, 67d].

Let us consider PAs which allow us to perform the most natural, to some extent, continuation of the power series. Let us formulate the definition. Let

$$F(\varepsilon) = \sum_{i=0}^{\infty} C_i \varepsilon^i , \qquad F_{mn}(\varepsilon) = \frac{\sum_{i=0}^{m} a_i \varepsilon^i}{\sum_{i=0}^{n} b_i \varepsilon^i},$$

where the coefficients a_i, b_i are determined from the following condition: the first $(m+n)$ components of the expansion of the rational function $F_{mn}(\varepsilon)$ into the Maclaurin series coincide with the first $(m+n+1)$ components of the series for $F(\varepsilon)$. Then F_{mn} is called the $[m/n]$ Padé approximant. The set of F_{mn} functions for different m and n forms the Padé table.

The diagonal PAs $(m = n)$ are the mostly widely used in practice. Let us notice that the PA is unique when m and n are specified. To construct the PAs, it is necessary to solve a system of linear algebraic equations (for optimal methods for the determination of PA coefficients see [42, 43, 99]). The PAs have found wide utilization in a series of branches of mathematics and physics, and particulary for enlarging the domain of applicability of series of perturbation methods. The PA performs meromorphic continuation of the function given in the form of power series, and for this reason it allow us to achieve success in the cases where analytic continuatioin cannot be applied. If the PA sequence converges to a given function, then the roots of its denominators tend to singular points. It allows us to determine the singularities with a sufficiently great number of series components, and then to perform the analytic continuation.

The data concerning convergence of the PA could have applications in practice only as options which would enhance the reliability of the results. Indeed, in practice it is possible to construct only a limited number of PAs, while all convergence theorems require information about an infinite number of them.

Gonchar's theorem [39d] states that if none of the diagonal PAs ($[n/n]$) has any pole in a circle of radius R, then the sequence $[n/n]$ converges uniformly in the circle to the initial function f. Futher, the lack of poles in the sequence $[n/n]$ in the circle of radius R implies the convergence of an initial Taylor series in the circle.

As the diagonal PAs are invariant with respect to the fractional-linear transformations $z \rightarrow z/(\alpha z + \beta)$, then the theorem is valid only for the open circle containing the expansion point and for any domain being a union of such circles.

The theorem has one important consequence for continuous fractions, namely: the holomorphity of all suitable fractions of an initial continuous fraction inside a domain Ω implies uniform convergence of the fraction inside Ω.

An essential disadvantage in practice is the necessity of verifying all diagonal PAs. The point is that if inside a circle of radius R only some subsequence of the diagonal sequence PA has no pole, then its uniform convergence to the initial holomorphic function, in the given circle, is guaranteed only for $r < r_0$, where $0.583R < r_0 < 0.584R$. There exists a counterexample showing that in general $r < 0.8R$.

Since in practice only a finite number of components of the series of the perturbation theory is known and there are no estimations of the convergence rate, then the above theorems could only increase the likelihood of the results obtained. This likelihood is also augmented by known "experimental results" since the practice of PA application shows that the convergence of PA series is usually wider than the convergence domain of the initial series.

Let us note that widely applied continued fractions form a particular case of PAs. In fact, the suitable fractions, representing the sequence of approximations of the continued fraction, coincide with the following PA sequence: $[0/0]$, $[1/0]$, $[1/1]$, $[2/1]$, $[2/2]$, ... Therefore we shall not separate the case of the application of continued fractions.

The following circumstances are essential. In the perturbation theory asymptotic series, divergent for all values of the parameter $\varepsilon \neq 0$, are very often obtained. This does not permit us to evaluate the value of the sought function with arbitrary precision for any ε. At the same time a transformation with a PA (or into a continuous fraction) gives an expression suitable over in a wide range of problems. The approach is strictly mathematically proved for those series where $(-1)^n C_n$ (C_n is the n-th coefficient of the series) is the n-th moment of some mass distribution, but numerous applications of similar approaches also show their applicability to more general cases.

2.12.2 Using One-Point Padé Approximants in Dynamics

We shall consider the Duffing equation which can be studied with different methods, which allows us to compare their efficiency. We shall apply the perturbation method combined with a PA to the problem.

The equation is written in the form

$$\ddot{u} + u + u^3 = 0. \tag{2.12.1}$$

The vibration frequency ω has its exact value

$$\omega = \frac{\pi\sqrt{1 + A^2}}{2\sqrt{2}K(\Theta)}, \tag{2.12.2}$$

where

$$\Theta = \text{arctg}\sqrt{\frac{A^2}{2 + A^2}}, \qquad K(\Theta) = \int\limits_0^{\pi/2} \frac{d\psi}{\sqrt{1 - A^2(2 + A^2)^{-1}\sin^2\psi}},$$

$K(\Theta)$ is an elliptic integral of type I.

The asymptotic expansion of ω in terms of small A^2 (where A is the amplitude of vibrations) has the form

$$\omega = 1 + \frac{3}{8}A^2 - \frac{21}{256}A^4 + \frac{81}{2048}A^6 - \frac{6549}{262144}A^8$$
$$+ \frac{37737}{2094152}A^{10} - \frac{9636183}{67108864}A^{12} + \dots. \tag{2.12.3}$$

First, we shall restrict ourselves to the first three components of the series (2.12.3) and we shall construct the PA [2/2]

$$\omega_2 = \frac{32 + 19A^2}{37 + 7A^2}. \tag{2.12.4}$$

Taking into consideration the components $\sim A^6$ of the frequency expansion, we have

$$\omega_4 = \frac{1 + 1.13A^2 + 0.261A^4}{1 + 0.756A^2 + 0.0599A^4}. \tag{2.12.5}$$

Continuing the process we obtain a sequence of diagonal PAs in the form

$$\omega_{2n} = \sum_{i=0}^{2n} \alpha_i \varepsilon^{2i} \left(\sum_{i=0}^{2n} \beta_i \varepsilon^{2i}\right)^{-1}.$$

Along with the diagonal PAs we shall study an element of the Padé table of order [2/4]

$$\omega = \frac{1 + 0.513A^2}{1 + 0.138A^2 + 0.030A^4},$$

constructed with the first three components of the series (2.12.3).

The result of the frequency calculations according to (2.12.3)–(2.12.5) are graphically presented in Fig. 2.22. The curves 1, 2, 3 correspond to the sums of three, seven and eleven components of the series (2.12.3). Curves 4, 5, 6, correspond to the Padé approximants of order [2/3], [3/3], [4/4]. The exact

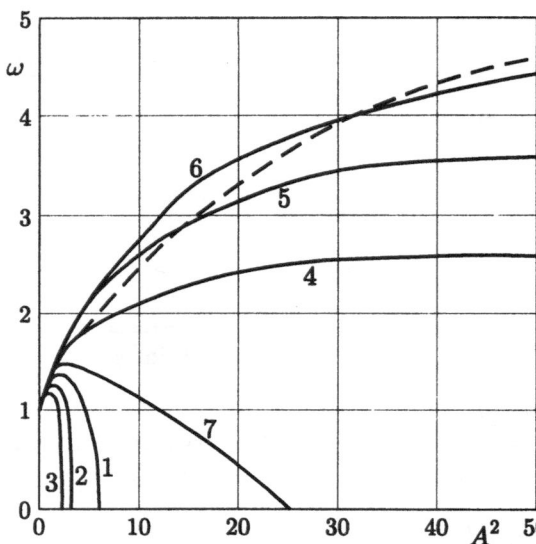

Fig. 2.22. Frequency of amplitude dependence for the Duffing equation constructed with the perturbation method and Padé approximants

solution is represented by the dashed curve. The nondiagonal approximant is represented by curve 7.

It can be seen that the best approximation is achieved with the diagonal PAs.

Lately solitons and solutions close to them have been widely used in mechanics. These are essentially nonlinear solutions which cannot be constructed using the quasi-linear approach when any number of components is conserved. It is still more interesting to note that the PA allows the construction of solutions of that type, beginning from local (quasi-linear) expansions. Moreover, the term "padeon" has appeared.

A model example is presented by the boundary problem

$$y'' - y + 2y^3 = 0, \quad y(0) = 1, \quad y(\infty) = 0,$$

which has the exact solution ("soliton")

$$y = \cosh^{-1}(x). \tag{2.12.6}$$

A solution in the form of the Dirichlet series

$$y = Ce^{-x}\left[1 - 0.25C^2 e^{-2x} + 0.0625C^4 e^{-4x} + \ldots\right], \quad C = \text{const},$$

after rearrangement into the PA and determination of C becomes the exact solution (2.12.6).

PAs often give a good result even for a small number of components of the perturbation series. Obviously, however, the efficiency of the PA increases when the number of approximations increases. So, in [4, 5] many components of the expansion series of the amplitude ε^2 of the period of the Van-der-Pol equation have been constructed by PAs which has led to the discovery of the singularities of the sought period as a function of ε^2 and then, using analytic continuation, the construct of a solution applicable throughout the range of ε^2. At present there is the a possibility of obtaining the approximations of a higher-order with computers. It can be imagined that in the case where a complicated problem of the construction of the approximation of a higher-order in the perturbation methods is solved, then it is desirable to try to apply PAs and other methods of convergence acceleration.

At the same time it must be noticed that iterative methods are essentially simpler to realize by means of computer technology. PAs can be used to improve these methods. Let the iterative process have the form

$$T(u_0) = 0, \quad u_n = T_1(u_{n-1}), \quad n = 1, 2, \ldots.$$

We introduce the function $S_n(\varepsilon)$:

$$S_n(\varepsilon) = u_0 + (u_1 - u_0)\varepsilon + (u_2 - u_1)\varepsilon^2 + \ldots + (u_n - u_{n-1})\varepsilon^n. \quad (2.12.7)$$

For $\varepsilon = 0$ we have $S_n(\varepsilon) \approx u_0$, for $\varepsilon = 1$ $S_n(\varepsilon) \approx u_n$. Then, we rearrange the series (2.12.7) with the PA and suppose $\varepsilon = 1$:

$$u \approx S_n = \frac{u_0 + \sum_{i=1}^m \alpha_i}{1 + \sum_{j=1}^p \beta_j}, \quad m + p = n. \quad (2.12.8)$$

Let us consider, as an example, the problem of big deflections of round isotropic plate of radius R, with a free opening of radius R_0 and a rigidly restrained external outline on which a superficial pressure of constant intensity is acting. The problem solution was found in [50d] using the method of finite central differences for the Young modulus $E = 62.4$ kg/m^3 and the Poisson coefficient $\nu = 0.335$, $R_0 R^{-1} = 0.1$.

The method of succesive approximations applied for the solution of the system of nonlinear algebraic equations, for comparatively big loads, converges for some 150–200 iterations, and the convergence to the solution has an oscillating nature.

Table 2.3. Radial forces T in a round isotropic plate – iteration procedure

Approxim. number	T	Approxim. number	T
0	5.27286	10	4.13072
1	1.09640	145	3.02320
2	4.81246	146	3.11416
3	1.45039	147	3.02603
4	4.55120	148	3.11236
5	1.67086	149	3.02680
6	4.37191	150	3.11063
7	1.82867	151	3.02849
8	4.23735	152	3.10890
9	1.94992	—	—

Table 2.3 gives the result of computations of the dimensionless radial force $T = N_r R^2 D^{-1}$ for $\rho = R^{-1}$, where r is the polar coordinate; $q^* = 0.5qR^4(Dh)^{-1} = 35$ is the intensity of the external load.

Applying the method of generalized summing the situation can be improved (Table 2.4). Let us present the proposed method. The PA (2.12.8), taking into account four approximations, will have the form

$$T = \frac{5.319 - 284.883\varepsilon - 27.606\varepsilon^2}{1 - 52.762\varepsilon - 47.992\varepsilon^2}. \tag{2.12.9}$$

Table 2.4. Radial forces T in a round isotropic plate – using of Padé approximants

Approxim. number	T	Approxim. number	T
0	—	6	3.0656
1	2.6955	7	3.0760
3	3.0140	8	3.0791
4	3.0941	9	3.0789
5	3.0656	—	3.0789
		Solution [50d]	3.0789

When $\varepsilon = 1$, the formula (2.12.9) gives $T = 3.079$. The boundary problem considered above demonstrates the high efficiency of Padé approximants to accelerate the convergence of iterative processes.

PAs can be used for a heuristic evaluation of the domain of applicability of the perturbation theory series. The ε values, up to which the difference between calculations according to the segment of the perturbation series and its diagonal PA does not exceed a given value (e.g. 5%), can be considered as approximative values for the domain of applicability of the initial series.

A transformation to a rational functional allows us to describe nontrivial behaviour at infinity and to take into consideration the singular points of the solutions. We shall consider, as an example, the problem of the flow around a thin elliptical airfoil ($|x| \leq 1$, $|y| \leq \varepsilon$, $\varepsilon \ll 1$) by a plane stream of perfect liquid incoming with velocity v. The first few components of the asymptotic expansion of the relative stream velocity q^* on the airfoil surface are:

$$q^* = \frac{q}{v} = 1 + \varepsilon - \frac{1}{2}\varepsilon^2\frac{x^2}{1-x^2} - \frac{1}{2}\varepsilon^3\frac{x^2}{1-x^2} + \ldots \tag{2.12.10}$$

The written solution diverges for $x = 1$. After replacing expansion (2.12.10) by the PA, the singularity for $x = 1$ disappears:

$$q^* = \frac{(1-x^2)(1+\varepsilon)}{1-x^2+0.5\varepsilon^2x^2} + O(\varepsilon^4). \tag{2.12.11}$$

Fig. 2.23 presents for $\varepsilon = 0.5$: 1 – the exact solution, dashed line – the solution (2.12.10), 2 – PA (2.12.11), and the point line – the solution according to the

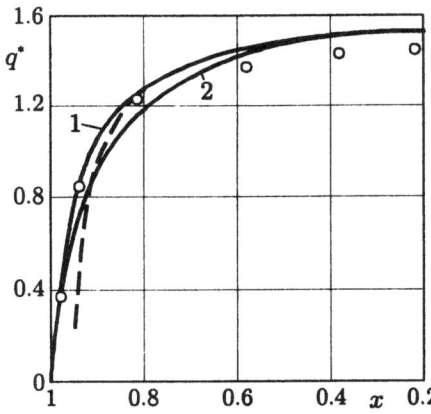

Fig. 2.23. Compansion of the PA approach and Lighthill method

Lighthill method [108, 154], which gives in this case worse results than the PA.

2.12.3 Matching Limit Expansions

From the physical point of view, every nontrivial asymptotic usually has an inverse. In other words, if an asymptotic for $\varepsilon \to 0$ ($\varepsilon \to \infty$) exists, the asymptotic for $\varepsilon \to \infty$ ($\varepsilon \to 0$) can be constructed. Then there appears one of the principal sharpest problems for the asymptotic approach – namely the construction of solutions appropriate for $0 \ll \varepsilon \ll \infty$. This may be solved both on the level of solutions and on the level of equations. In particular, one can try to synthesize the limit equations with the purpose of obtaining a "complex" relationship allowing for a smooth transition from $\varepsilon \to 0$ to $\varepsilon \to \infty$. For a synthesis of solutions one can utilize two-point PAs (TPPAs) [72, 81, 85, 87, 117, 127, 128]. The definition of TPPAs is given below. Let

$$F(\varepsilon) \approx \sum_{i=0}^{\infty} a_i \varepsilon^i \quad \text{for } \varepsilon \to 0; \tag{2.12.12}$$

$$F(\varepsilon) \approx \sum_{i=0}^{\infty} b_i \varepsilon^i \quad \text{for } \varepsilon \to \infty. \tag{2.12.13}$$

The following function will be called the TPPA

$$\phi_{mn}(\varepsilon) = \left(\sum_{k=0}^{m} \alpha_k \varepsilon^k \right) \Big/ \left(\sum_{k=0}^{n} \beta_k \varepsilon^k \right),$$

where the coefficients α_k, β_k are defined so that the first p coefficients of the proper part of the Laurent expansion of ϕ_{mn} coincide with the coefficients (2.12.12), and the $(m + n + 2 - p)$ coefficients of the main part coincide with the coefficients (2.12.13).

Let us investigate the model problem of vibration of a chain consisting of n masses m, joined with springs of rigidity α. The detailed model is a finite difference approximation of the longitudinal vibration of a rod.

The deflection of the k-th particle (y_k) complies with the equation

$$m\ddot{y}_k = \alpha\left[(y_{k+1} - y_k) - (y_k - y_{k-1})\right] , \qquad k = 1, 2, \ldots, n. \qquad (2.12.14)$$

At the ends of the chain the boundary conditions are given by

$$y_k = 0 \quad \text{for} \quad k \leq 0 \quad \text{and} \quad k > n.$$

There are n possible proper forms of vibrations:

$$y_k = A_s \sin\frac{ks\pi}{n+1} \cos(\omega_s t + \phi_s), \qquad s = 1, 2, \ldots, n, \qquad (2.12.15)$$

and the appropriate frequencies of free vibrations are given by

$$\omega_s = 2\sqrt{\frac{\alpha}{m}} \sin\frac{s\pi}{2(n+1)}. \qquad (2.12.16)$$

Let us construct the asymptotic expansions of the frequency ω_s in the vicinities of the points $s = 0$ and $s = 2(n+1)$. We substitute variables in the expression for ω_s putting

$$\bar{x} = x(0.5\pi - x)^{-1}, \qquad x = s\pi[2(n+1)]^{-1}.$$

In the same way, instead of the segment $[0, 2(n+1)]$ for s, we obtain the semi-interval $\bar{x} \in [0, \infty)$.

Enumerating the expansions for $\bar{x} \to 0$ and $\bar{x} \to \infty$, we obtain

$$\sin\frac{\pi\bar{x}}{2(1+\bar{x})} = \frac{\pi}{2}\left[\bar{x} - \bar{x}^2 + \left(1 - \frac{\pi^2}{12}\right)\bar{x}^3\right.$$
$$\left. - \left(1 - \frac{\pi^2}{8}\right)\bar{x}^4 + \ldots\right], \qquad \bar{x} \to 0, \qquad (2.12.17)$$

$$\sin\frac{\pi\bar{x}}{2(1+\bar{x})} = 1 - \frac{\pi^2}{8}\bar{x}^{-2} + \left(1 - \frac{\pi^2}{12}\right)\bar{x}^{-3}$$
$$- \left(1 - \frac{\pi^2}{8}\right)\bar{x}^{-4} + \ldots , \qquad \bar{x} \to \infty. \qquad (2.12.18)$$

A solution, appropriate for $0 \leq \bar{x} \leq \infty$, can be obtained with the TPPA method

$$\omega_s = P\sqrt{\frac{\alpha}{m}}\left[\frac{1.57\bar{x} + 0.81\bar{x}^2}{1 + 1.57\bar{x} + 0.81\bar{x}^2}\right]. \qquad (2.12.19)$$

The results of frequency calculations according to (2.12.16)–(2.12.19) are presented in Fig. 2.24. The exact solution (2.12.16) is designated by 1, the expansions (2.12.17) and (2.12.18) by 2 and 3. The rearranged Padé solution coincides very well with the exact solution over the considered interval. An analysis of the diagrams shows that the TPPA has enabled us to construct an approximative solution appropriate for any frequency of vibrations.

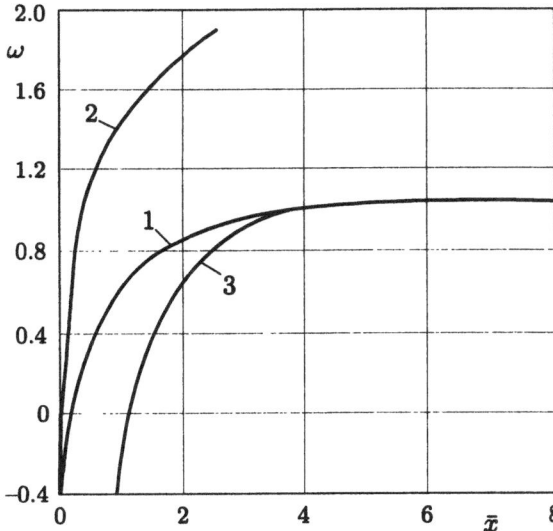

Fig. 2.24. Two-point PA in the theory of the oscillations of chain

An important TPPA application may be the inverse Laplace transform. Indeed, having a given transform, it is possible to investigate the inverse transform behaviour for $t \to 0$ and $t \to \infty$. The problem is the inverse transform description for $0 \ll t \ll \infty$ [30d, 68d]. It is proposed in the monograph [68d] that only those components which give an asymptotic for $p \to 0$ and $p \to \infty$ (whare p is the transform parameter) and also the principal singularities of the expression should be left under the integral sign of Mellin's integral. If the simplified integral can be calculated, then an approximate analytic solution is obtained.

In spite of the unquestionable utility of this approach, convincingly proved in the above monograph, the problem on the whole remains unsolved.

The PAs have been applied to the inverse Laplace transform to widen the domain of applicability of power expansions. One of the TPPA application methods for the solution of the inverse Laplace transform is the $F(p)$ transform expansion in Taylor and Laurent series in the vicinities of the points $p = 0$ and $p = \infty$, which is then followed by the replacement of $F(p)$ by a rational function according to this scheme. Then the transition to the inverse transform is realized according to well-known rules. But the aim is achieved more quickly by applying the TPPA directly to the asymptotics of the inverse transform $f(t)$. Here is an example:

$$F(p) = K_0(p)e^{-p},$$

where K_0 is the McDonald function,

$$f(t) = \frac{1}{\sqrt{2t}} - \frac{\sqrt{t}}{4\sqrt{2}} + \dots \qquad \text{for} \quad t \to 0;$$
$$f(t) = t^{-1} - t^{-2} + \dots \qquad \text{for} \quad t \to \infty.$$

The exact value of $f(t)$ is

$$f(t) = [t(t+2)]^{-1/2}, \qquad t > 0. \tag{2.12.20}$$

The TPPA has the form

$$f(t) \approx (t + \sqrt{2t})^{-1}. \tag{2.12.21}$$

Figure 2.25 presents the solutions (2.12.20) and (2.12.21) (curves 1 and 2 as appropriate). If the asymptotics are not of the power form, the difficulties are also surmountable. Sometimes the asymptotic for $p \to \infty$ may be represented as a sum of exponential functions or of sines and power series [85]. In other cases, it is necessary to introduce nonpower functions into the fractional-rational expressions and expand the latter into power series for $t \to 0$ [11].

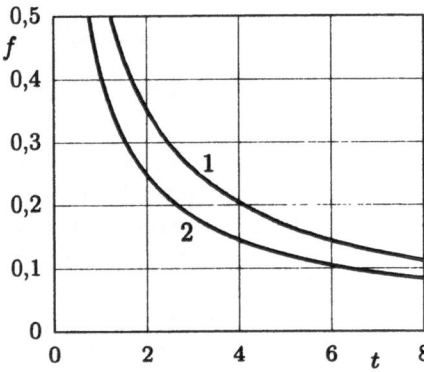

Fig. 2.25. Matching limiting solution in the theory of the Laplace transform

Another interesting example is the Van der Pol equation. We give some necessary preliminary information according to [94]. The Van der Pol oscillator is governed by the equation

$$\ddot{x} + k\dot{x}(X^2 - 1) + X = 0.$$

The solution tends in time to an oscillation with a particular amplitude which does not depend on the initial conditions. The period of this limit oscillation T is of interest and is plotted in Fig. 2.26 as a function of the strength of the nonlinear friction, k. The continuous line gives the numerical results obtained by means of the Runge-Kutta method. The dashed curves give the second-order perturbation approximations

$$T = 2\pi(1 + \frac{1}{16}k^2) + O(k^4) \quad \text{as } k \to 0 \tag{2.12.22}$$

$$T = k(3 - 2\ln 2) + 7.0143k^{-1/3} + O(k^{-1}\ln k) \quad \text{as } k \to \infty. \tag{2.12.23}$$

The TPPA formula uses two terms of the expansion (2.12.22) and the first term of the expansion (2.12.23):

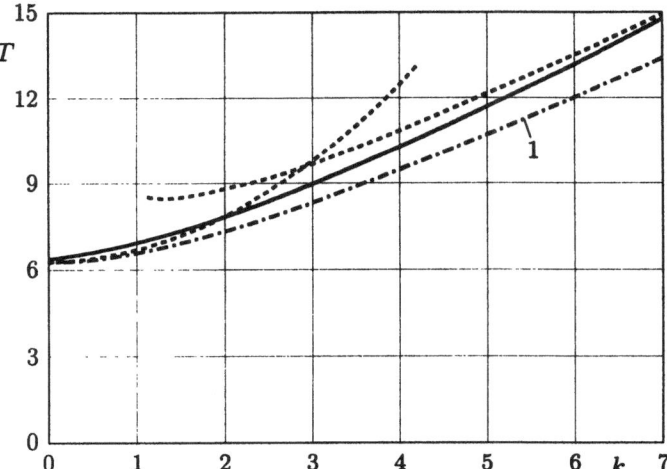

Fig. 2.26. Various solution for the period of the Van der Pol equation

$$T = \frac{6.2832 + 1.5294k + 0.3927k^2}{1 + 0.2433k}$$

and shows good agreement with the numerical results for all values of k (curve 1 in Fig. 2.26).

2.12.4 Matching Local Expansions in Nonlinear Dynamics[1]

Interesting results were obtained by the use of two-point Padé approximants in the theory of normal vibrations in nonlinear finite-dimensional systems [118, 56d].

Consider a conservative system

$$m_i \ddot{x}_i + \Pi_{x_i} = 0, \quad \dot{x}_i = \frac{dx_i}{dt}, \quad \Pi_z = \frac{\partial \Pi}{\partial z}, \quad i = 1, 2, \dots, n, \qquad (2.12.24)$$

where $\Pi = \Pi(x)$ is the potential energy, assumed to be a positive definite function; and $x = (x_1, x_2, ..., x_n)^T$. The power series expansion for $\Pi(x)$ begins with terms having a power of at least 2. Without reducing the degree of generalization, assume that $m_i = 1$, since this can always be ensured by dilatation of the coordinates.

The energy integral for system (2.12.24) is

$$\frac{1}{2} \sum_{k=1}^{n} \dot{x}_k^2 + \Pi(x_1, x_2, ..., x_n) = h, \qquad (2.12.25)$$

h being the system energy. Assume that within configuration space, bounded by the closed maximum equipotential surface $\Pi = h$, the only equilibrium position is $x_i = 0$ $(i = 1, 2, ..., n)$.

---

[1] By courtesy of Yu.V. Mikhlin

In order to determine the trajectories of normal vibrations, the following relationships can be used [135]:

$$2x_i'' \frac{h - \Pi}{1 + \sum_{k=2}^n x_k'^2} + x_i'(-\Pi) = -\Pi_{x_i} \quad (i = 1, 2, 3, \ldots, n; \; x \equiv x_i). (2.12.26)$$

These are obtained either as Euler equations for the variational principle in Jacobi form or by elimination of time from the equations of motion (2.12.24) with consideration for the energy integral (2.12.25).

An analytical extension of the trajectories on the maximum isoenergy surface $\Pi = h$ is possible if the boundary conditions, i.e. the conditions of orthogonality of a trajectory to the surface, are satisfied [135]:

$$x_i'[-\Pi_x(X, \dot{x}_2(X), \ldots, x_n(X)] = -\Pi_{x_i}(X, x_2(X), \ldots, x_n(X)), \quad (2.12.27)$$

$(X, x_2(X), ..., x_n(X))$ being the trajectory return points lying on the $\Pi = h$ surface where all velocities are equal to zero. If the trajectory $x_i(x)$ is defined, the law of motion with respect to time can be found using

$$\ddot{x} + \Pi_x(x_1, x_2(x), \ldots, x_n(x)) = 0,$$

for which the periodic solution $x(t)$ is obtained by inversion of the integral.

Now consider the problem of normal vibrational behaviour in certain nonlinear systems when the amplitude (or energy) of the vibrations is varied from zero to an extremely large value. Assume that in the system

$$\ddot{z} + \Pi_{z_i}(z_1, z_2, \ldots, z_n) = 0 \tag{2.12.28}$$

the potential energy $\Pi(z_1, z_2, ..., z_n)$ is a positive definite polynomial of $z_1, ..., z_n$ having a minimum power of 2 and a maximum power of $2m$. On choosing a coordinate, say z_1, substitute $z_i = cx_i$ where $c = z_1(0)$. Obviously, $x_1(0) = 1$. Furthermore, without loss of generality, assume $\dot{x}_1(0) = 0$. Then

$$\ddot{x}_i + V_{x_i}(c, x_1, x_2, \ldots, x_n) = 0, \tag{2.12.29}$$

where $V = \sum_{k=0}^{2m-1} c^k V^{(k+2)}(x_1, x_2, \ldots, x_n)$, $V^{(r+1)}$ contains terms of the power $(r + 1)$ of the variables in the potential

$$V(c, x_1, x_2, \ldots, x_n) = \Pi(z_1(x_1), z_2(x_2), \ldots, z_n(x_n)).$$

It is assumed below that the amplitude of vibration $c = z(0)$ is the independent parameter.

At small amplitudes a homogenous linear system with a potential energy $V^{(2)}$ is selected as the initial one while, at large amplitudes, a homogenous nonlinear system with a potential energy $V^{(2m)}$ is selected. Both linear and nonlinear homogenous systems allow normal vibrations of the type $x_i = k_i x_1$, where the constants k_i are determined from the algebraic equations

$$k_i V_{x_i}^{(r)}(1, k_2, \ldots, k_n) = V_{x_i}^{(r)}(1, k_2, \ldots, k_n).$$

A number of vibrations of this type can be greater than the number of degrees of freedom in the nonlinear case.

In the vicinity of a linear system at small values of c, trajectories of the normal vibrations $x_i^{(1)}(x)$ can be determined as a power series of x and c (assuming that $x_1 = x$), while in the vicinity of a homogenous nonlinear system (at large values of c), $x_i^{(2)}(x)$ can be determined as a power series of x and c^{-1}.

The construction of the series is described in [55d].

The amplitude values (at $\dot{x} = \dot{x}_i = 0$) define the normal vibration mode completely. Therefore, for the sake of simplicity, only the expansions of $\rho_i^{(1)} = x_i^{(1)}(1)$ and $\rho_i^{(2)} = x_i(1)$ in terms of the powers of c wil be discussed below:

$$\rho_i^{(1)} = \sum_{j=0}^{\infty} \alpha_j^{(i)} c^j, \quad \rho_i^{(2)} = \sum_{j=0}^{\infty} \beta_j^{(i)} c^{-j}. \tag{2.12.30}$$

In order to join together the local expansions (2.12.30) and to investigate the behaviour of the normal vibration trajectories at arbitrary values of c, fractional rational TPPAs are used:

$$P_s^{(i)} = \frac{\sum_{j=0}^{s} a_j^{(i)} c^j}{\sum_{j=0}^{s} b_j^{(i)} c^j} \tag{2.12.31}$$

or

$$P_s^{(j)} = \frac{\sum_{j=0}^{s} a_j^{(i)} c^{j-s}}{\sum_{j=0}^{s} b_j^{(i)} c^{j-s}}. \tag{2.12.32}$$

Compare expressions (2.12.31) and (2.12.32) with the expansions (2.12.30).

By preserving only the terms with the order of $c^r (-s \le r \le s)$ and equating the coefficients at equal powers of c, $n-1$ systems of $2(s+1)$ linear algebraic equations will be obtained for the determination of $a_j^{(i)}$, $b_j^{(i)}$ ($j = 0, 1, 2, \ldots$).

Since the determinants of these systems $\Delta_s^{(i)}$ are generally not equal to zero, the systems of algebraic equations have a single exact solution, $a_j^{(i)} = b_j^{(i)} = 0$.

Select a TPPA corresponding to the preserved terms in (2.12.30) having the nonzero coefficients $a_j^{(1)}$, $b_j^{(1)}$. Assume that $b_0^{(i)} \ne 0$, for otherwise as $c \to 0$ $x_i^{(1)} \to \infty$. Without loss of generality, it can also be assumed that $b_0^{(i)} = 1$. Now the systems of algebraic equations for the determination of $a_j^{(i)}$, $b_j^{(i)}$ become overdetermined. All the unknown coefficients $a_0^{(1)}, \ldots, a_n^{(1)}$, $b_1^{(1)}, \ldots, b_n^{(1)}$ ($i = 2, 3, \ldots, n$) are determined from $(2s+1)$ equations while the "error" of this approximate solution can be obtained by a substitution of all coefficients in the remaining equation. Obviously, the "error" is determined by the value of $\Delta_s^{(i)}$, since at $\Delta_s^{(i)} = 0$ nonzero solutions and, consequently, the exact Padé approximants will be obtained in the given approximation in terms of c.

Hence, the following is the necessary condition for convergence of a succession of the TPPA (2.12.31), at $s \to \infty$, to fractional rational functions:

$$P^{(i)} = \frac{\sum_{j=0}^{\infty} a_j^{(i)} c^j}{\sum_{j=0}^{\infty} b_j^{(i)} c^j} \quad (b_0^{(i)} = 1),$$

(2.12.33)

namely,

$$\lim_{s \to \infty} \Delta_s^{(i)} = 0 \quad (i = 2, 3, \ldots, n).$$

(2.12.34)

Indeed, if conditions (2.12.34) are not satisfied, nonzero values of the coefficients $A_i^{(i)}$, $b_i^{(i)}$ in (2.12.33) will obviously not be obtained.

Conditions (2.12.34) are necessary but not sufficient for the convergence of the approximants (2.12.31) to the functions (2.12.33); nevertheless, the role of conditions (2.12.34) is determined by the following consideration.

Since in the general case there is more than one quasilinear local expansion and essentially nonlinear local expansions are alike, the numbers of expansions of the respective type being not necessarily equal, it is the convergence conditions (2.12.34) that allow one to establish a relation between the quasilinear and essentially nonlinear expansions, that is, to decide which of them corresponds to the same solution and which to different ones.

For a concrete analysis based on the above technique, consider a conservative system with two degrees of freedom, whose potential energy contains the terms of the 2nd and 4th powers of the variables z_1, z_2. Substituting $z_1 = cx$, $z_2 = cy$, where $c = z_1(0)$, $(x(0) = 1)$, one obtains

$$V = c^2 \left(d_1 \frac{x^2}{2} + d_2 \frac{y^2}{2} + d_3 xy \right) + c^4 \left(\gamma_1 \frac{x^4}{4} + \gamma_2 x^3 y \right.$$
$$\left. + \gamma_3 \frac{x^2 y^2}{2} + \gamma_4 xy^3 + \gamma_5 \frac{y^4}{4} \right) \equiv c^2 V^{(2)} + c^4 V^{(4)}.$$

The equation for determining the trajectory $y(x)$ is of the form

$$2y''(h - V) + (1 + y'^2)(-y'V_x + V_y) = 0,$$

(2.12.35)

while the boundary conditions (2.12.27) can be written

$$(-y'V_x + V_y)|_{h=V} = 0.$$

For definiteness, let $d_1 = d_2 = 1 + \gamma$; $d_3 = -\gamma$; $\gamma_1 = 1$; $\gamma_2 = 0$; $\gamma_3 = 3$; $\gamma_4 = 0.2091$; $\gamma_5 = \gamma$. Write the equations of motion for such a system:

$$\ddot{x} + x + \gamma(x - y) + c^2(x^3 + 3xy^2 + 0.2091y^3) = 0,$$
$$\ddot{y} + y + \gamma(y - x) + c^2(2y^3 + 3x^2 y + 0.6273y^2 x) = 0.$$

(2.12.36)

In the linear limiting case ($c = 0$) two rectilinear normal modes of vibrations $y = k_0 x$, $k_0^{(1)} = 1$; $k_0^{(2)} = -1$ are obtained, while a nonlinear system (where the equations of motion contain only the third power terms with respect to x, y) admits four such modes: $k_0^{(3)} = 1.496$; $k_0^{(4)} = 0$; $k_0^{(5)} = -1.279$; $k_0^{(6)} = -5$.

In order to determine nearly rectilinear curvilinear trajectories of normal vibrations, (2.12.35) is used along with the boundary conditions.

By matching the local expansions the following Padé approximants are obtained

$$
\begin{array}{ccc}
 & I-IV & II-V \\
\gamma = 2 & \rho = \dfrac{1+1.20c^2}{1+1.61c^2+0.72c^4} & \rho = \dfrac{-1-1.11c^2-0.275c^4}{1+1.00c^2+0.215c^4} \\[3mm]
\gamma = 0.5 & \rho = \dfrac{1+1.06c^2}{1+2.06c^2+3.20c^4} & \rho = \dfrac{-1-2.76c^2-1.36c^4}{1+2.31c^2+1.04c^4} \\[3mm]
\gamma = 0.2 & \rho = \dfrac{1+1.70c^2}{1+3.96c^2+13.29c^4} & \rho = \dfrac{-1-6.41c^2-9.03c^4}{1+5.30c^2+7.02c^4}\,.
\end{array}
\tag{2.12.37}
$$

The two additional modes of vibration exist only in a nonlinear system; as ν increases (the amplitude c decreases), they vanish at a certain limiting point. For the analysis of these vibration modes, assume a new variable $\sigma = (\rho - 1.496)/(\rho - 5)$.

By using the variable σ, two expansions in terms of positive and negative powers were obtained; therefore, fractional rational representations can be introduced as above. By comparing these expansions, the following TPPAs are obtained

$$
\begin{array}{cc}
 & III-VI \\
\gamma = 2 & \nu = \dfrac{8.874\sigma+1.126\sigma^2}{1+4.300\sigma+2.836\sigma^2+0.549\sigma^3} \\[3mm]
\gamma = 0.5 & \nu = \dfrac{35.497\sigma+5.108\sigma^2}{1+3.021\sigma-0.794\sigma^2+0.622\sigma^3} \\[3mm]
\gamma = 0.2 & \nu = \dfrac{88.986\sigma+1.470\sigma^2}{1-0.143\sigma+3.747\sigma^2+0.072\sigma^3}\,.
\end{array}
\tag{2.12.38}
$$

Now proceed to the determination of the limiting point. Obviously, it can be found from

$$\frac{\partial \nu}{\partial \sigma} = 0.$$

From (2.12.38):

at $\gamma = 2$ the limiting point is $\nu \cong 1.21$, $c \cong 0.91$;
at $\gamma = 0.5$ the limiting point is $\nu \cong 11.10$, $c \cong 0.30$;
at $\gamma = 0.2$ the limiting point is $\nu \cong 23.93$, $c \cong 0.20$.

Hence, as $\gamma \to 0$ the limiting point is characterized by the amplitude $c \to 0$. Therefore, the two additional vibration modes in a nonlinear system can exist at rather small amplitudes of vibrations. Note that the quasilinear analysis does not allow one to find these solutions even at small amplitudes.

In the limit, when $\gamma = 0$, a linear system decomposes into two independent oscillators having identical frequencies and admits any rectilineal modes of normal vibrations. Obviously, the full system (2.12.36) at $\gamma = 0$ admits four modes of vibrations (in the nonlinear case) $y_2 = ky_1$, $k = \{1.496, 0, -1.279, -5\}$.

Thus, fractional rational Padé approximants allow us to estimate the nonlocal behaviour of normal vibrations in nonlinear finite-dimensional systems. For system (2.12.36) the evolution of the modes of normal vibrations is shown in Fig. 2.27 using parameters $\zeta = \ln(1 + c^2 h^2)$ and $\varphi = \text{arctg}\, \rho$ (the picture shows periodicity in φ, the period being 2π). The solid lines correspond to analytical solutions ((2.12.37) and (2.12.38) were employed), while the dotted lines were obtained in computer check calculations for $\gamma = 2$, carried out by A. L. Zhupiev. The analytical solutions and numerical computations show good agreement. For solution II, the TPPA relationship and the numerical calculations gave, in the scale selected, the same curve (Fig. 2.27).

2.12.5 Generalizations and Problems

Evidently, the TPPA is not a panacea. For example, one of the "bottlenecks" of the TPPA method is related to the presence of logarithmic components in numerous asymptotic expansion. Van Dyke [154] writes: "A technique analogous to rational functions is needed to improve the utility of series containing logarithmic terms. No striking results have yet been achieved".

This problem is most essential for the TPPA, because, as a rule, one of the limits $\varepsilon \to 0$ or $\varepsilon \to \infty$ for a real mechanical problem gives expansions with logarithmic terms or other gives complicated functions.

It is worth noting that in some cases these obstacles may be overcome by using an approximate method of TPPA construction by taking as limit points not $\varepsilon = 0$ and $\varepsilon = \infty$, but some small and large (but finite) values [69d].

On the other hand, in [61, 58d] were proposed so-called quasifractional approximants. Let us suggest that we have a perturbation approach in powers of ε for $\varepsilon \to 0$ and the asymptotic expansions $F(x)$, containing, for example, a logarithm for $\varepsilon \to \infty$. By definition QA is the ratio R with the unknown coefficients α_1, β_1, containing both the powers of ε and $F(x)$. The coefficients α_1, β_1 are chosen in such a way that (a) the expansion of R in powers of ε matches the corresponding perturbation expansion; and (b) the asymptotic behaviour of R for $\varepsilon \to \infty$ coincides with $F(x)$.

The main advantage of the TPPA and QA is simplicity of algorithms and the possibility of using only a few terms of the expansions. Besides, it is possible to take into account the known singularities of the defined functions.

On the other hand, one of the important problems of the TPPA and QA is to control the correctness of the realized matching. Sometimes we can use numerical methods [100] or a procedure of recalculation of the matching

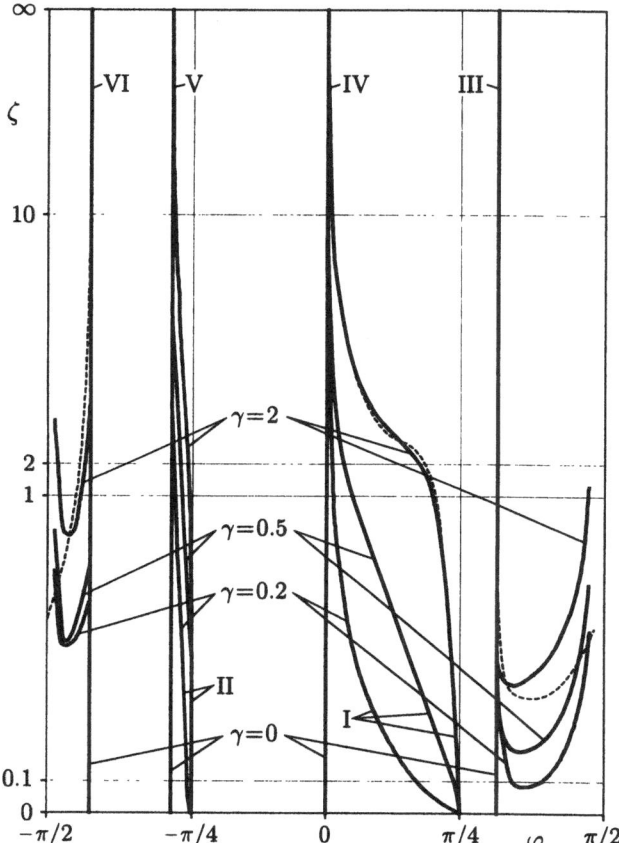

Fig. 2.27. The evolution of normal vibration modes. Local normal mode expansions are marked by I, II (quasilinear case) and III, IV, V and VI (essentially nonlinear case)

parameters [64d]. Along with the comparison of the known numerical or analytical solutions, numerical or experimental results, it is possible to verify the modified expansions by their mutual correspondence. To estimate the error of the obtained TPPA, the Newton–Kantorovich method is used, and then one can utilize the well-developed mathematical techniques concerning the effective estimators. Also one may use one-point approximants for the expansion for $\varepsilon \to 0$ and $\varepsilon \to \infty$ for comparing with the TPPA. But in general the question is open.

It is known that the PA posseses the property of self-correction of the error [111, 113] and may be used for the solution of ill-posed problems. In other words, errors of the nominator and denominator mutually vanish. This effect is closely connected with the fact that errors in the coefficients of the PA don't spread arbitraly, but "mistaken" coefficients are created in the new good approximations to the solution. But we do not know whether this property exists for the TPPA.

With reference to this, we must say that many results in the theory of the one-point PA were obtained on the basis of numerical experiments. For example, many different methods for accelerating convergence of sequences and series were tested and compared in a wide range of test problems, including both linearly and logarithmically convergent series, monotonic and alternating series [147]. This paper gives detailed comparisons of all the tested methods on the basis of the number of correct digits as a function of the number of terms of the series used. Such computations would be very useful for the theory of the TPPA.

3. Continuous Systems

3.1 Continuous Approximation for a Nonlinear Chain

Let us consider the system in which the masses, interconnected by a weightless beam, interact with nonlinearly elastic supports distributed equidistantly along the length (Fig. 3.1). The corresponding equation of the free motion in the absence of friction is written in the form

$$m \sum_{j=-\infty}^{\infty} \frac{\partial^2 w}{\partial t^2} \delta(x - jl) + EJ\frac{\partial^4 w}{\partial x^4} - S\frac{\partial^2 w}{\partial x^2}$$

$$+ \sum_{j=-\infty}^{\infty} q(w)\delta(x - jl) = 0; \tag{3.1.1}$$

here m is the magnitude of each concentrated mass, $w(x,t)$ is the transversal displacement, l is the spacing between supports (masses), δ is the Dirac delta function, $q(w) = aw + bw^3$, and S is the stretching force.

In a linearized system ($b = 0$), harmonic vibrations and waves conserving their shape in time play a fundamental role. But other types of waves, e.g., localized wave packets, inevitably "spread out" because of dispersion. In a nonlinear system, for $EJ = 0$ (the masses are connected by a string), the situation turns out to be reversed. Quasiharmonic waves are distorted, but there exist localized solutions of soliton type with a time frequency that exceeds the highest frequency of the natural vibrations of the linearized system.

We demonstrate the possibility of constructing such solutions in the general case ($EJ \neq 0$).

We write the equations of motion using a finite difference approximation of the elastic forces in the beam

$$\frac{d^2 v_j}{d\tau^2} + \alpha(6v_j - 4v_{j+1} - 4v_{j-1} + v_{j+2} + v_{j-2})$$

$$+ \beta(2v_j - v_{j+1} - v_{j-1}) + v_j + v_j^3 = 0, \tag{3.1.2}$$

where

$$v_j = w_j\sqrt{\frac{b}{a}}, \quad \tau = \sqrt{\frac{a}{m}}t, \quad \alpha = \frac{EJ}{al^3}, \quad \beta = \frac{S}{al}.$$

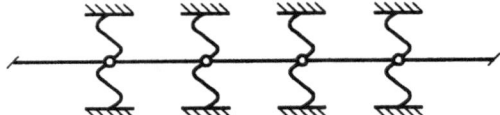

Fig. 3.1. Weightless beam with discrete masses on nonlinearly elastic supports

System (3.1.2) has a stationary solution in the form of a "sawtooth" standing wave

$$v_j = (-1)^j V(\tau),$$

where the function $V(\tau)$ satisfies the differential equation

$$\frac{d^2 V}{d\tau^2} + 4(4\alpha + \beta)V + V + V^3 = 0.$$

We shall seek localized solutions of soliton type in the form

$$v_j(\tau) = (-1)^j V_j(\tau), \tag{3.1.3}$$

assuming that the functions $V_{j(\tau)}$ vary smoothly with the index j. Substitution of (3.1.3) into (3.1.2) leads, after some transformations, to the system

$$\frac{d^2 V_j}{d\tau^2} + \alpha(6V_j - 4V_{j+1} - 4V_{j-1} + V_{j+2} + V_{j-2}) \tag{3.1.4}$$

$$+ (8\alpha + \beta)(V_{j+1} + V_{j-1} - 2V_j) + 4(4\alpha + \beta)V_j + V_j + V_j^3 = 0.$$

Using the continuum approximation to replace the set of functions $V_{j(\tau)}$ by the function $V(\xi, \tau)$ of two variables, we obtain the equation

$$\frac{\partial^2 V}{\partial \tau^2} + \alpha \frac{\partial^4 V}{\partial \xi^4} + (8\alpha + \beta)\frac{\partial^2 V}{\partial \xi^2} + 4(4\alpha + \beta)V + V + V^3 = 0, \tag{3.1.5}$$

where $\xi = x/l$.

Seeking solutions of soliton type with small amplitudes, we take as the first approximation a function that is harmonic in time:

$$V(\xi, \tau) = A(\xi) \sin \omega\tau. \tag{3.1.6}$$

Applying the procedure of the Galerkin method to (3.1.5) and taking (3.1.6) into account, we obtain the equation

$$\alpha \frac{\partial^4 A}{\partial \xi^4} + (8\alpha + \beta)\frac{\partial^2 A}{\partial \xi^2} + (\omega^2 - \omega_1^2)A + \frac{3}{4}A^3 = 0, \tag{3.1.7}$$

where $\omega^2_1 = 1 + 4(4\alpha + \beta)$.

Let $\omega^2 - \omega^2_1 = \epsilon^2 > 0$; then for the existence of the desired soliton solution it is necessary for the quantity ϵ to be a small parameter, with $A \sim \epsilon$, $\frac{d}{d\xi} \sim \epsilon$. Here in (3.1.7) the first term can be discarded, so that the soliton solution corresponds to the separatrix of the second-order differential equation

$$(8\alpha + \beta)\frac{\partial^2 A}{\partial \xi^2} - \epsilon A + \frac{3}{4}A^3 = 0$$

and has the form

$$V_j(\tau) = 4(-1)^j \frac{\varepsilon \sin \omega\tau}{\mathrm{ch}(\varepsilon\xi_j/\sqrt{4/3\,\alpha + \beta})}. \tag{3.1.8}$$

For ω^2 close to ω_1^2, the function (3.1.8) is evidently a spatially localized perturbation (Fig. 3.2) that performs periodic pulsations in time (an envelope soliton or "pulson"). As numerical investigations show, pulsons occur not only under special initial conditions but also as a result of "self-localization" of nonlocalized perturbations such as a sawtooth standing wave. Thus, a high-frequency vibration can precede the "soliton" stage of a dynamical process.

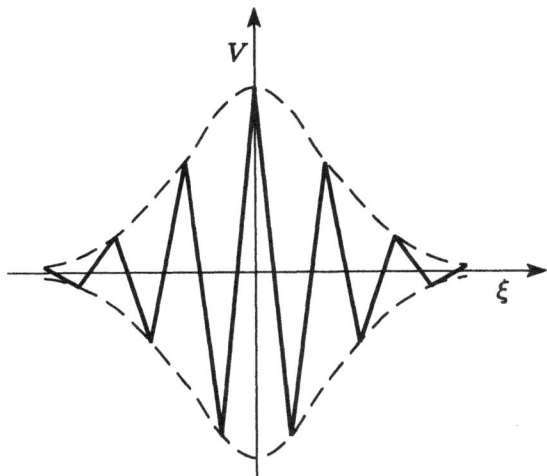

Fig. 3.2. Spatial localization of oscillations

The most important condition for the appearance of the soliton mode in real systems is the presence of a "quasilimiting" frequency in the spectrum of the natural vibrations of the linearized system, above which the pulson frequencies are located.

Now we consider a simple formal way of constructing a continuum approximation for the problem of high-frequency oscillations of a chain of masses and we also formulate continuum equations which are able to describe rather satisfactorily both low- and high-frequency oscillations.

The oscillations of a chain of masses coupled by nonlinear springs are described by the equations

$$m\frac{\mathrm{d}^2 u_k}{\mathrm{d}t^2} + c(2u_k - u_{k+1} - u_{k-1}) + c_1 u_k^3 = 0. \tag{3.1.9}$$

The well-known continuum approximation of set (3.1.9) for the case of low-frequency oscillations has the form

$$m\frac{\partial^2 u}{\partial t^2} - h^2 c\frac{\partial^2 u}{\partial x^2} + c_1 u^3 = 0, \tag{3.1.10}$$

where h is the distance between the masses.

We write (3.1.9) in the form [153]

$$m\frac{\partial^2 u}{\partial t^2} + 4c\sin^2\left(-\frac{ih}{2}\frac{\partial}{\partial x}\right)u + c_1 u^3 = 0. \tag{3.1.11}$$

The low-frequency approximation is now obtained by expanding the operator $\sin^2(-1/2ih\partial/\partial x)$ in a Taylor: series

$$\sin^2\left(-\frac{ih}{2}\frac{\partial}{\partial x}\right) = -\frac{h^2}{4}\frac{\partial^2}{\partial x^2} + \cdots,$$

and the high-frequency continuum approximation is obtained by expanding that operator in the vicinity of the identity transformation:

$$\sin^2\left(-\frac{ih}{2}\frac{\partial}{\partial x}\right) = 1 + \frac{h^2}{4}\frac{\partial^2}{\partial x^2} + \cdots.$$

In the latter case the equation for the function $u(x,t)$ is as follows:

$$m\frac{\partial^2 u}{\partial t^2} + 4cu + ch^2\frac{\partial^2 u}{\partial x^2} + c_1 u^3 = 0. \tag{3.1.12}$$

For the displacements of the masses in the chain we obtain

$$u(x+h,t) = -u(x,t) + \frac{h}{2}\frac{\partial u(x,t)}{\partial x} + \cdots.$$

For the Toda chain [153] the short-wavelength continuum approximation is

$$m\frac{\partial^2 u}{\partial t^2} + \left(4c + \frac{\partial^2}{\partial x^2}\right)e^{-bu} = 0.$$

The existence of simple expressions for continuum approximations for low- and high-frequency oscillations makes it possible to construct composite equations which quite satisfactorily describe the processes for arbitrary oscillation frequencies. For instance, using (3.1.11) and (3.1.12), we can construct the composite equation

$$m\left(1 - \frac{h}{4}\frac{\partial^2}{\partial x^2}\right)\frac{\partial^2 u}{\partial t^2} - ch^2\frac{\partial^2 u}{\partial x^2} + c_1\left(1 - \frac{h^2}{4}\frac{\partial^2}{\partial x^2}\right)u^3 = 0. \tag{3.1.13}$$

To illustrate the efficiency of such a combination, we show in Fig. 3.3 the results of determining the frequencies of a linear ($c_1 = 0$) chain of n masses. The exact values can be determined from the formula

$$\omega_k = 2\sqrt{\frac{c}{m}}\sin\frac{k\pi}{2(n+1)}.$$

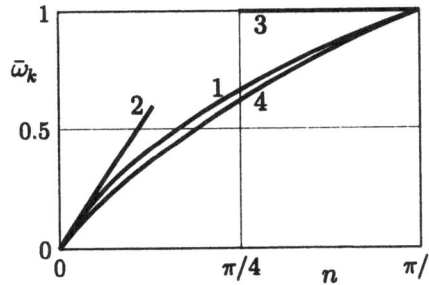

Fig. 3.3. Various approximation for the frequencies of the chain

We plot in Fig. 3.3 the quantities $\overline{\omega}_k = 0.5\omega_k(mc^{-1})^{1/2}$; the numbers 1, 2, 3, 4 indicate the exact solution (the discrete values of $\overline{\omega}_k$ are connected by a solid line) and the solutions obtained on the basis of (3.1.10), (3.1.11), and (3.1.13), respectively.

The proposed method can be easily adopted to a lattice of higher dimensionality, and it allows the construction of relations with a higher degree of accuracy.

3.2 Homogenization Procedure in the Nonlinear Dynamics of Thin-Walled Structures

At present, the method of homogenization is used to great advantage for solving variable-coefficient partial differential equations in such disciplines as the theory of composites [44, 50, 63, 101, 139] or the design of reinforced, corrugated, perforated, etc., shells [20, 23, 60, 64, 65, 8d, 9d, 12d, 13d]. An original nonhomogenous medium or structure is reduced to a homogeneous one (generally speaking, to an anisotropic one) with certain effective characteristics. The homogenization method allows one not only to obtain effective characteristics but also to investigate the nonhomogenous distribution of mechanical stresses in different materials and structures which is of great significance for evaluating their strength. Then, the main idea of the method is based on the separation of "fast" and "slow" variables. As a start, a certain periodic boundary problem is formulated ("cell" or "local" problem) and its solution, assuming periodic continuation of boundary conditions, is obtained. For that purpose the local coordinates ("fast" variables, in the case of using a multiscaling method) are introduced. After that the averaging itself upon local ("fast") coordinates is performed.

3.2.1 Nonhomogeneous Rod

As an example we treat here the case of axial oscillations of a rod with periodic cross-section $a(x/\varepsilon)$ in a nonlinear resisting medium with periodic properties $b(x/\varepsilon)$. The governing equation may be written in the following form:

$$\frac{\partial}{\partial x}\left[a\left(\frac{x}{\varepsilon}\right)\frac{\partial u}{\partial x}\right] + b\left(\frac{x}{\varepsilon}\right)u^3 - c\left(\frac{x}{\varepsilon}\right)\frac{\partial^2 u}{\partial t^2} = 0, \qquad (3.2.1)$$

where a, b, c are periodic functions,

$$a\left(\frac{x+\varepsilon}{\varepsilon}\right) = a\left(\frac{x}{\varepsilon}\right); \quad b\left(\frac{x+\varepsilon}{\varepsilon}\right) = b\left(\frac{x}{\varepsilon}\right); \quad c\left(\frac{x+\varepsilon}{\varepsilon}\right) = c\left(\frac{x}{\varepsilon}\right).$$

We introduce the "fast" variable $y = x/\varepsilon$. Then, the differential operator $\partial/\partial x$, applied to the function $u(x,y,t)$, becomes

$$\frac{\partial}{\partial x} + \varepsilon^{-1}\frac{\partial}{\partial y}. \qquad (3.2.2)$$

Let us consider the following ansatz for the solution of (3.2.1):

$$u = u_0(x,y,t) + \varepsilon u_1(x,y,t) + \varepsilon^2 u_2(x,y,t) = \ldots . \qquad (3.2.3)$$

Now we substitute expressions (3.2.2) and (3.2.3) in (3.2.1) and identify the powers of ε:

$$\varepsilon^{-2} \; : \; \frac{\partial}{\partial y}\left[a(y)\frac{\partial u_0}{\partial y}\right] = 0; \qquad (3.2.4)$$

$$\varepsilon^{-1} \; : \; \frac{\partial}{\partial y}\left[a(y)\frac{\partial u_1}{\partial x}\right] + a(y)\frac{\partial^2 u_0}{\partial x \partial y} + \frac{\partial}{\partial y}\left[a(y)\frac{\partial u_1}{\partial y}\right] = 0; \qquad (3.2.5)$$

$$\varepsilon^0 \; : \; \frac{\partial}{\partial y}\left[a(y)\frac{\partial u_2}{\partial y}\right] + a(y)\frac{\partial^2 u_0}{\partial x^2} + \frac{\partial}{\partial y}\left[a(y)\frac{\partial u_1}{\partial x}\right]$$
$$+ a(y)\frac{\partial^2 u_1}{\partial x \partial y} + b(y)u_0^3 - c(y)\frac{\partial^2 u_0}{\partial t^2} = 0; \qquad (3.2.6)$$

$\cdots$ $\cdots\cdots\cdots\cdots\cdots\cdots\cdots$

We use the technique described in [50, 139].

From (3.2.4) and the conditions of periodicity we have

$$u_0 = u_0(x,t).$$

Then, (3.2.5) (the so-called "cell" or "local" problem) may be rewritten as

$$\frac{\partial}{\partial y}\left[a(y)\frac{\partial u_1}{\partial y}\right] = -\frac{du_0}{dy}\frac{\partial u_0}{\partial x}.$$

After integration one obtains

$$\frac{\partial u_1}{\partial y} = -\frac{\partial u_0}{\partial x} + \frac{c(x,t)}{a(y)}. \qquad (3.2.7)$$

The function $c(x,t)$ is defined from the conditions of periodicity for u_1:

$$u_1(x,y+1) = u_1(x,y)$$

and may be written as

$$C(x,t) = \hat{a}\frac{\partial u_0}{\partial x},$$

$$\hat{a} = \left[\int\limits_0^1 a^{-1}\,dy\right]^{-1}.$$

Excluding the function $\partial u_1/\partial y$ from (3.2.6), one obtains

$$\frac{\partial}{\partial y}\left[a\frac{\partial u_2}{\partial y}\right] + \frac{\partial}{\partial y}\left[a\frac{\partial u_1}{\partial x}\right] + \hat{a}\frac{\partial^2 u_0}{\partial x^2} + bu_0^3 - c\frac{\partial^2 u_0}{\partial t^2} = 0. \qquad (3.2.8)$$

Now we apply the homogenization operator

$$\int\limits_0^1 (\ldots)\,dy$$

to equation (3.2.8). The first two terms vanish due to the periodicity of the corresponding functions, and finally we have

$$\hat{a}\frac{\partial^2 u_0}{\partial x^2} + \bar{b}u_0^3 - \bar{c}\frac{\partial^2 u_0}{\partial t^2} = 0,$$

where

$$\bar{b} = \int\limits_0^1 b(y)\,dy; \quad \bar{c} = \int\limits_0^1 c(y)\,dy.$$

This homogenized equation has only constant coefficients and its solution is simpler than in the case of the governing equation (3.2.1).

For the function u_1 one may obtain from (3.2.7)

$$u_1 = \frac{\partial u_0}{\partial x}\left[\hat{a}\int\limits_0^y a^{-1}\,dy - y\right],$$

$$u_1(x, y+1, t) = u_1(x, y, t).$$

Now we must make two very important remarks.
First of all, we have

$$u = u_0 + o(\varepsilon),$$

but

$$\frac{\partial u}{\partial x} = \frac{\partial u_0}{\partial x} + \frac{\partial u_1}{\partial y} + o(\varepsilon).$$

So, we cannot obtain the correct expression for the derivative in the framework of the homogenized problem solution.

Secondly, the solution of the local (cell) problem for a quasilinear differential operator (when the highest derivatives are linear) may be obtained from the linear boundary value problem.

3.2.2 Stringer Plate

The governing object is depicted in Fig. 3.4.

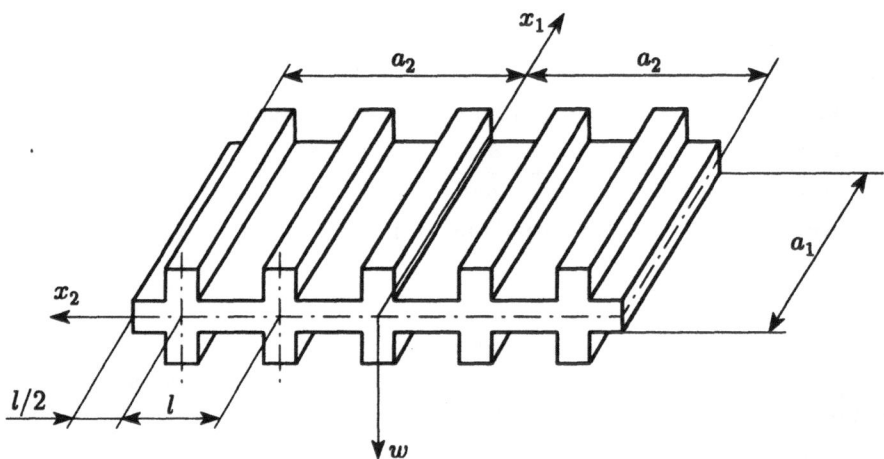

Fig. 3.4. Stringer plate

We use Berger's equation ([51], see also Sect. 3.3.1 of this book) and add some terms to it, taking into account the rib discreteness:

$$D\Delta\Delta W - N\Delta W + E_c I_c \sum_{k=-M_1}^{M_1} \delta(y - kl) W_{xxxx}$$

$$= -\left[\rho h - E_c I_c \sum_{k=-M_1}^{M^1} \delta(y - kl)\right] W_{tt}.$$

Here

$$N h^2 a_1 a_2 = 3D \int_0^a \int_0^b [(W_x)^2 + (W_y)^2]\, dx\, dy,$$

$$M_1 = 0.5(M - 1), \quad l = 2a_2 M^{-1}.$$

The transition conditions from one part of the plate between ribs to the other may be written as

$$W^+ = W^-, \quad W_y^+ = W_y^-, \quad W_{yy}^+ = W_{yy}^-,$$

$$D(W_{yyy}^+ - W_{yyy}^-) = E_c I_c W_{xxxx} + \rho_c F_c W_{tt}, \quad (\ldots)^{\pm} = \lim_{y \to kl \pm 0} (\ldots).$$

We suppose that a typical period of the solution in the y direction (L) is much smaller than the distance between the ribs l $(\varepsilon = iL^{-1} \ll 1)$. Then, we use a multiscale approach. Let us introduce "fast" η_1 ($\eta_1 = l^{-1}y$) and "slow" η ($\eta = L^{-1}y$). Then

$$\frac{\partial}{\partial y} = L^{-1}\frac{\partial}{\partial \eta} + l^{-1}\frac{\partial}{\partial \eta_1}.$$

The normal displacement W will be represented as the expansions

$$W = W_0(\xi, \eta) + \varepsilon^4 W_1(\xi, \eta, \eta_1) + \dots,$$

where $\xi = L^{-1}x$.

After substituting the above expressions into the governing relations and separating them with respect to ε, one obtains

$$-W_{1\eta_1\eta_1\eta_1\eta_1} = \Delta\Delta W_0 - N_0 L^2 D^{-1}\Delta W_0 + \rho h L^4 D^{-1}W_{0tt}, \tag{3.2.9}$$

$$N_0 L^2 h^2 b_1 b_2 = 3D \int_{-b_2}^{b_2} \int_0^{b_1} [(W_{0\xi})^2 + (W_{0\eta})^2]\,\mathrm{d}\xi\,\mathrm{d}\eta, \quad b = L^{-1}a_1, \quad b = L^{-1}a_2,$$

$$W_1\big|_{\eta_1=0} = W_1\big|_{\eta_1=1}, \quad W_{1\eta_1}\big|_{\eta_1=0} = W_{1\eta_1}\big|_{\eta_1=1},$$

$$W_{1\eta_1\eta_1}\big|_{\eta_1=0} = W_{1\eta_1\eta_1}\big|_{\eta_1=1},$$

$$W_{1\eta_1\eta_1\eta_1}\big|_{\eta_1=0} - W_{1\eta_1\eta_1\eta_1}\big|_{\eta_1=1} = (Dl)^{-1}(E_c I_c W_{0\xi\xi\xi} + \rho F_c L^4 W_{0tt}).$$

The conditions for a nontrivial solution of the boundary value problem (3.2.9) are

$$\Delta\Delta W_0 - N_0 L^{-2} D^{-1}\Delta W_0 + \rho h L^4 D^{-1}W_{0tt}$$
$$= (Dl)^{-1}(E_c I_c W_{0tt} = (Dl)^{-1}(E_c I_c W_{0\eta\eta\eta\eta} + \rho_c F_c L^4 W_{0tt}.$$

The exact solution of the cell problem (3.2.9) is

$$W_1 = (24DL)^{-1}(E_c I_c W_{0\xi\xi\xi\xi} + \rho_c F_c W_{0tt})\eta_1^2(\eta_1^2 - 1)^2.$$

The function W_1 in the general case does not satisfy the boundary conditions, and leads to the appearance of boundary layers near the ends $x = 0$, a_1. Let us suppose that the ends $x = 0$, a_1 are clamped. To obtain the boundary layer function W_{kp}, we introduce the "fast" variable $\eta_1 = l_{-1}x$ (then $\partial/\partial x = L^{-1}\partial/\partial\xi_1$) and the expansions

$$W_{kp} = \varepsilon^4 W_{1kp}(\xi, \eta, \xi_1, \eta_1, t) + \dots.$$

After separating, one obtains

$$\Delta_1\Delta_1 W_{1kp} = 0, \quad \text{where } \Delta_1(\dots) = (\dots)_{\xi_1\xi_1} + (\dots)_{\eta_1\eta_1};$$
$$W_1\big|_{\eta_1} = W_{1\eta}\big|_{\eta_1} = 0; \quad W_{1\eta}\big|_{\eta_1=0} = W_1\big|_{\eta_1=1} = 0;$$
$$\text{for } \xi_1 = 0 \quad W_{1kp} = -W_1, \quad W_{1kp\xi_1} = 0,$$
$$\text{for } \xi_1 \to \infty \quad W_{1kp}, W_{1kp\xi_1} \to 0;$$
$$\text{for } \xi_1 = l^{-1}a_1 \quad W_{1kp} = -W_1, \quad W_{1kp\xi_1} = 0,$$
$$\text{for } \xi_1 \to -\infty \quad W_{1kp}, W_{1kp\xi_1} \to 0.$$

This boundary value problem may be solved routinely by Kantorovich's variational procedure [48d].

Now let us compare the asymptotic solution with the exact one, which may be obtained for the static problem. In the nonlinear theory the exact solution may be constructed very rarely, and it is wonderful that we may do it for our very complicated problem. We choose the governing equation in the following form:

$$D\nabla^4 W - N\nabla^2 W + E_c I_c \sum_{k=-M}^{M} \delta(x_2 - kl)\frac{\partial^4 W}{\partial x_1^4} = Q(x_1, x_2)$$

$$\equiv \sum_{s=1}^{\infty}\sum_{p=1}^{\infty} q_{sp} \sin\left(\frac{\pi s x_1}{a_1}\right) \cos\left(\frac{0.5\pi(2p+1)x_2}{a_2}\right). \qquad (3.2.10)$$

The plate is simply supported:

$$W = \frac{\partial^2 W}{\partial x_1^2} = 0 \qquad \text{for } x_1 = 0, a_1;$$

$$W = \frac{\partial^2 W}{\partial x_2^2} = 0 \qquad \text{for } x_2 = \pm a_2. \qquad (3.2.11)$$

One can obtain a solution of the nonlinear boundary value problem (3.2.10), (3.2.11) in the form

$$W(x_1, x_2) = \sum_{m=1}^{\infty}\sum_{n=1}^{\infty} w_{mn}(x_1, x_2),$$

where

$$w_{mn}(x_1, x_2) = \left\{ f_n \cos\left(\frac{0.5\beta_n x_2}{a_2}\right) + \sum_{j=1}^{\infty} f_{nj}^{(+)} \cos\left(\frac{0.5\beta_n x_2}{a_2}\right.\right.$$

$$\left. + \frac{2\pi j}{l}x_2\right) + \sum_{j=1}^{\infty} f_{nj}^{(-)} \cos\left(\frac{0.5\beta_n x_2}{a_2} - \frac{2\pi j}{l}x_2\right)\right\}$$

$$\cdot \sin\left(\frac{\alpha_m x_1}{a_1}\right), \qquad \alpha_m = \pi m, \quad \beta_n = \pi(2n+1).$$

Substituting into (3.2.10) and splitting it into cosines, one obtains an infinite recurrent system of nonlinear algebraic equations:

$$(\alpha_m^2 + \mu^2\beta_n^2)(\alpha_m^2 + \mu^2\beta_n^2 + A)\xi_n + \alpha_m^4\gamma K = P_{mn}; \qquad (3.2.12)$$

$$\left[\alpha_m^2 + \mu^2(\alpha_{ni}^{(+)})^2\right]\left[\alpha_m^2 + \mu^2(\alpha_{ni}^{(+)})^2 + A\right]\xi_{ni}^{(+)} + \alpha_m^4\gamma K = 0;$$

$$\left[\alpha_m^2 + \mu^2(\alpha_{ni}^{(-)})^2\right]\left[\alpha_m^2 + \mu^2(\alpha_{ni}^{(-)})^2 + A\right]\xi_{ni}^{(-)} + \alpha_m^4\gamma K = 0; \qquad (3.2.13)$$

where $i = 1, 2, 3, \ldots$

$$K = \xi_n + \sum_{j=1}^{\infty} [\xi_{nj}^{(+)} + \xi_{nj}^{(-)}]; \qquad (3.2.14)$$

$$A = 3\{(\alpha_m^2 + \mu^2\beta_n^2)\xi_n^2 + \sum_{j=1}^{\infty} \left[\alpha_m^2 + \mu^2(\alpha_{nj}^{(+)})^2\right] \left(\xi_{nj}^{(+)}\right)^2$$

$$+ \sum_{j=1}^{\infty} \left[\alpha_m^2 + \mu^2(\alpha_{nj}^{(-)})^2\right] \left(\xi_{nj}^{(-)}\right)^2\};$$

$$\gamma = \frac{E_r I}{D a_2} \sum_{k=-M_1}^{M_1} \cos^2\left(\frac{\pi(2n+1)k}{M}\right);$$

$$\alpha_{ni}^{(+)} = \beta_n + 2\pi M i; \quad \alpha_{ni}^{(-)} = \beta_n - 2\pi M i.$$

Then, one can rewrite the system in the form

$$\xi_n = \frac{P_{mn} - \gamma\alpha_m^4 K}{(\alpha_m^2 + \mu^2\beta_n^2)(\alpha_m^2 + \mu^2\beta_n^2 + A)};$$

$$\xi_{ni}^{(+)} = -\frac{\gamma\alpha_m^4 K}{\left[\alpha_m^2 + \mu^2(\alpha_{ni}^{(+)})^2\right]\left[\alpha_m^2 + \mu^2(\alpha_{ni}^{(+)})^2 + A\right]};$$

$$\xi_{ni}^{(-)} = -\frac{\gamma\alpha_m^4 K}{\left[\alpha_m^2 + \mu^2(\alpha_{ni}^{(-)})^2\right]\left[\alpha_m^2 + \mu^2(\alpha_{ni}^{(-)})^2 + A\right]}, \quad i = 1, 2, \ldots.$$

Substituting expressions for ξ_n, $\xi_{ni}^{(+)}$, $\xi_{ni}^{(-)}$ $(i = 1, 2, ...)$ into (3.2.14), one can obtain K as a function of A:

$$K = P_{mn}S(1 + \gamma\alpha_m^4 S + S^{(+)} + S^{(-)})^{-1}, \qquad (3.2.15)$$

where

$$S = [(\alpha_m^2 + \mu^2\beta_n^2)(\alpha_m^2 + \mu^2\beta_n^2 + A)]^{-1};$$

$$S^{(+)} = \sum_{j+1}^{\infty} \left[\alpha_m^2 + \mu^2(\alpha_{ni}^{(+)})^2\right]^{-1}\left[\alpha_m^2 + \mu^2(\alpha_{ni}^{(+)})^2 + A\right]^{-1};$$

$$S^{(-)} = \sum_{j+1}^{\infty} \left[\alpha_m^2 + \mu^2(\alpha_{ni}^{(-)})^2\right]^{-1}\left[\alpha_m^2 + \mu^2(\alpha_{ni}^{(-)})^2 + A\right]^{-1}.$$

Taking into account formulae (3.2.13)–(3.2.15) one obtains (3.2.12) as a transcendental equation (with respect to the unknown A) that may be solved routinely by numerical methods.

Then we will obtain K (using formula (3.2.15)) and the amplitudes ξ_n, $\xi_{ni}^{(+)}$, $\xi_{ni}^{(-)}$ $(i = 1, 2, ...)$.

For the numerical investigation we choose a square plate loaded by the lateral load

$$Q = q_{10} \sin\left(\frac{\pi x_1}{a_1}\right) \cos\left(\frac{0.5\pi x_2}{a_2}\right).$$

We also suppose $\nu = 0.2$, $E_c IM/(Da_2) = 200$; $q_{10} = 8 \cdot 10^3 Dha_1^{-4}$, $x_1 = 0.5a_1$, $\bar{x}_2 = a_2^{-1}x_2$, $\overline{M}_2 = (Dh/a_2)M_2$.

The numerical results are plotted in Fig. 3.5.

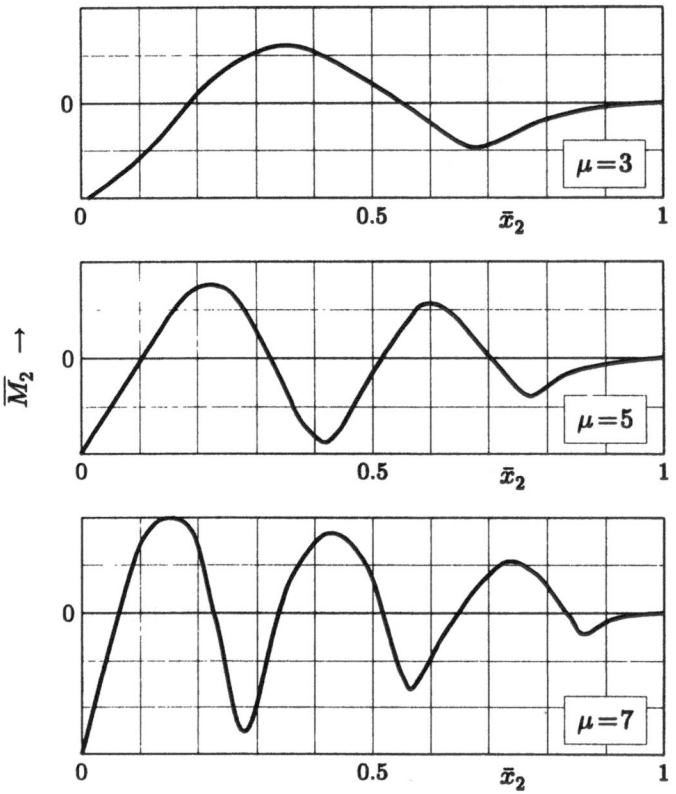

Fig. 3.5. Bending moments in the Stringer plate in the perpendicular direction to the ribs

3.2.3 Perforated Membrane

Consider the problem of transverse oscillations of a rectangular membrane weakened by a double periodic system of regularly spaced identical circular holes of radius a. The ratio ε of the period of perforations to the characteristic size of the region Ω is a small quantity. The outer contour $\partial\Omega$ of the membrane is rigidly clamped, while the edges of the apertures $\partial\Omega_i$ are free. Let us begin from the linear case. In mathematical language we have the boundary value problem

$$c^2 \left(\frac{\partial^2 u}{\partial x^2} + \frac{\partial^2 u}{\partial y^2} \right) = \frac{\partial^2 u}{\partial t^2} \quad \text{in} \quad \Omega, \tag{3.2.16}$$

$$u = 0 \quad \text{on} \quad \partial\Omega, \tag{3.2.17}$$

$$\frac{\partial u}{\partial n} = \quad \text{on} \quad \partial\Omega_i, \tag{3.2.18}$$

where u is the transverse displacement of the points of the membrane, $c^2 = p/\rho$, p is the tension in the membrane, and ρ is the density.

We take a solution for the characteristic oscillations of the membrane in the form $u(x, y, t) = u(x, y)e^{i\omega t}$, where $\lambda = \omega^2/c^2$ and ω is the circular frequency. Then, instead of (3.2.16) we obtain

$$\frac{\partial^2 u}{\partial x^2} + \frac{\partial^2 u}{\partial y^2} + \lambda u = 0. \tag{3.2.19}$$

We have presented the solution of the problem (3.2.19), (3.2.17), (3.2.18) just posed as an asymptotic series in the powers of a small parameter

$$u = u_0(x, y) + \varepsilon \left(u_{10}(x, y) + u_1(x, y, \xi, \eta) \right)$$
$$+ \varepsilon^2 \left(u_{20}(x, y) + u_2(x, y, \xi, \eta) \right) + \cdots,$$

where $\xi = x/\varepsilon$ and $\eta = y/\varepsilon$ are the "fast" variables.

The functions $u_0, u_{10}, u_{20}, \ldots$ depend only on the "slow" variables, and the other variables $u_i (i = 1, 2\ldots)$ are periodic together with their derivatives with respect to the "fast" variables and have a period equal to that of the structure. Similarly, we expand the frequency

$$\lambda = \lambda_0 + \varepsilon\lambda_1 + \varepsilon^2\lambda_2 + \cdots.$$

After separating ε we obtain from (3.2.19) an infinite system

$$\varepsilon^{-1} : \quad \frac{\partial^2 u_1}{\partial \xi^2} + \frac{\partial^2 u_1}{\partial \eta^2} = 0, \tag{3.2.20}$$

$$\varepsilon^0 : \quad \frac{\partial^2 u_0}{\partial x^2} + \frac{\partial^2 u_0}{\partial y^2} + 2\frac{\partial^2 u_1}{\partial x \partial \xi} + 2\frac{\partial^2 u_1}{\partial y \partial \eta}$$
$$+ \frac{\partial^2 u_2}{\partial \xi^2} + \frac{\partial^2 u_2}{\partial \eta^2} + \lambda_0 u_0 = 0, \tag{3.2.21}$$

$$\varepsilon^1 : \quad \frac{\partial^2 u_1}{\partial x^2} + \frac{\partial^2 u_1}{\partial y^2} + 2\frac{\partial^2 u_2}{\partial x \partial \xi} + 2\frac{\partial^2 u_2}{\partial y \partial \eta} + \frac{\partial^2 u_3}{\partial \xi^2} + \frac{\partial^2 u_3}{\partial \eta^2}$$
$$+ \frac{\partial^2 u_{10}}{\partial x^2} + \frac{\partial^2 u_{10}}{\partial y^2} + \lambda_0(u_1 + u_{10}) + \lambda_1 u_0 = 0. \tag{3.2.22}$$

The corresponding boundary conditions (3.2.18) assume the form

$$\varepsilon^0 : \quad \frac{\partial u_1}{\partial \overline{n}} + \frac{\partial u_0}{\partial n} = 0; \quad \text{on} \quad \partial\Omega_i \tag{3.2.23}$$

$$\varepsilon^1 : \quad \frac{\partial u_2}{\partial \overline{n}} + \frac{\partial u_1}{\partial n} + \frac{\partial u_{10}}{\partial n} = 0, \quad \text{on} \quad \partial\Omega_i$$

$$\cdots \quad \cdots\cdots\cdots$$

where $\partial/\partial\overline{n}$ is the derivative with respect to the "fast" variables.

The solution of the boundary problem for a complex multiply connected region now breaks up into three stages. The first stage is the solution of the "cell problem" (3.2.20), (3.2.23).

On the opposite sides of the "cell" the function u_1 must satisfy the periodicity conditions

$$u_1|_{\xi=b} = u_1|_{\xi=-b}, \quad u_1|_{\eta=b} = u_1|_{\eta=-b},$$
$$\frac{\partial u_1}{\partial \xi}\bigg|_{\xi=b} = \frac{\partial u_1}{\partial \xi}\bigg|_{\xi=-b}, \quad \frac{\partial u_1}{\partial \eta}\bigg|_{\eta=b} = \frac{\partial u_1}{\partial \eta}\bigg|_{\eta=-b}. \tag{3.2.24}$$

Using Galerkin's variational method to solve problems (3.2.20), (3.2.23) and (3.2.24), we represent u_1 as

$$u_1 = \sum_{m=0}^{\infty}\sum_{n=0}^{\infty}\left(A_{1mn}\sin\frac{m\pi\xi}{b}\cos\frac{n\pi\eta}{b} + A_{2mn}\cos\frac{m\pi\xi}{b}\sin\frac{n\pi\eta}{b}\right).$$

After performing the necessary operations, we obtain

$$A_{1mn} = a\frac{\partial u_0}{\partial x}A_{mn}^*, \quad A_{2mn} = a\frac{\partial u_0}{\partial y}A_{mn}^*,$$

where A_{mn}^* is a constant determined from the vanishing of the variation of Galerkin functional.

The second stage of the solution of the problem is the construction of the averaged relations. Applying the averaging operator to (3.2.21)

$$\tilde{\Phi}(x,y) = \frac{1}{|\Omega_i^*|}\int\int_{\Omega_i}\Phi(x,y,\xi,\eta)\,d\xi\,d\eta,$$

where Ω_i^* is a "cell" without holes, we obtain the averaged equation with a boundary condition on the outer contour of the membrane:

$$\frac{\partial^2 u_0}{\partial x^2} + \frac{\partial^2 u_0}{\partial y^2} + B\lambda_0 u_0 = 0 \quad \text{in} \quad \Omega^*, \qquad u_0 = 0 \quad \text{on} \quad \partial\Omega,$$

where Ω^* is a membrane without perforations,

$$B = \left(1 - \frac{\pi}{4}\frac{a^2}{b^2}\right)\left(1 - \frac{\pi}{4}\frac{a^2}{b^2} - \frac{\pi}{2}\frac{a^2}{b^2}\sum_{m=0}^{\infty}\sum_{n=0}^{\infty}A_{mn}^*\frac{J_1\left(\frac{a}{b}\pi\sqrt{m^2+n^2}\right)}{\sqrt{m^2+n^2}}\right)^{-1},$$

and J_1 is the Bessel function of order 1.

At the third stage we find the first correction to the frequency λ_1. To do this we must determine the function u_2 as a solution of the boundary value problem

$$\frac{\partial^2 u_2}{\partial \xi^2} + \frac{\partial^2 u_2}{\partial \eta^2} = -\frac{\partial^2 u_0}{\partial x^2} - \frac{\partial^2 u_0}{\partial y^2} - 2\frac{\partial^2 u_1}{\partial x\partial\xi} - 2\frac{\partial^2 u_1}{\partial y\partial\eta} - \lambda_0 u_0 \quad \text{in} \quad \Omega_i,$$
$$\frac{\partial u_2}{\partial \overline{n}} + \frac{\partial u_1}{\partial n} + \frac{\partial u_{10}}{\partial n} = 0 \quad \text{on} \quad \partial\Omega_i,$$

with the periodicity conditions similar to (3.2.24). Proceeding as in the determination of the function u_1, we can obtain

$$u_2 = \sum_{m=0}^{\infty} \sum_{n=0}^{\infty} \left(C_{1mn} \sin \frac{m\pi\xi}{b} \cos \frac{n\pi\eta}{b} \right.$$
$$\left. + C_{2mn} \cos \frac{m\pi\xi}{b} \sin \frac{n\pi\eta}{b} \right) + \varphi(\xi, \eta),$$

where

$$C_{1mn} = A_{1mn} \left(\frac{\partial u_0}{\partial x} \rightleftharpoons \frac{\partial u_{10}}{\partial x} \right), \quad C_{2mn} = A_{2mn} \left(\frac{\partial u_0}{\partial y} \rightleftharpoons \frac{\partial u_{10}}{\partial y} \right),$$

and $\varphi(\xi, \eta)$ is a function satisfying the condition $\varphi(-\xi, -\eta) = \varphi(\xi, \eta)$. The form of the function $\varphi(\xi, \eta)$ is unimportant, since it makes no contribution to the averaged equation and, consequently, is not used in the determination of λ_1. After averaging (3.2.22) we obtain

$$\frac{\partial^2 u_{10}}{\partial x^2} + \frac{\partial^2 u_{10}}{\partial y^2} + B(\lambda_0 u_{10} + \lambda_1 u_0) = 0.$$

To determine λ_1 we multiply the equation just obtained termwise by u_0 and integrate over the region Ω^* [120, 122]. If $u_{10} = \tilde{u}_1 = 0$ on $\partial\Omega$, then the differential operator

$$L(u_{10}) = \frac{\partial^2 u_{10}}{\partial x^2} + \frac{\partial^2 u_{10}}{\partial y^2}$$

is self-adjoint. Then $\lambda_1 = 0$ and the expansion of the characteristic frequencies begins with λ_2 – a term of order ε^2. In that case, if $\tilde{u}_1$ does not satisfy the boundary condition on the contour of the membrane and consequently $u_{10} \neq 0$ on $\partial\Omega$, we obtain a nonzero first correction to the characteristic frequency.

Now let us investigate the nonlinear but quasilinear (the terms with derivatives in the governing equation are linear) case – a membrane on nonlinear support. The governing equation is

$$c^2 \left(\frac{\partial^2 u}{\partial x^2} + \frac{\partial^2 u}{\partial y^2} \right) + c_1^2 u^3 = \frac{\partial^2 u}{\partial t^2},$$

and the boundary conditions are given by (3.2.17), (3.2.18). Here c_1^2 is the rigidity of the nonlinear support. After splitting into ε, one obtains the cell problem in the form (3.2.20), (3.2.23). Then its solution coincides with the solution of the linear problem, and the homogenized nonlinear equation may be written in the form

$$c^2 \left(\frac{\partial^2 u_0}{\partial x^2} + \frac{\partial^2 u_0}{\partial y^2} \right) + Bc_1^2 u_0^3 = B\frac{\partial^2 u_0}{\partial t^2} \quad \text{in } \partial\Omega^*,$$
$$u_0 = 0 \quad \text{on } \partial\Omega.$$

3.2.4 Perforated Plate

We use the averaging method for the computation of densely perforated plates. As has been mentioned above, having the solution of the static local problem, this approach gives the possibility to obtain, without basic difficulties, the solution of dynamical and quasilinear problems.

We consider the problem of the bending of a rectangular plate, weakened by a doubly periodic system of holes. Let Ω be the domain occupied by the plate, let the exterior contour be $\partial\Omega$ and let $\partial\Omega_i$ be the boundary of the hole. The periodic ε of the structure is the same in both directions and small in comparison with the characteristic dimension of the domain Ω ($\varepsilon \ll 1$). The boundaries $\partial\Omega_i$ of the holes are free, and the exterior contour $\partial\Omega$ of the domain is fastened in a definite manner. We have the boundary value problem

$$\frac{\partial^4 u}{\partial x^4} + 2\frac{\partial^4 u}{\partial x^2 \partial y^2} + \frac{\partial^4 u}{\partial y^4} = f \quad \text{in } \Omega; \tag{3.2.25}$$

$$M_{\mathrm{r}} = 0, \qquad V_{\mathrm{r}} = 0 \quad \text{on } \partial\Omega_i, \tag{3.2.26}$$

where

$$M_{\mathrm{r}} = \nu\Delta u + (1-\nu)\left(\cos^2\alpha\frac{\partial^2 u}{\partial x^2} + \sin^2\alpha\frac{\partial^2 u}{\partial y^2} + \sin 2\alpha\frac{\partial^2 u}{\partial x \partial y}\right);$$

$$V_{\mathrm{r}} = \cos\alpha\frac{\partial}{\partial x}\Delta u + \sin\alpha\frac{\partial}{\partial y}\Delta u$$

$$+ (1-\nu)\frac{\partial}{\partial s}\left[\cos 2\alpha\frac{\partial^2 u}{\partial x \partial y} + \frac{1}{2}\sin 2\alpha\left(\frac{\partial^2 u}{\partial y^2} - \frac{\partial^2 u}{\partial x^2}\right)\right];$$

u is the normal deflection, and α is the angle between the exterior normal n to the contour and the x axis.

We represent the solution of the problem in the form of a series of powers of a small parameter ε:

$$u = u_0 + \varepsilon u_1 + \varepsilon^2 u_2 + \dots, \tag{3.2.27}$$

where $u_i = u_i(x, y, \xi, \eta)$ $(i = 0, 1, 2, \dots)$, $\xi = x/\varepsilon, \eta = y/\varepsilon$ are the "fast" variables.

Taking into account the relations

$$\frac{\partial}{\partial x} = \frac{\partial}{\partial x} + \frac{1}{\varepsilon}\frac{\partial}{\partial \xi}; \quad \frac{\partial}{\partial y} = \frac{\partial}{\partial y} + \frac{1}{\varepsilon}\frac{\partial}{\partial \eta}$$

the initial equation and each of the boundary conditions splits with respect to ε into an infinite system of equations

$$\frac{\partial^4 u_0}{\partial \xi^4} + 2\frac{\partial^4 u_0}{\partial \xi^2 \partial \eta^2} + \frac{\partial^4 u_0}{\partial \eta^4} = 0, \quad \text{in } \Omega_i;$$

$$\nu\left(\frac{\partial^2 u_0}{\partial \xi^2} + \frac{\partial^2 u_0}{\partial \eta^2}\right) + (1-\nu)\left(\cos^2\alpha\frac{\partial^2 u_0}{\partial \xi^2} + \sin^2\alpha\frac{\partial^2 u_0}{\partial \eta^2}\right.$$
$$\left. + \sin 2\alpha\frac{\partial^2 u_0}{\partial \xi\partial\eta}\right) = 0, \quad \text{on } \partial\Omega_i; \tag{3.2.28}$$

$$\left[(1-\nu)\frac{\sin 2\alpha\cos\alpha}{2} + \cos\alpha\right]\frac{\partial^3 u_0}{\partial\xi^3} + \left[(1-\nu)\frac{\sin 2\alpha\cos\alpha}{2}\right.$$
$$\left. + \sin\alpha\right]\frac{\partial^3 u_0}{\partial\eta^3} + \left[(1-\nu)\left(-\cos 2\alpha\sin\alpha - \frac{1}{2}\sin 2\alpha\cos\alpha\right)\right.$$
$$\left. + \sin\alpha\right]\frac{\partial^3 u_0}{\partial\xi^2\partial\eta}\left[(1-\nu)\left(\cos 2\alpha\cos\alpha - \frac{1}{2}\sin 2\alpha\sin\alpha\right)\right.$$
$$\left. + \sin\alpha\right]\frac{\partial^3 u_0}{\partial\xi\partial\eta^2} = 0, \quad \text{on } \partial\Omega_i;$$

$$\frac{\partial^4 u_1}{\partial\xi^4} + 2\frac{\partial^4 u_1}{\partial\xi^2\partial\eta^2} + \frac{\partial^4 u_1}{\partial\eta^4} = 0, \quad \text{in } \Omega_i; \tag{3.2.29}$$

$$\nu\left(\frac{\partial^2 u_1}{\partial \xi^2} + \frac{\partial^2 u_1}{\partial \eta^2}\right) + (1-\nu)\left(\cos^2\alpha\frac{\partial^2 u_1}{\partial \xi^2} + \sin^2\alpha\frac{\partial^2 u_1}{\partial \eta^2}\right.$$
$$\left. + \sin 2\alpha\frac{\partial^2 u_1}{\partial \xi\partial\eta}\right) = 0, \quad \text{on } \partial\Omega_i;$$

$$\left[(1-\nu)\frac{\sin 2\alpha\cos\alpha}{2} + \cos\alpha\right]\frac{\partial^3 u_1}{\partial\xi^3} + \left[(1-\nu)\frac{\sin 2\alpha\cos\alpha}{2}\right.$$
$$\left. + \sin\alpha\right]\frac{\partial^3 u_1}{\partial\eta^3} + \left[(1-\nu)\left(-\cos 2\alpha\sin\alpha - \frac{1}{2}\sin 2\alpha\cos\alpha\right)\right.$$
$$\left. + \sin\alpha\right]\frac{\partial^3 u_1}{\partial\xi^2\partial\eta} + \left[(1-\nu)\left(\cos 2\alpha\cos\alpha - \frac{1}{2}\sin 2\alpha\sin\alpha\right)\right.$$
$$\left. + \cos\alpha\right]\frac{\partial^3 u_1}{\partial\xi\partial\eta^2} = 0, \quad \text{on } \partial\Omega_i;$$

$$\frac{\partial^4 u_2}{\partial\xi^4} + 2\frac{\partial^4 u_2}{\partial\xi^2\partial\eta^2} + \frac{\partial^4 u_2}{\partial\eta^4} = 0, \quad \text{in } \Omega_i; \tag{3.2.30}$$

$$\nu\left(\frac{\partial^2 u_2}{\partial \xi^2} + \frac{\partial^2 u_0}{\partial x^2} + \frac{\partial^2 u_2}{\partial \eta^2} + \frac{\partial^2 u_0}{\partial y^2}\right) + (1-\nu)\left[\cos^2\alpha\left(\frac{\partial^2 u_2}{\partial\xi^2}\right.\right.$$
$$\left. + \frac{\partial^2 u_0}{\partial x^2}\right) + \sin^2\alpha\left(\frac{\partial^2 u_2}{\partial\eta^2} + \frac{\partial^2 u_0}{\partial y^2}\right) + \sin 2\alpha\left(\frac{\partial^2 u_2}{\partial\xi\partial\eta}\right.$$
$$\left.\left. + \frac{\partial^2 u_0}{\partial x\partial y}\right)\right] = 0, \quad \text{on } \partial\Omega_i; \tag{3.2.31}$$

$$\left[(1-\nu)\frac{\sin 2\alpha \sin \alpha}{2} + \cos \alpha\right]\frac{\partial^3 u_2}{\partial \xi^3} + \left[(1-\nu)\frac{\sin 2\alpha \cos \alpha}{2}\right.$$

$$\left.+\sin \alpha\right]\frac{\partial^3 u_2}{\partial \eta^3} + \left[(1-\nu)\left(-\cos 2\alpha \sin \alpha - \frac{1}{2}\sin 2\alpha \cos \alpha\right)\right.$$

$$\left.+\sin \alpha\right]\frac{\partial^3 u_2}{\partial \xi^2 \partial \eta} + \left[(1-\nu)\left(\cos 2\alpha \cos \alpha - \frac{1}{2}\sin 2\alpha \sin \alpha\right)\right.$$

$$\left.+\cos \alpha\right]\frac{\partial^3 u_2}{\partial \xi \partial \eta^2} = 0, \quad \text{on } \partial \Omega_i; \tag{3.2.32}$$

$$\frac{\partial^4 u_3}{\partial \xi^4} + 4\frac{\partial^4 u_2}{\partial x \partial \xi^3} + 2\frac{\partial^4 u_3}{\partial \xi^2 \partial \eta^2} + 4\frac{\partial^4 u_2}{\partial y \partial \xi^2 \partial \eta} + 4\frac{\partial^4 u_2}{\partial x \partial \xi \partial \eta^2}$$

$$+\frac{\partial^4 u_3}{\partial \eta^4} + 4\frac{\partial^4 u_2}{\partial y \partial \eta^3} = 0, \quad \text{in } \Omega_i; \tag{3.2.33}$$

$$\nu\left(\frac{\partial^2 u_3}{\partial \xi^2} + +2\frac{\partial^2 u_2}{\partial x \partial \xi} + \frac{\partial^2 u_1}{\partial x^2} + \frac{\partial^2 u_3}{\partial \eta^2} + 2\frac{\partial^2 u_2}{\partial y \partial \eta} + \frac{\partial^2 u_1}{\partial y^2}\right)$$

$$+(1-\nu)\left[\cos^2 \alpha \left(\frac{\partial^2 u_3}{\partial \xi^2} + 2\frac{\partial^2 u_2}{\partial x \partial \xi} + \frac{\partial^2 u_1}{\partial x^2}\right)\right.$$

$$+\sin^2 \alpha \left(\frac{\partial^2 u_3}{\partial \eta^2} + 2\frac{\partial^2 u_2}{\partial y \partial \eta} + \frac{\partial^2 u_1}{\partial y^2}\right) \tag{3.2.34}$$

$$\left.+\sin 2\alpha \left(\frac{\partial^2 u_3}{\partial \xi \partial \eta} + \frac{\partial^2 u_2}{\partial x \partial \eta} + \frac{\partial^2 u_2}{\partial y \partial \xi} + \frac{\partial^2 u_1}{\partial x \partial y}\right)\right] = 0, \quad \text{on } \partial \Omega_i;$$

$$\left[(1-\nu)\frac{\sin 2\alpha \sin \alpha}{2} + \cos \alpha\right]\left(\frac{\partial^3 u_3}{\partial \xi^3} + 3\frac{\partial^3 u_2}{\partial x \partial \xi^2} + \frac{\partial^3 u_0}{\partial x^3}\right)$$

$$+\left[(1-\nu)\frac{\sin 2\alpha \cos \alpha}{2} + \sin \alpha\right]\left(\frac{\partial^3 u_3}{\partial \eta^3} + 3\frac{\partial^3 u_2}{\partial y \partial \eta^2} + \frac{\partial^3 u_0}{\partial y^3}\right)$$

$$+\left[(1-\nu)\left(-\cos 2\alpha \sin \alpha - \frac{1}{2}\sin 2\alpha \cos \alpha\right) + \sin \alpha\right]$$

$$\cdot \left(\frac{\partial^3 u_3}{\partial \xi^2 \partial \eta} + \frac{\partial^3 u_2}{\partial y \partial \xi^2} + 2\frac{\partial^3 u_2}{\partial y \partial \xi \partial \eta} + \frac{\partial^3 u_0}{\partial x^2 \partial y}\right)$$

$$+\left[(1-\nu)\left(\cos 2\alpha \cos \alpha - \frac{1}{2}\sin 2\alpha \sin \alpha\right) + \cos \alpha\right]$$

$$\cdot \left(\frac{\partial^3 u_3}{\partial \eta^2 \partial \xi} + \frac{\partial^3 u_2}{\partial x \partial \eta^2} + 2\frac{\partial^3 u_2}{\partial y \partial \xi \partial \eta} + \frac{\partial^3 u_0}{\partial y^2 \partial x}\right) + \left[-\frac{1-\nu}{2}\frac{\partial}{\partial s}(\sin 2\alpha)\right]$$

$$\cdot \left(\frac{\partial^2 u_2}{\partial \xi^2} + \frac{\partial^2 u_0}{\partial x^2}\right) + \frac{1-\nu}{2}\frac{\partial}{\partial s}(\sin 2\alpha)\left(\frac{\partial^2 u_2}{\partial \eta^2} + \frac{\partial^2 u_0}{\partial y^2}\right)$$

$$+(1-\nu)\frac{\partial}{\partial s}(\cos 2\alpha)\left(\frac{\partial^2 u_2}{\partial \xi \partial \eta} + \frac{\partial^2 u_0}{\partial x \partial y}\right) = 0, \quad \text{on } \partial \Omega_i; \tag{3.2.35}$$

$$\frac{\partial^4 u_4}{\partial \xi^4} + 4\frac{\partial^4 u_3}{\partial x \partial \xi^3} + 6\frac{\partial^4 u_2}{\partial x^2 \partial \xi^2} + \frac{\partial^4 u_0}{\partial x^4} + 2\frac{\partial^4 u_4}{\partial \xi^2 \partial \eta^2} + 4\frac{\partial^4 u_3}{\partial y \partial \xi^2 \partial \eta}$$

$$+4\frac{\partial^4 u_3}{\partial x \partial \xi \partial \eta^2} + 2\frac{\partial^4 u_2}{\partial y^2 \partial \xi^2} + 2\frac{\partial^4 u_2}{\partial x^2 \partial \eta^2} + 8\frac{\partial^4 u_2}{\partial x \partial y \partial \xi \partial \eta} + 2\frac{\partial^4 u_0}{\partial x^2 \partial y^2}$$

$$+\frac{\partial^4 u_4}{\partial \eta^4} + 4\frac{\partial^4 u_3}{\partial y \partial \eta^3} + 6\frac{\partial^4 u_2}{\partial y^2 \partial \eta^2} + \frac{\partial^4 u_0}{\partial y^4} = f, \quad \text{in } \partial \Omega_i, \tag{3.2.36}$$

. . . .

where Ω_i is a characteristic cell of the structure.

Thus, the solution of the formulated problem (3.2.25), (3.2.26) for a composite multiply connected domain splits into a series of steps in domains with a simpler geometry, from which we can distinguish two fundamental problems: a local problem ("a problem on the cell") which consists in the solving of the biharmonic equation in the domain Ω_i with the given boundary conditions on the contour of the hole and the periodic continuation conditions on the opposite sides of the "cells"

$$u_1|_{\xi=b} = u_1|_{\xi=-b} \quad : \quad u_1|_{\eta=b} = u_1|_{\eta=-b}$$

$$\frac{\partial u_1}{\partial \xi}|_{\xi=b} = \frac{\partial u_1}{\partial \xi}|_{\xi=-b} \quad : \quad \frac{\partial u_1}{\partial \eta}|_{\eta=b} = \frac{\partial u_1}{\partial \eta}|_{\eta=-b}$$

$$\frac{\partial^2 u_1}{\partial \xi^2}|_{\xi=b} = \frac{\partial^2 u_1}{\partial \xi^2}|_{\xi=-b} \quad : \quad \frac{\partial^2 u_1}{\partial \eta^2}|_{\eta=b} = \frac{\partial^2 u_1}{\partial \eta^2}|_{\eta=-b} \tag{3.2.37}$$

$$\frac{\partial^3 u_1}{\partial \xi^3}|_{\xi=b} = \frac{\partial^3 u_1}{\partial \xi^3}|_{\xi=-b} \quad : \quad \frac{\partial^3 u_1}{\partial \eta^3}|_{\eta=b} = \frac{\partial^3 u_1}{\partial \eta^3}|_{\eta=-b};$$

and a global problem which consists in the solving of an averaged equation of the form (3.2.36) in the domain Ω^* without perforations and with the initial boundary conditions on the contour of the plate.

As follows from relations (3.2.28), (3.2.37) and (3.2.29), (3.2.37), the functions u_0, u_1 do not depend on the fast variables, i.e.,

$$u_0 = u_0(x, y); \quad u_1 = u_1(x, y). \tag{3.2.38}$$

Consequently, the solution of the problem is represented in the form of a sum of a certain smooth function and a small fast oscillating correction; moreover, the expansion starts with the $u_2 \sim O(\varepsilon^2)$ term. After the successive solving of the "cell problems" (3.2.30)–(3.2.32), (3.2.37) and (3.2.33)–(3.2.35), (3.2.37) and the determination of the functions u_2, u_3, we determine the principal part of the solution, i.e., the function u_0. Applying to (3.2.36) the averaging operator (...)

$$(\tilde{\dots}) = \frac{1}{|\Omega_i^*|} \iint\limits_{\Omega_i} (\dots) \, d\xi \, d\eta,$$

we obtain the averaged equation in the form

$$\left(\frac{\partial^4 u_0}{\partial x^4} + 2\frac{\partial^4 u_0}{\partial x^2 \partial y^2} + \frac{\partial^4 u_0}{\partial y^4} - f\right)\frac{|\Omega_i|}{|\Omega_i^*|} \tag{3.2.39}$$

$$+ \frac{1}{|\Omega_i^*|}\iint\limits_{\Omega_i}\left(\frac{\partial^4 u_3}{\partial x \partial\xi^3} + \frac{\partial^4 u_3}{\partial y \partial\xi^2\partial\eta} + \frac{\partial^4 u_3}{\partial x\partial\xi\partial\eta^2} + \frac{\partial^4 u_3}{\partial y\partial\eta^3} + 3\frac{\partial^4 u_2}{\partial x^2\partial\xi^2}\right.$$

$$\left.+ \frac{\partial^4 u_2}{\partial y^2\partial\xi^2} + 3\frac{\partial^4 u_2}{\partial y^2\partial\eta^2} + \frac{\partial^4 u_2}{\partial x^2\partial\eta^2} + 4\frac{\partial^4 u_2}{\partial x\partial y\partial\xi\partial\eta}\right)\,d\xi\,d\eta = 0.$$

We obtain the solution of the "cell problems" (3.2.30)–(3.2.32), (3.2.37) and (3.2.33)–(3.2.35), (3.2.37) for a plate with a square net of perforation holes of radius a by making use of the Bubnov–Galerkin method, modified for the case of natural boundary conditions.

We consider them successively. We represent

$$u_2 = \sum_{m=0}^{\infty}\sum_{n=0}^{\infty}\left(A_{1mn}\sin\frac{m\pi\xi}{b}\sin\frac{n\pi\eta}{b} + A_{2mn}\cos\frac{m\pi\xi}{b}\cos\frac{n\pi\eta}{b}\right.$$

$$\left.+ A_{3mn}\sin\frac{m\pi\xi}{b}\cos\frac{n\pi\eta}{b} + A_{4mn}\cos\frac{m\pi\xi}{b}\sin\frac{n\pi\eta}{b}\right), \tag{3.2.40}$$

where $A_{1mn}, A_{2mn}, A_{3mn}, A_{4mn}$ are constants, defined from the conditions of the vanishing of the variation of the Galerkin functional.

The selection of the function u_2 in the form (3.2.40) allows us to satisfy the periodic continuation conditions (3.2.37); then the variation of the Galerkin functional becomes

$$\iint\limits_{\Omega_i}\Delta^2 u_2\delta u_2\,ds - \int\limits_{\partial\Omega_i}[M_r(u_2) - f]\frac{\partial\delta u_2}{\partial n}\,dl + \int\limits_{\partial\Omega_i}\tilde{V}_r(u_2)\delta u_2\,dl = 0, \tag{3.2.41}$$

where by $M_r - f$, $\tilde{V}_r$ we have denoted expressions (3.2.31), (3.2.32), respectively.

As one can see from (3.2.41), by virtue of the symmetry of the considered domain the constants A_{3mn}, A_{4mn} are equal to zero. The unknowns A_{1mn}, A_{2mn} are determined after carrying out the standard procedure of the Galerkin method:

$$A_{1mn} = b^2\frac{\partial^2 u_0}{\partial x\partial y}A_{1mn}^*; \quad A_{2mn} = b^2\frac{\partial^2 u_0}{\partial x^2}\overline{A}_{2mn}^* + b^2\frac{\partial^2 u_0}{\partial y^2}\overline{\overline{A}}_{2mn}^*, \tag{3.2.42}$$

where A_{1mn}^*, $\overline{A}_{2mn}^*$, $\overline{\overline{A}}_{2mn}^*$ are numerical coefficients.

By means of the same scheme, after transforming the right-hand sides of the equations and the boundary conditions, we obtain the solution of problem (3.2.33)–(3.2.35), (3.2.37). From similar considerations, we represent

$$u_3 = \sum_{m=0}^{\infty}\sum_{n=0}^{\infty}\left(B_{1mn}\sin\frac{m\pi\xi}{b}\cos\frac{n\pi\eta}{b} + B_{2mn}\cos\frac{m\pi\xi}{b}\sin\frac{n\pi\eta}{b}\right)$$

$$+ u_2(u_0 \to u_1); \tag{3.2.43}$$

and with the aid of the Galerkin method we find

$$B_{1mn} = b^3 \left(\overline{B^*}_{1mn} \frac{\partial^3 u_0}{\partial x^3} + \overline{\overline{B^*}}_{1mn} \frac{\partial^3 u_0}{\partial x \partial y^2} \right);$$

$$B_{2mn} = b^3 \left(\overline{B^*}_{2mn} \frac{\partial^3 u_0}{\partial y^3} + \overline{\overline{B^*}}_{2mn} \frac{\partial^3 u_0}{\partial x^2 \partial y} \right);$$

$\overline{B^*}_{1mn}, \overline{\overline{B^*}}_{1mn}, \overline{B^*}_{2mn}, \overline{\overline{B^*}}_{2mn}$ being numerical coefficients.

This approach to solving local problems with the aid of the modified Bubnov–Galerkin method turns out to be especially efficient for the determination of the global characteristics, displacements and averaged coefficients, since for the determination of the latter one can use integral representations.

As an example we consider a plate for which $a/b = 1/3$. If in expansions (3.2.40), (3.2.43) we restrict ourselves to one-term approximations, then for the coefficients we obtain

$$A^*_{111} = 0.0102; \quad \overline{A^*}_{210} = \overline{\overline{A^*}}_{201} = -0.0156;$$

$$\overline{A^*}_{210} = \overline{A^*}_{201} = -0.0090;$$

$$\overline{B^*}_{110} = \overline{B^*}_{201} = 0.0074; \quad \overline{\overline{B^*}}_{110} = \overline{\overline{B^*}}_{201} = 0.0014;$$

$$\overline{B^*}_{111} = \overline{B^*}_{211} = 0.0042; \quad \overline{\overline{B^*}}_{111} = \overline{\overline{B^*}}_{211} = 0.0055.$$

Then, after some transformations of (3.2.39), we obtain the averaged equation in the form

$$\tilde{A} \frac{\partial^4 u_0}{\partial x^4} + 2\tilde{B} \frac{\partial^4 u_0}{\partial x^2 \partial y^2} + \tilde{A} \frac{\partial^4 u_0}{\partial y^4} = \tilde{f},$$

where $\tilde{A}, 2\tilde{B}$ are the averaged coefficients

$$\tilde{A} = 0.860; \quad 2\tilde{B} = 1.690.$$

In the case of one-term approximations, comparison with the known values [40d] shows the satisfactory accuracy of the results.

3.3 Averaging Procedure in the Nonlinear Dynamics of Thin-Walled Structures

3.3.1 Berger and Berger-Like Equations for Plates and Shells

In 1955 Berger proposed approximate nonlinear equations for the deformation of rectangular and circular plates, neglecting the second invariant of the strain tensor in the potential energy expression (the "Berger hypothesis" [51]). Berger's equations have become widely used due to their simplicity and visualization. Later Berger's results were generalized for shallow shell and sandwich plate problems.

Similar equations were applied to dynamic problems. The adequacy and applicability of the "Berger hypothesis" were frequently and widely discussed in scientific papers. It has been shown that the "Berger hypothesis" leads to insufficient results when applied to orthotropic plates; there is no obvious pattern of generalization to shallow shell equations (for example the direct application to dynamic equations of a shallow shell was shown in [57] to be erroneous).

Various approaches were proposed to verify the "Berger hypothesis", including extravagant ones (propositions to regard the $(1 - \nu)$ term as a small parameter, and to neglect the second invariant of the stress tensor instead of the strain tensor in the potential energy terms).

Here we describe a noncontradictory derivation procedure of Berger-type equations in the application to rectangular and circular isotropic plates, and isotropic and sandwich shallow shells. It is shown that the second invariant of the strain tensor is small in a random way and this takes place only for isotropic single layered and transversally-isotropic three-layered plates; logically sequential procedures for the composition of Berger-type simplified theories require us to apply the homogenization approach.

First of all, let us consider several intuitive considerations.

The applicability of the "Berger hypothesis" to isotropic rectangular plates was justified by considerable amount of numerical analysis and appears to be beyond doubt. In other words, the contribution of the second invariant J_2 of the strain tensor to the potential is undoubtedly smaller than that of the first invariant J_1.

Taking into account

$$J_1 = \varepsilon_1 + \varepsilon_2; \qquad J_2 = \varepsilon_1\varepsilon_2 - 0.25\varepsilon_{12}^2;$$
$$\varepsilon_1 = u_x + 0.5w_x^2; \qquad \varepsilon_2 = v_y + 0.5w_y^2; \qquad \varepsilon_{12} = u_y + v_x + w_x w_y$$

the corresponding inequality for a rectangular plate $0 \leq x \leq a$, $0 \leq y \leq b$ may be written as

$$\left| \int_0^a \int_0^b (A + B_1 + C)\mathrm{d}x\,\mathrm{d}y \right| \gg (1 - \nu) \left| \int_0^a \int_0^b (A - B_2)\mathrm{d}x\,\mathrm{d}y \right|; \qquad (3.3.1)$$

$$A = 2u_x v_y + w_y^2 u_x + w_x^2 v_y,$$
$$B_1 = B_{11} + B_{12}, \qquad B_2 = 0.5B_{11} + B_{22},$$
$$B_{11} = u_x^2 + u_y^2, \qquad B_{12} = u_x w_x^2 + v_y w_y^2,$$
$$B_{22} = (u_y + v_x)w_x w_y, \qquad C = 0.25(w_x^2 + w_y^2)^2.$$

The main difference between the left and right hand parts of (3.3.1) is connected with the C-term. Let us consider the eigenvalue problem assuming that the displacements and bending moments are equal to zero along the plate boundary. Applying Galerkin's procedure for the one-term approximation $(u, v, w) = A_i(t) \times \sin(m\pi x/a) \cdot \sin(n\pi y/b)$ one can see that the $(A + B_1)$

and $(A - B_2)$ terms contribute equally (at least, by order of magnitude) to the potential energy, except for the special case $a = b$, $m = n$. Hence, the C-term contribution to the potential energy must prevail, as for the $m, n \gg 1$ case. Then, due to differentiations, the magnitude of the C-term becomes significant. Moreover, the C-term contains a slowly varying part instead of the rapidly varying B_{12}, B_{22} terms, and the integrals of the former ones become small. These considerations have led us to the decision to use the homogenization method (the nonlinear WKB-method [160]), based on the high variability of the solution along spatial coordinates, for the purpose of composition of Berger-type equations.

The nondimensional equations of motion of a rectangular plate may be written as

$$(12(1 - \nu^2))^{-1}\varepsilon\nabla^2\nabla^2\tilde{w} + \varepsilon(\tilde{F}_{\xi\xi}\tilde{w}_{\eta\eta} - 2\tilde{F}_{\xi\eta}\tilde{w}_{\xi\eta} + \tilde{F}_{\eta\eta}\tilde{w}_{\xi\xi}) + \tilde{w}_{\tau\tau} = 0,$$

$$\nabla^2\nabla^2\tilde{F} + \varepsilon(\tilde{w}_{\xi\xi}\tilde{w}_{\eta\eta} - \tilde{w}_{\xi\eta}^2) = 0,$$

$$\tilde{F}_{\eta\eta} = (1 - \nu^2)^{-1}(\tilde{u}_\xi + 0.5\varepsilon\tilde{w}_\xi^2 + v(\tilde{v}_\eta + 0.5\tilde{w}_\eta^2)),$$

$$\tilde{F}_{\xi\xi} = (1 - \nu^2)^{-1}(\tilde{v}_\eta + 0.5\xi\tilde{w}_\eta^2 + \nu(\tilde{u}_\xi + 0.5\tilde{w}_\xi^2)),$$

$$\tilde{F}_{\xi\eta} = -0.5(1 + \nu)^{-1}(\tilde{u}_\eta + \tilde{v}_\xi + \varepsilon\tilde{w}_\eta\tilde{w}_\eta).$$

where $\varepsilon = h/a$; $(\xi, \eta) = (x/y)a$; $\tilde{F} = F/Eha$; $(\tilde{u}, \tilde{v}, \tilde{w}) = (u, v, w)/h$; $\tau = dt\sqrt{\rho(1 - nu^2)/E}$; $\nabla^2 = \partial^2/\partial\xi^2 + \partial^2/\partial\eta^2$.

The most natural way of introducing "rapid variability" into the nonlinear system requires one to include the "rapid" variable $\varepsilon^\alpha\theta(\xi, \eta)$, regarding it as an independent variable. The value of α would be specified during the limiting $(\varepsilon \to 0)$ system derivation process. Now, following the multiple scale method, we obtain (the notation ξ, η describes the "slow" variables, as before)

$$\frac{\partial}{\partial\xi} = \frac{\partial}{\partial\xi} + \varepsilon^\alpha\theta_\eta\frac{\partial}{\partial\theta}; \qquad \frac{\partial}{\partial\eta} = \frac{\partial}{\partial\eta} + \varepsilon^\alpha\theta_\eta\frac{\partial}{\partial\theta}.$$

We suggest that the functions $\tilde{F}, \tilde{w}, \tilde{u}, \tilde{v}$ are sums of "slow" (i.e. depending upon the "slow" variables only) and "rapid" periodic components of the unknown period $\theta_0(\xi, \eta)$ [12, 3d, 4d]:

$$\tilde{F} = F^0(\xi, \eta) + \varepsilon\beta_1 F^1(\xi, \eta, \varepsilon^\alpha\theta),$$

$$\tilde{w} = w^0(\xi, \eta) + \varepsilon\beta_2 w^1(\xi, \eta, \varepsilon^\alpha\theta),$$

$$\tilde{u} = u^0(\xi, \eta) + \varepsilon\beta_3 u^1(\xi, \eta, \varepsilon^\alpha\theta),$$

$$\tilde{v} = v^0(\xi, \eta) + \varepsilon\beta_4 v^1(\xi, \eta, \varepsilon^\alpha\theta).$$

The following relations are to be used, too:

$$F^0 \sim \varepsilon\gamma_1 w^0; \quad w^0 \sim \varepsilon\gamma_2; \quad u^0 \sim \varepsilon\gamma_3; \quad v^0 \sim \varepsilon\gamma_4; \quad \partial/\partial\tau(\ldots) \sim \varepsilon^\delta(\ldots).$$

There are asymptotic integration parameters $\beta_i, \gamma_1, \gamma_2, \gamma_3, \gamma_4, \delta$ describing the relative orders of magnitude of the "slow" and "rapid" components: F^0 and w^0, u^0, v^0 and ε. The noncontradictory choice of its values, being routine work, has to be managed while satisfying the conditions of the noncontradictive character of the limiting ($\varepsilon \to 0$) systems. The nontrivial limiting systems may be obtained from (3.3.1), assuming

$$\alpha = -0.5, \quad \beta_1 = 0, \quad \beta_2 < 0, \quad \beta_3, \beta_4 \geq -0.5,$$
$$\gamma_1 = 1, \quad \gamma_2 = 0, \quad \gamma_3, \gamma_4 > 0, \quad \delta = 0$$

and may be written as

$$(12(1 - \nu^2))^{-1} w_{\theta\theta\theta\theta}^1 (\theta_\xi^2 + \theta_\eta^2)^2 +$$
$$+\underline{(F_{\xi\xi}^0 \theta_\eta^2 - 2F_{\xi\eta}^0 \theta_\xi \theta_\eta + F_{\eta\eta}^0 \theta_\xi^2) w_{\theta\theta}^1} + w_{\tau\tau}^2 = 0, \tag{3.3.2}$$

$$F_{\theta\theta\theta\theta}^1 (\theta_\xi^2 + \theta_\eta^2) = 0, \tag{3.3.3}$$

$$\varepsilon^{-1} F_{\theta\theta}^1 \theta_\eta^2 + F_{\eta\eta}^0 = 0.5(1 - \nu^2)^{-1} (w_\theta^1)^2 (\theta_\xi^2 + \nu\theta_\eta^2), \tag{3.3.4}$$

$$\varepsilon^{-1} F_{\theta\theta}^1 \theta_\eta^2 + F_{\xi\xi}^0 = 0.5(1 - \nu^2)^{-1} (w_\theta^1)^2 (\theta_\eta^2 + \nu\theta_\xi^2), \tag{3.3.5}$$

$$\varepsilon^{-1} F_{\theta\theta}^1 \theta_\xi \theta_\eta + F_{\xi\eta}^0 = -0.5(1 + \nu)^{-1} (w_\theta^1)^2 \theta_x i \theta_\eta. \tag{3.3.6}$$

The underlined term in (3.3.2) may be derived using a θ-averaging procedure, $\underline{(\ldots)} = \theta_0^{-1} \int_0^\theta (\ldots) d\theta$

$$F_{\theta\theta}^1 = 0, \quad (\bar{F}_{\xi\xi}^0, \bar{F}_{\eta\eta}^0, \bar{F}_{\xi\eta}^0) = (F_{\xi\xi}^0, F_{\eta\eta}^0, F_{\xi\eta}^0),$$
$$\underline{(\ldots)} = 0.5(1 - \nu^2)^{-1} (w_\theta^1)^2 (\theta_\xi^2 + \theta_\eta^2).$$

Using the previously introduced variables we get

$$(w_\theta^1)^2 (\theta_\xi^2 + \theta_\eta^2) = \frac{1}{ab} \int_0^a \int_0^b (w_x^2 + w_y^2) dx \, dy + O(\varepsilon);$$

equation (3.3.2) becomes the Berger equation

$$D\nabla^2 \nabla^2 w + N\nabla^2 w + \rho h w_{tt} = 0;$$
$$\nabla^2 = \frac{\partial^2}{\partial x^2} + \frac{\partial^2}{\partial y^2}; \quad D = \frac{Eh^2 B}{12}; \quad B = \frac{Eh}{1 - \nu^2}, \tag{3.3.7}$$

$$N = \frac{B}{2ab} \int_0^a \int_0^b (w_x^2 + w_y^2) \, dx \, dy.$$

The strain compability equation becomes linear:

$$\nabla^4 F = 0. \tag{3.3.8}$$

One could easily obtain Berger's equations for the viscoelastic plate,

$$\frac{h^2}{12} \Gamma \nabla^2 \nabla^2 w - \Gamma N \nabla^2 w + \rho h w_{tt} = 0$$

where $\Gamma\psi = \psi \int_0^\tau R(t - \tau_1) d\tau_1$, and R is the relaxation kernel.

Circural plates are to be considered separately for centre-holed and continuous plates. In the first case, Cartesian coordinates may be used, regarding (3.3.7), (3.3.8). In the second ($r_0 = 0$) case, r^{-1} varying coefficients are to be taken into account. Finally, one obtains (3.3.7), where

$$\nabla^2 = \frac{\partial^2}{\partial r^2} + \frac{1}{r}\frac{\partial}{\partial\theta} + \frac{1}{r^2}\frac{\partial^2}{\partial\theta^2},$$

$$N = \frac{B}{2\pi R^2} \int_0^{2\pi}\int_0^R (w_r^2 + w_\theta^2)\, dr\, d\theta.$$

Let us consider a shallow shell, the curvatures of which are k_1, k_2, and the in-plane dimensions are a and b. Assuming $a \sim b$, $k_i \sim 1$ and following the procedure described herein, we obtain Berger's equations

$$D\nabla^4 w + h\nabla_k F + N\nabla^2 w + \frac{B}{ab}\left(w_{xx} \int_0^a\int_0^b (k_1 + \nu k_2)w\, dy\, dx \right.$$

$$\left. + w_{yy}\int_0^a\int_0^b (k_2 + \nu k_1)w\, dy\, dy \right) + \rho h w_{tt} = 0,$$

$$\nabla^4 F + E\nabla_k w = 0, \tag{3.3.9}$$

$$\nabla_k = k_1\frac{\partial^2}{\partial x^2} + k_2\frac{\partial^2}{\partial y^2}.$$

Two points of special value may be outlined. Firstly, equations (3.3.9) allow all the possible natural limiting passags: a Berger plate; a Kirchhoff nonlinear bar, a shallow arc and, finally (the problem which the Berger hypothesis approach failed to overcome), linear shallow shell equations. Secondly, the "Berger's hypothesis" applied to (3.3.9) appears to be invalid (the second invariant of the strain tensor energy term is not smaller in order of magnitude compared with the first invariant term).

Let us look for approximate equations of transversally-isotropic sandwich shells. Introducing

$$D_0 = D\theta_0; \qquad \mu = \frac{h^2}{R^2\beta}; \qquad \rho_1 = \sum_{k=1}^3 \rho_k h_k,$$

where the parameters θ, μ depend on the bending stiffness of the load-carrying layers, and on the sandwich shear-resistance (θ, θ_0, β), can be calculated using the formulas given in [2] (ρ_k, h_k denote the density and thickness of the k-th layer).

Assuming $\theta_0 \sim 1$, $\mu \sim h/R$, $\theta \sim 1$, approximate equations can written as

$$D_0(1 - \underline{\theta\mu R^2\nabla^2})\nabla^2\nabla^2\chi + h\nabla_k F + N\nabla^2 w$$

$$+\frac{B}{ab}\left(w_{xx}\int_0^a\int_0^b(k_1 + \nu k_2)w\,dy\,dx\right.$$

$$\left.+w_{yy}\int_0^a\int_0^b(k_2 + \nu k_1)w\,dy\,dx\right) + \rho_1 w_{tt} = 0, \tag{3.3.10}$$

$$\nabla^4 F + E\nabla_k w = 0,$$
$$\left.\begin{array}{c}0.5(1 - \nu)\mu R^2\nabla^2\psi = \psi \\ w = (1 - \mu R^2\nabla^2)\chi\end{array}\right\}. \tag{3.3.11}$$

Considering the case $\theta < 1$, the underlined term in (3.3.10) must be omitted as well.

Limiting systems, governing static and dynamic behaviour of nonlinear bars, and linear plates are of no interest to us and are omitted.

This investigation can be concluded as follows:

1. "Berger's hypothesis" in its initial formulation appears to be true for isotropic single-layered and transversally isotropic multi-layered plates only.
2. As a matter of fact, Berger's equations represent the first approximation a of homogenization procedure (the nonlinear WKB method) when the rapid variability of the solution with respect to spatial coordinates is assumed.

3.3.2 "Method of Freezing" in the Nonlinear Theory of Viscoelasticity

The classical averaging method (in the form of the "method of freezing" [37d]) is a very usefull approach for solving the integro-differential equations of nonlinear dynamics.

Let us consider for example the equation of the nonlinear oscillation of the viscoelastic rectangular plate ($0 \le x \le a$, $0 \le y \le b$)

$$\Gamma\nabla^4 w - 12\Gamma h^{-2}J\nabla^2 w + phD^{-1}w_{tt} = 0. \tag{3.3.12}$$

Here $\Gamma_\varphi = \varphi + \int_0^t R(t - t_1)\varphi(t_1)\,dt_1$; $J = \frac{1}{2ab}\int_0^a\int_0^b(w_x^2 + w_y^2)\,dx\,dy$; and R is the kernel of relaxation.

If the plate is simply supported, the spatial and time variables may be separated:

$$w(x, y, t) = A(t)\sin\frac{m\pi x}{a}\sin\frac{n\pi y}{b}.$$

Then, for the amplitude $A(t)$ one obtains a very complicated integro-diffe-rential equation

$$D_1 \Gamma(A + 3h^{-2}A^3) + A_{tt} = 0. \tag{3.3.13}$$

Here

$$D_1 = \frac{D\pi^4 \left[\left(\frac{m}{a}\right)^2 + \left(\frac{n}{b}\right)^2\right]^2}{ph}.$$

For low-frequency oscillations the function $A(t_1)$ is changed slowly with re-spect to the relation kernel R, so we may "freeze" $A(t_1)$ at the point $t = t_1$ [37d] and replace the integral

$$J_1 = \int_0^t A(t_1)R(t - t_1)\,dt_1$$

by the following one

$$J_1 \approx A(t)J;$$

$$J = \int_0^t R(t - t_1)\,dt_1.$$

Then (3.3.13) turns out to be the ordinary nonlinear differential equation with variable coefficients

$$D_1 J(A + 3h^{-2}A^3) + A_{tt} = 0,$$

which can be solved by using the averaging procedure [141]. It is possible also to use the second procedure of freezing, applying to J the averaging operator Π [37d]

$$\overline{(\ldots)} \equiv \Pi(\ldots) \equiv \lim_{T \to \infty} \frac{1}{T} \int_0^T (\ldots)\,dt.$$

As a result, we obtain the following equation with constant coefficients:

$$D_1 \overline{J}(A + 3h^{-2}A^3) + A_{tt} = 0,$$

which may be solved exactly or by the perturbation or averaging procedure [141].

3.4 Bolotin-Like Approach for Nonlinear Dynamics

3.4.1 Straightforward Bolotin Approach

Bolotin [56] proposed an effective asymptotic method for the investigation of linear continuous elastic system oscillations with complicated boundary conditions. Bolotin's method is also called the dynamic edge effect method. The main idea of this approach is to separate the continuous elastic system into two parts. In one of them – the so-called interior zone – solutions may be expressed by trigonometric functions with unknown constants. In the second part – the dynamic edge effect zone – Bolotin used exponential functions. The matching procedure (along the edges or unknown interior lines) permits one to obtain the unknown constants, and the complete solution of the dynamics problem may be written in a relatively simple form. This approximate solution is very good for high-frequency oscillations, but even for low-frequency oscillation cases the error is not excessive (see references [62, 73, 74] and the references quoted therein).

These considerations are devoted to nonlinear oscillations of shallow cylindrical shells and rectangular plates.

As the governing equations we use the approximate nonlinear equations obtained in Sect. 3.3.1:

$$DV^4 w - hR^{-1}\frac{\partial^2 F}{\partial x_1^2} - NV^2 w + N_1\frac{\partial^2 w}{\partial x_1^2} + N_2\frac{\partial^2 w}{\partial x_2^2} + \rho h\frac{\partial^2 w}{\partial t^2} = 0, \quad (3.4.1)$$

$$\nabla^4 F + ER^{-1}\frac{\partial^2 w}{\partial x_1^2} = 0,$$

where

$$N = \frac{6D}{h^2 a_1 a_2}\int_0^{a_2}\int_0^{a_1}\left[\left(\frac{\partial w}{\partial x_1}\right)^2 + \left(\frac{\partial w}{\partial x_2}\right)^2\right]dx_1\,dx_2,$$

$$N_2 = \frac{12D}{h^2 a_1 a_2 R}\int_0^{a_2}\int_0^{a_1}w\,dx_1\,dx_2, \qquad N_1 = N_2\nu.$$

System (3.4.1) may be rewritten in the mixed form

$$\frac{\partial^2 u_1}{\partial x_1^2} + 0.5(1-\nu)\frac{\partial^2 u_1}{\partial x_2^2} + 0.5(1+\nu)\frac{\partial^2 u_2}{\partial x_1\partial x_2} - \nu R^{-1}\frac{\partial w}{\partial x_1} = 0, \quad (3.4.2)$$

$$\frac{\partial^2 u_2}{\partial x_2^2} + 0.5(1-\nu)\frac{\partial^2 u_2}{\partial x_1^2} + 0.5(1+\nu)\frac{\partial^2 u_1}{\partial x_1\partial x_2} - \nu R^{-1}\frac{\partial w}{\partial x_2} = 0, \quad (3.4.3)$$

$$\nabla^4 w - D^{-1}\left(NV^2 w - N_1\frac{\partial^2 w}{\partial x_1^2} - N_2\frac{\partial^2 w}{\partial x_2^2}\right)$$

$$-12h^{-2}R^{-1}\left(\frac{\partial u_2}{\partial x_2} + \nu\frac{\partial u_1}{\partial x_1} - R^{-1}w\right) + \rho h^2 D^{-1}\frac{\partial^2 w}{\partial t^2} = 0. \quad (3.4.4)$$

Let us consider the boundary conditions as follows (clamped edges)

$$u_1 = u_2 = w = \frac{\partial w}{\partial x_1} = 0 \qquad \text{for } x_1 = 0, a_1, \qquad (3.4.5)$$

$$u_1 = u_2 = w = \frac{\partial w}{\partial x_2} = 0 \qquad \text{for } x_2 = 0, a_2. \qquad (3.4.6)$$

Here we shall investigate normal modes of nonlinear oscillations. For continuous systems this means that the dependences on the spatial and time variables may be separated in an exact or in an approximate way [159].

Let us represent the interior solution of (3.4.1) in the form

$$w(x_1, x_2, t) = w_0 = f_1 \cos k_1(x_1 - x_{10}) \sin k_2(x_2 - x_{20})\xi_1(t), \qquad (3.4.7)$$

$$F(x_1, x_2, t) = F_0 = f_2 \cos k_1(x_1 - x_{10}) \sin k_2(x_2 - x_{20})\xi_2(t), \qquad (3.4.8)$$

where $k_1(k_2)$ and $x_{10}(x_{20})$ are unknown constants; $k_1(k_2)$ is the wavelength and $x_{10}(x_{20})$ is the phase shift in the $x_1(x_2)$ direction.

Substituting expressions (3.4.7) and (3.4.8) into the initial relations, one can obtain an ordinary differential equation for the time function ξ_1 and a relation between the functions ξ_1 and ξ_2:

$$\frac{\partial^2 \xi_1}{\partial t^2} + \omega^2 \left(1 + \gamma_1 \xi_1 + \gamma_2 \xi_1^2\right) \xi_1 = 0, \qquad (3.4.9)$$

$$\xi_2 = E f_1 k_1^2 (R f_2)^{-1} (k_1^2 + k_2^2)^{-2} \xi_1, \qquad (3.4.10)$$

where

$$\omega^2 = D\rho^{-1}h^{-2}\Omega, \qquad \Omega = (k_1^2 + k_2^2)^2 + 12(1 - \nu^2)(hR)^{-2}k_1^4(k_1^2 + k_2^2)^{-2},$$

$$\gamma_1 = -12A_3 f_1(\Omega R h^2 a_1 a_2)^{-1}(\nu k_1^2 + k_2^2),$$

$$\gamma_2 = 1.5 f_1^2(\Omega h^2 a_1 a_2)^{-1}(k_1^2 + k_2^2)\left[k_1^2(a_1 - A_1)(a_2 - A_2)\right.$$
$$\left. + k_2^2(a_1 + A_1)(a_2 + A_2)\right],$$

$$A_1 = 0.5k_1^{-1}\left[\sin 2k_1(a_1 - x_{10}) + \sin 2k_1 x_{10}\right],$$

$$A_2 = 0.5k_2^{-1}\left[\sin 2k_2(a_2 - x_{20}) + \sin 2k_2 x_{20}\right],$$

$$A_3 = -(k_1 k_2)^{-1}\left[\sin k_1(a_1 - x_{10}) + \sin k_1 x_{10}\right]$$
$$\cdot \left[\cos k_2(a_2 - x_{20}) - \cos k_2 x_{20}\right].$$

Let us designate the solution of (3.4.9) satisfying the initial conditions $\xi = 0$, $d\xi/dt = 1$ for $t = 0$ as $\varphi(t)$. This solution may be represented by elliptical functions [55].

The radial displacement w in the interior zone may be expressed in the form

$$w_0 = f_1 \cos k_1(x_1 - x_{10}) \sin k_2(x_2 - x_{20})\varphi(t). \qquad (3.4.11)$$

Using (3.4.2), (3.4.3) one finds u_{10} and u_{20} in the interior zone to be

$$u_{10} = f_3 \sin k_1(x_1 - x_{10}) \sin k_2(x_2 - x_{20})\varphi(t), \qquad (3.4.12)$$

$$u_{20} = f_4 \cos k_1(x_1 - x_{10}) \cos k_2(x_2 - x_{20})\varphi(t), \qquad (3.4.13)$$

where

$$f_3 = \frac{f_1 k_1(\nu k_1^2 - k_2^2)}{R(k_1^2 + k_2^2)^2}, \qquad f_4 = \frac{-f_1 k_2\left(k_2^2 + (2 + \nu)k_1^2\right)}{R(k_1^2 + k_2^2)^2}.$$

The constants k_1, k_2, x_{10} and x_{20} are unknown, and the boundary conditions are not yet satisfied. Consequently, one proceeds to construct corrective solutions in the narrow zone near the edges.

Let us introduce the new variables u_{1b}, u_{2b} and w_b – the components of the corrective solutions localized near the boundaries. The shell displacements can thus be expressed in the forms

$$u_1 = u_{10} + u_{1b}, \quad u_2 = u_{20} + u_{2b}, \quad w = w_0 + w_b. \tag{3.4.14}$$

Substitution of expressions (3.4.14) into (3.4.2)–(3.4.4) yields

$$\frac{\partial^2(u_{10} + u_{1b})}{\partial x_1^2} + 0.5(1 - \nu)\frac{\partial^2(u_{10} + u_{1b})}{\partial x_2^2}$$
$$+0.5(1 + \nu)\frac{\partial^2(u_{20} + u_{2b})}{\partial x_1 \partial x_2} - \frac{\nu}{R}\frac{\partial(w_0 + w_b)}{\partial x_1} = 0, \tag{3.4.15}$$

$$\frac{\partial^2(u_{20} + u_{2b})}{\partial x_2^2} + 0.5(1 - \nu)\frac{\partial^2(u_{20} + u_{2b})}{\partial x_1^2}$$
$$+0.5(1 + \nu)\frac{\partial^2(u_{10} + u_{1b})}{\partial x_1 \partial x_2} - \frac{1}{R}\frac{\partial(w_0 + w_b)}{\partial x_2} = 0, \tag{3.4.16}$$

$$\nabla^4(w_0 + w_b) - \frac{6}{h^2 a_1 a_2}\left\{\nabla^2(w_0 + w_b)\int_0^{a_2}\int_0^{a_1}\left[\left(\frac{\partial(w_0 + w_b)}{\partial x_1}\right)^2\right.\right.$$
$$+\left(\frac{\partial(w_0 + w_b)}{\partial x_2}\right)^2\right]dx_1\, dx_2 - 2\left[\nu\frac{\partial^2(w_0 + w_b)}{\partial x_1^2}\right.$$
$$\left.+\frac{\partial^2(w_0 + w_b)}{\partial x_2^2}\right]\int_0^{a_2}\int_0^{a_1}(w_0 + w_b)\,dx_1\, dx_2\right\} - \frac{12}{h^2 R}\left[\nu\frac{\partial(u_{10} + u_{1b})}{\partial x_1}\right.$$
$$\left.+\frac{\partial(u_{20} + u_{2b})}{\partial x_2} - \frac{w_0 + w_b}{R}\right] + \frac{\rho h^2}{D}\frac{\partial^2(w_0 + w_b)}{\partial t^2} = 0. \tag{3.4.17}$$

Equations (3.4.15) and (3.4.17) are very complicated and cannot be solved in an explicit way without asymptotic simplifications. First of all, we must separate the interior and the corrective solutions. For this purpose one can use energy estimations. Thus, let us estimate the integral coefficients in (3.4.17) for large parameters $k_1 \sim k_2 \gg 1$

$$\int_0^{a_2}\int_0^{a_1}\left(\frac{\partial w_0}{\partial x_1}\right)^2 dx_1\, dx_2 \sim k_1^2, \qquad \int_0^{a_2}\int_0^{a_1}\left(\frac{\partial w_0}{\partial x_2}\right)^2 dx_1\, dx_2 \sim k_2^2,$$

$$\int_0^{a_2}\int_0^{a_1}\left(\frac{\partial w_b}{\partial x_1}\right)^2 dx_1\, dx_2 \sim k_1, \qquad \int_0^{a_2}\int_0^{a_1}\left(\frac{\partial w_b}{\partial x_2}\right)^2 dx_1\, dx_2 \sim k_2,$$

$$\int_0^{a_2}\int_0^{a_1}\left(\frac{\partial w_0}{\partial x_1}\right)\left(\frac{\partial w_b}{\partial x_1}\right) dx_1\, dx_2 \sim k_1,$$

$$\int_0^{a_2}\int_0^{a_1} \left(\frac{\partial w_0}{\partial x_2}\right)\left(\frac{\partial w_b}{\partial x_2}\right) dx_1\, dx_2 \sim k_2.$$

If we eliminate all terms of lower order in (3.4.17), we obtain the simplified equation

$$\nabla^4(w_0 + w_b) - \frac{6}{h^2 a_1 a_2}\left\{\nabla^2(w_0 + w_b)\int_0^{a_2}\int_0^{a_1}\left[\left(\frac{\partial w_0}{\partial x_1}\right)^2\right.\right.$$

$$+ \left(\frac{\partial w_o}{\partial x_2}\right)^2\Bigg] dx_1\, dx_2 - 2\left[\nu\frac{\partial^2(w_0 + w_b)}{\partial x_1^2} + \frac{\partial^2(w_0 + w_b)}{\partial x_2^2}\right]$$

$$\cdot \int_0^{a_2}\int_0^{a_1} w_0\, dx_1\, dx_2\Bigg\} - \frac{12}{h^2 R}\left[\nu\frac{\partial(u_{10} + u_{1b})}{\partial x_1} + \frac{\partial(u_{20} + u_{2b})}{\partial x_2}\right.$$

$$\left. - \frac{w_0 + w_b}{R}\right] + \frac{\rho h^2}{D}\frac{\partial^2(w_0 + w_b)}{\partial t^2} = 0. \tag{3.4.18}$$

Substituting equations (3.4.11)–(3.4.13) for the interior solution into system (3.4.15) and (3.4.18), one obtains approximate equations for the corrective solutions

$$\frac{\partial^2 u_{1b}}{\partial x_1^2} + 0.5(1 - \nu)\frac{\partial^2 u_{1b}}{\partial x_2^2} + 0.5(1 + \nu)\frac{\partial^2 u_{2b}}{\partial x_1 \partial x_2} - \frac{\nu}{R}\frac{\partial w_b}{\partial x_1} = 0, \tag{3.4.19}$$

$$\frac{\partial^2 u_{2b}}{\partial x_2^2} + 0.5(1 - \nu)\frac{\partial^2 u_{2b}}{\partial x_1^2} + 0.5(1 + \nu)\frac{\partial^2 u_{1b}}{\partial x_1 \partial x_2} - \frac{\nu}{R}\frac{\partial w_b}{\partial x_2} = 0, \tag{3.4.20}$$

$$\nabla^4 w_b - \frac{\gamma_1 \Omega}{k_2^2 + \nu k_1^2}\varphi(t)\left(\frac{\partial^2 w_b}{\partial x_2^2} + \nu\frac{\partial^2 w_b}{\partial x_1^2}\right) - \frac{\gamma_2 \Omega}{k_1^2 + k_2^2}\varphi^2(t)\nabla^2 w_b$$

$$- \frac{12}{h^2 R}\left[\nu\frac{\partial u_{1b}}{\partial x_1} + \frac{\partial u_{2b}}{\partial x_2} - \frac{w_b}{R}\right] + \frac{\rho h^2}{D}\frac{\partial^2 w_b}{\partial t^2} = 0. \tag{3.4.21}$$

Equations (3.4.19)–(3.4.10), describing the corrective solutions, are linear differential equations with time-dependent coefficients.

Spatial and time variables cannot be separated exactly in (3.4.19)–(3.4.21), but one can use the variational Kantorovitch method [48d]. Let us briefly describe this method.

First of all, let us represent the solution of (3.4.19)–(3.4.21) in the form satisfying the condition of periodicity:

$$u_{1b}(x_1, x_2, t) \approx U_1(x_1, x_2)\varphi(t),$$
$$u_{2b}(x_1, x_2, t) \approx U_2(x_1, x_2)\varphi(t), \tag{3.4.22}$$
$$w_b(x_1, x_2, t) \approx W(x_1, x_2)\varphi(t).$$

Now, one can substitute expressions (3.4.22) into (3.4.19)–(3.4.21), multiply these equations by $\varphi(t)$ and integrate over a period. Then one obtains

$$d_{11}U_1 + d_{12}U_2 + d_{13}W = 0, \qquad d_{21}U_1 + d_{22}U_2 + d_{23}W = 0,$$
$$d_{31}U_1 + d_{32}U_2 + d_{33}W = 0, \qquad\qquad\qquad\qquad (3.4.23)$$

where

$$d_{11} = \frac{\partial^2}{\partial x_1^2} + 0.5(1-\nu)\frac{\partial^2}{\partial x_2^2}, \qquad d_{12} = d_{21} = 0.5(1-\nu)\frac{\partial^2}{\partial x_1 \partial x_2},$$

$$d_{13} = -\frac{\nu}{R}\frac{\partial}{\partial x_1}, \quad d_{22} = 0.5(\nu-1)\frac{\partial^2}{\partial x_1^2} + \frac{\partial^2}{\partial x_2^2}, \quad d_{23} = -\frac{1}{R}\frac{\partial}{\partial x_2},$$

$$d_{31} = -c_0\nu R\frac{\partial}{\partial x_1}, \qquad d_{32} = -c_0\nu R\frac{\partial}{\partial x_2},$$

$$d_{33} = h^2\nabla^4 + c_1\left(\nu\frac{\partial^2}{\partial x_1^2} + \frac{\partial^2}{\partial x_2^2} + \nu k_1^2 + k_2^2\right)$$
$$\qquad - c_2\left(\nabla^2 + k_1^2 + k_2^2\right) + c_0 - h^2\Omega,$$

$$c_0 = \frac{12}{R^2}, \qquad c_1 = -\frac{\gamma_1\lambda\Omega h^2}{\nu k_1^2 + k_2^2}\int_0^\tau \varphi^2(t)\,dt,$$

$$c_2 = \frac{\gamma_2\lambda\Omega h^2}{k_1^2 + k_2^2}\int_0^\tau \varphi^3(t)\,dt, \qquad \lambda^{-1} = \int_0^\tau \varphi(t)\,dt.$$

Now the partial differential equations (3.4.23) have constant coefficients, and one can use the operational method of the solution of differential equations with constant coefficients. In accordance with the main idea of this method one can operate with the derivatives $\partial/\partial x_i$ as with the constants and use the methods of linear algebra [82].

Then (3.4.23) may be reduced to a single equation for the function Φ,

$$D^*\Phi = 0, \qquad\qquad\qquad (3.4.24)$$

where

$$U_1 = 0.5(1-\nu)D_{13}^*\Phi, \qquad U_2 = 0.5(1-\nu)D_{23}^*\Phi, \qquad (3.4.25)$$
$$W = 0.5(1-\nu)D_{33}^*\Phi. \qquad\qquad\qquad (3.4.26)$$

Here D^* is the determinant of system (3.4.23), and D_{i3}^* $(i = 1, 2, 3)$ are the minors of the determinant D^*.

Let us now consider the edge effect at the $x_1 = 0$ zone edge. In this zone we represent U_1, U_2 and W in the form

$$U_1 = \theta_1(x_1)\sin k_2(x_2 - x_{20}), \qquad U_2 = \theta_2(x_1)\cos k_2(x_2 - x_{20}), \quad (3.4.27)$$
$$W = \theta(x_1)\sin k_2(x_2 - x_{20}). \qquad\qquad\qquad (3.4.28)$$

For θ_1, θ_2 and θ one can obtain a system similar to system (3.4.23) (where $\partial/\partial x_2 \to k_2$).

The characteristic equation for system (3.4.23) is

$$(p^2 + k_1^2)(h^2 p^6 + a_{11}p^4 + a_{12}p^2 + a_{13}) = 0, \qquad (3.4.29)$$

where

$$a_{11} = -h^2(k_1^2 - 4k_2^2) + \alpha_1,$$

$$a_{12} = k_2^2 \left[h^2(2k_1^2 + 5k_2^2) + \frac{c_0(1 - \nu^2)(2k_1^2 + k_2^2)}{(k_1^2 + k_2^2)^2} - 2\alpha_1 \right],$$

$$a_{13} = -k_2^4 \left[h^2(k_1^2 + 2k_2^2) + \frac{c_0(1 - \nu^2)}{(k_1^2 + k_2^2)^2} - \alpha_1 \right],$$

$$\alpha_1 = \nu c_1 - c_2.$$

Equation (3.4.29) has two imaginary roots ($p_7 = +ik_1$ and $p_8 = -ik_1$, $i^2 = -1$) belonging to the interior solutions (3.4.11)–(3.4.13) and one must eliminate them from (3.4.29).

The next six roots are

$$p_{4,1} = \pm \left(-2r \cos \left(\frac{\omega_1}{3} \right) + \frac{a_{11}}{3} \right)^{0.5}, \tag{3.4.30}$$

$$p_{5,2} = \pm \left(2r \cos \left(\frac{\pi - \omega_1}{3} \right) + \frac{a_{11}}{3} \right)^{0.5}, \tag{3.4.31}$$

$$p_{6,3} = \pm \left(2r \cos \left(\frac{\pi + \omega_1}{3} \right) + \frac{a_{11}}{3} \right)^{0.5}, \tag{3.4.32}$$

where

$$\omega_1 = \arccos(qr^{-3}), \qquad q = \frac{a_{11}^3}{27} - \frac{a_{11}a_{12}}{6} + \frac{a_{13}}{2},$$

$$r = \text{sign}\,(q) \frac{(3a_{12} - a_{11}^2)^{0.5}}{3}.$$

Then, near the boundary $x_1 = 0$, one has

$$\Phi = \sum_{k=1}^{6} C_{1k} \exp(p_k x_1),$$

(where C_{1k} are arbitrary constants), and the corrective solution displacement may be written in the form

$$u_{1b}^{(1)} = c_0 R \sum_{k=1}^{6} C_{1k} p_k (\nu p_k^2 + k_2^2) \exp(p_k x_1) \sin k_2 (x_2 - x_{20}) \varphi(t),$$

$$u_{2b}^{(1)} = c_0 R k_2 \sum_{k=1}^{6} C_{1k} [k_2^2 - (2 + \nu) p_k^2] \exp(p_k x_1) \sin k_2 (x_2 - x_{20}) \varphi(t),$$

$$w_b^{(1)} = \sum_{k=1}^{6} C_{1k} (p_k^2 - k_2^2)^2 \exp(p_k x_1) \sin k_2 (x_2 - x_{20}) \varphi(t).$$

As described above, one can thus easily deal with the edge effect at the $x_2 = 0$ boundary.

Let us rewrite boundary conditions (3.4.5) and (3.4.6) in the form

$$u_{10} + u_{1b}^{(1)} = 0, \qquad u_{20} + u_{2b}^{(1)} = 0, \tag{3.4.33}$$

$$w_0 + w_b^{(1)} = 0, \qquad \frac{\partial w_0}{\partial x_1} + \frac{\partial w_b^{(1)}}{\partial x_1} = 0 \quad \text{for } x_1 = 0, \tag{3.4.34}$$

$$u_{10} + u_{1b}^{(2)} = 0, \qquad u_{20} + u_{2b}^{(2)} = 0, \tag{3.4.35}$$

$$w_0 + w_b^{(2)} = 0, \qquad \frac{\partial w_0}{\partial x_2} + \frac{\partial w_b^{(2)}}{\partial x_2} = 0 \quad \text{for } x_2 = 0. \tag{3.4.36}$$

These conditions must be supplemented by

$$u_{1b}^{(1)}, u_{2b}^{(1)}, w_b^{(1)} \to 0 \quad \text{for } x_1 \to \infty, \tag{3.4.37}$$

$$u_{1b}^{(2)}, u_{2b}^{(2)}, w_b^{(2)} \to 0 \quad \text{for } x_2 \to \infty. \tag{3.4.38}$$

Then, the arbitrary constants may be determined from conditions (3.4.33), (3.4.35), (3.4.37) and (3.4.38).

Using conditions (3.4.34) and (3.4.36), one has

$$x_{10} = k_1^{-1} \operatorname{arctg} \left\{ -\sum_{k=1}^{3} C_{1k} p_k (p_k^2 - k_2^2)^2 \left[k_1 \sum_{k=1}^{3} C_{1k}(p_k^2 - k_2^2)^2 \right]^{-1} \right\},$$

$$x_{20} = k_2^{-1} \operatorname{arctg} \left\{ k_2 \sum_{k=1}^{3} C_{2k}(s_k^2 - k_1^2)^2 \left[k_1 \sum_{k=1}^{3} C_{2k}(s_k^2 - k_1^2)^2 \right]^{-1} \right\}. \tag{3.4.39}$$

The oscillation forms can be separated into symmetry types. For the type symmetric in both directions one has

$$\frac{\partial w_0}{\partial x_1} = u_{10} = 0 \text{ for } x_1 = 0.5a_1, \qquad \frac{\partial w_0}{\partial x_2} = u_{20} = 0 \text{ for } x_2 = 0.5a_2. \tag{3.4.40}$$

For the type antisymmetric in both directions one has

$$w_0 = u_{20} = 0 \text{ for } x_1 = 0.5a_1, \qquad w_0 = u_{10} = 0 \text{ for } x_2 = 0.5a_2. \tag{3.4.41}$$

Substituting the displacement into (3.4.40) and (3.4.41) and taking into account formulas (3.4.11)–(3.4.13), one obtains the transcendental equations

$$k_1(a_1 - 2x_{10}) = m\pi, \qquad k_2(a_2 - 2x_{20}) = n\pi, \qquad m, n = 1, 2, \ldots. \tag{3.4.42}$$

For $m = 2k$, $n = 2k + 1$, one has antisymmetric (in both directions) modes, and for $n = 2k$, $m = 2k + 1$ one has symmetric (in both directions) modes.

Equations (3.4.39) and (3.4.42) may be solved routinely.

In the limiting case $1/R \to 0$ one obtains the solution for the nonlinear oscillations of a rectangular plate from equations (3.4.1), (3.4.4)–(3.4.7), (3.4.11), (3.4.17), (3.4.18), (3.4.26), (3.4.28)–(3.4.30), (3.4.34) and (3.4.36)–(3.4.42).

Now we examine the accuracy of Bolotin's method for the nonlinear case. Let us consider a simply supported square plate and introduce the notation

$f^* = f_1/h$, $\omega^* = \omega/\omega_0$ (ω being the natural frequency of the square plate clamped along its edges and $\omega_0 = \pi(D/\rho h a^4)$ being the square of the fundamental frequency of linear oscillations of the simply supported square plate).

Amplitude-frequency dependencies for the nondimensional amplitude and frequency obtained by the present method (continuous lines) and by the method of approximate variables separation [158] (dashed lines) are shown in Fig. 3.6. The corresponding curves show satisfactory agreement. The discrepancy is not excessive, which confirms the acceptable accuracy of this method.

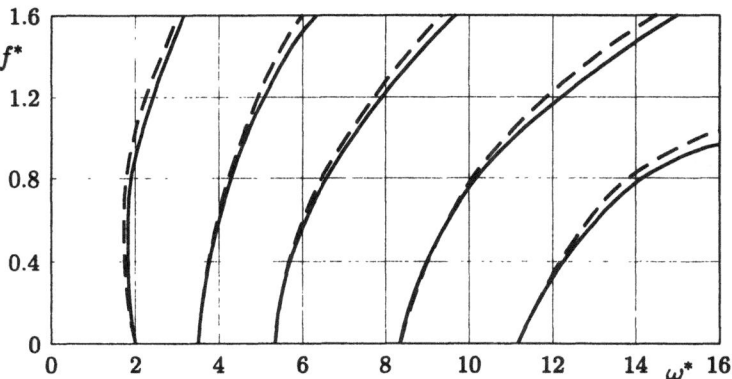

Fig. 3.6. Curves of the dimensionless amplitude f^* versus the dimensionless frequency ω^* for the first five modes of the simply supported square plate

3.4.2 Modified Bolotin Approach

Unfortunately, the above approach may be used only for rectangular regions.

Here we propose a modification of the dynamic edge effect method (DEEM) that can be used to find the natural vibration frequencies and modes of plates and shells of a nonrectangular form at high amplitudes. For example, we deal with sector plates.

To describe the motion of the plate, we proceed from the simplified equation proposed in Sect. 2.3.1:

$$\Delta^2 w - N\Delta w + \sigma^2 \frac{\partial^2 w}{\partial t^2} = 0$$

$$\Delta \equiv \frac{\partial^2}{\partial r^2} + \frac{\partial}{r\partial r} + \frac{\partial^2}{\partial \varphi^2}$$

$$N = \frac{12}{\theta} \int\limits_0^1 \int\limits_0^\theta \left[\left(\frac{\partial w}{\partial r}\right)^2 + \left(\frac{\partial w}{r\partial \varphi}\right)^2 \right] r \, dr \, d\varphi \qquad (3.4.43)$$

where r, ϕ are the polar coordinates: $r \in [0,1]$, $\phi \in [0,\theta]$, R is the sector radius, $w(r,\phi,t)$ is the deflection function as a fraction of r, and $\sigma = R^2(\rho h/D)^{1/2}$.

For the sake of argument, we shall assume elastic-restraint conditions on the contour of the plate:

$$w|_{\Gamma_1,\Gamma_2} = \left[u\frac{\partial w}{r\partial \varphi} - (1-u)\frac{\partial^2 w}{r\partial \varphi^2} \right]|_{\Gamma_1,\Gamma_2} = 0 \tag{3.4.44}$$

$$w|_{\Gamma_3} = \left\{ Q\frac{\partial w}{\partial r} - (1-Q)\left[\Delta w - (\nu - 1)\frac{\partial w}{r\partial r} \right] \right\}|_{\Gamma_3} = 0 \tag{3.4.45}$$

$$|w_r^{(i)}(0,\varphi,t)| < \infty \quad (i = 0,1),$$

where $\Gamma_1, \Gamma_2, \Gamma_3$ are the straight and circular parts of the contour, respectively, and u and Q are the reduced elastic parameters of the constraint ($u, Q \in [0,1]$).

Let the initial conditions be

$$w(r,\varphi,t) = \max w, \quad w_t'(r,\varphi,t) = 0. \tag{3.4.46}$$

We present the solution of (3.4.43) in the form

$$w(r,\varphi,t) = Az(r,\varphi)\eta(t), \tag{3.4.47}$$

where A is the amplitude as a fraction of h. For $\eta(t)$ we first choose an approximation that satisfies the initial conditions (3.4.46):

$$\eta(t) = \cos \omega t. \tag{3.4.48}$$

We substitute expressions (3.4.47) and (3.4.48) into (3.4.43) and separate variables by Kantorovich's approximate method [48d] integrating over time on the segment $[0, 2\pi/\omega]$:

$$(\Delta^2 - H\Delta - \lambda^2)z(r,\varphi) = 0 \tag{3.4.49}$$

$$H = \frac{\theta A}{\theta} \int_0^1 \int_0^\theta \left[\left(\frac{\partial z}{\partial r}\right)^2 + \left(\frac{\partial z}{r\partial \varphi}\right)^2 \right] r\, dr\, d\varphi, \quad \lambda = \sigma\omega. \tag{3.4.50}$$

We shall assume below that λ is large. We present the solution of (3.4.49) far from the edges $\phi = 0, \theta$ in the form

$$z(r,\varphi) = W(r)\Psi(\varphi), \quad \Psi_1 = \sin k(\varphi - \xi), \tag{3.4.51}$$

where k, ξ are the unknown wave number and phase. Substituting (3.4.51) into (3.4.49), we obtain

$$(\Phi^2 - H\Phi - \lambda^2)W = 0, \quad \Phi \equiv \frac{d^2}{dr^2} + \frac{d}{rdr} - \left(\frac{k}{r}\right)^2. \tag{3.4.52}$$

The solution of (3.4.52) that satisfies the bound (3.4.45) has the form (J_k and I_k are Bessel functions of the first kind for real and imaginary arguments)

$$W(r) = C_1 J_k(\alpha r) + C_2 I_k(\beta r),$$

$$\alpha = \left\{ -\frac{H}{2} + \left[\left(\frac{H}{2} \right)^2 + \lambda^2 \right]^{\frac{1}{2}} \right\}^{\frac{1}{2}}, \tag{3.4.53}$$

$$\beta = \left\{ \frac{H}{2} + \left[\left(\frac{H}{2} \right)^2 + \lambda^2 \right]^{\frac{1}{2}} \right\}^{\frac{1}{2}}.$$

From (3.4.53), using the representations (3.4.47) and (3.4.51) and the boundary conditions (3.4.44), we obtain the transcendental equation

$$\frac{\alpha J_{k+1}(\alpha)}{J_k(\alpha)} + \frac{\beta I_{k+1}(\beta)}{I_k(\beta)} = P(\alpha^2 + \beta^2),$$

$$P = \frac{1 - Q}{1 - \nu(1 - Q)}. \tag{3.4.54}$$

The first term in expression (3.4.53) corresponds to the contour state, while the second one desribes the dynamic edge effect (DEE) near the circular edge. Using the asymptotic formulas for the Bessel functions [93] (which are different for the cases $k^2 = o(\lambda)$ and $k^2 = o(\lambda)$), we can show that the energy contribution of the DEE tends to zero as $\lambda \to \infty$.

To find a DEE-type solution at the straight boundaries, we present the solution of (3.4.49) in the form

$$z(r, \varphi) = W_1(r)\Psi(\varphi), \quad W_1 = J_k(\alpha r). \tag{3.4.55}$$

Substituting (3.4.55) into (3.4.49), applying (3.4.52) and the Bessel equation, and retaining only the terms of $o(\lambda^2)$, we obtain

$$\left(\frac{\partial^2}{\partial \varphi^2} + k^2 \right) \left[\frac{\partial^2}{\partial \varphi^2} + k^2 - r^2(2\alpha^2 + H) \right] W_1 \Psi = 0. \tag{3.4.56}$$

The expression for H that retains only the term corresponding to the ground state and is derived using the Green's second formula has the form

$$H = \frac{9}{2} A^2 [J_k^2(\alpha) - J_{k-1}^2(\alpha) J_{k+1}^2(\alpha)] B \alpha^2 \tag{3.4.57}$$

$$B = 1 \quad \text{for } k = 0, \quad B = 0.5 \left(1 - \frac{\sin 2k\xi}{k\theta} \right) 2 \quad \text{for } k \neq 0.$$

Since the variables in (3.4.56) cannot be exactly separated, we use Kantorovich's method [48d]:

$$\left(\frac{\partial^2}{\partial \varphi^2} + k^2 \right) \left[\frac{\partial^2}{\partial \varphi^2} + k^2 - s(\alpha^2 + H) \right] \Psi = 0, \tag{3.4.58}$$

$$s = \int_0^1 W_1^2 r^3 \left(\int_0^1 W_1^2 r \, dr \right)^{-1} dr; \quad s \to \frac{2}{3} \quad \text{as } \lambda \to \infty \quad k^2 = o(\lambda).$$

The general solution of (3.4.58) has the form

$$\Psi(\varphi) = \Psi_1(\varphi) + \Psi_2(\varphi), \tag{3.4.59}$$

$$\Psi_2 = C_{11} \exp(g\varphi) + C_{21} \exp(-g\varphi), \tag{3.4.60}$$

where Ψ_2 is the DEE-type solution at the straight boundaries and

$$g = \left[s(\alpha^2 + H) - k^2 \right]^{\frac{1}{2}}. \tag{3.4.61}$$

The asymptotic equation follows from (3.4.55), (3.4.59), (3.4.60) and boundary conditions (3.4.44):

$$k\theta = 2k\xi + m\pi, \quad m = 1, 2, \ldots; \tag{3.4.62}$$

$$k\xi = \operatorname{arctg} \left[\frac{uk}{ug + (1 - u)(g^2 + k^2)} \right]. \tag{3.4.63}$$

From (3.4.62) and (3.4.63) as $\lambda \longrightarrow 0$

$$k\theta = m\pi, \quad m = 1, 2, \ldots. \tag{3.4.64}$$

The same result can be obtained directly from (3.4.56), using r as a parameter. This means that expression (3.4.62) becomes more nearly exact as λ increases.

Let us obtain the DEE-type solution at the straight boundaries under the assumption that $k^2 = o(\lambda)$. Substituting expression (3.4.55) into (3.4.49), eliminating $\lambda^2 W_1$ with the aid of (3.4.64), and retaining only the terms of $o(k^4)$, we obtain

$$\left(\frac{\partial^2}{\partial \varphi^2} + k^2 \right) \left(\frac{\partial^2}{\partial \varphi^2} - k^2 - Hr^2 \right) \Psi(\varphi) = 0. \tag{3.4.65}$$

Separating the variables in (3.4.65) by Kantorovich's method, we obtain in the same way as above an equation that agrees superficially with (3.4.63) except that g has the form

$$g = (k^2 + sH)^{\frac{1}{2}}. \tag{3.4.66}$$

The DEE-type solution is expressed by (3.4.60).

If $k^2 = o(\lambda)$, we use the asymptotic representations for the Bessel functions to obtain from (3.4.66)

$$\alpha = \operatorname{arctg} \left(\gamma P - \frac{\beta}{\alpha} \right) + \pi \left(\frac{k}{2} + \frac{1}{4} + n \right);$$

$$\gamma = \alpha + \frac{\beta^2}{\alpha}, \quad n = l, 1, 2, \ldots. \tag{3.4.67}$$

The quantity l can assume the values 0 or 1, depending on the value of p. For example, $l = 0$ if $P = 1$ and $l = 1$ if $P = 0$. We can express β and λ in terms of α: $\beta = (\alpha^2 + H)^{1/2}, \lambda = \alpha\beta$.

We now substitute expression (3.4.67) into (3.4.63) and, assuming the constant to be unknown, it is required that (3.4.61) and (3.4.66) agree, retaining the terms of orders k and k^2 in the expression for α. We then obtain for s

$$s = \left(\frac{8k}{\pi^2}\right)\left(k + 4\mathrm{arctg}\,\frac{\gamma P - \frac{\beta}{\alpha}}{\pi} + 4n + 1\right)^{-1}. \tag{3.4.68}$$

The unknowns k and α are found from the system of transcendental equations (3.4.54), (3.4.51) and are used to express the constants C_i and C_{i1}. The equations in which s is given by (3.4.68) are evidently to be preferred because they simplify the solution greatly without changing the essentials of the asymptotic method. Thus, (3.4.54) and (3.4.51) become independent if $P = 0$. It also becomes unnecessary to evaluate the integrals in (3.4.58).

To obtain an improved expression for $\eta(t)$, we represent the solution of (3.4.43) in the form

$$w(r, \varphi, \theta) = AJ_k(\alpha r)\sin k(\varphi - \xi)\eta(t), \tag{3.4.69}$$

where k and α are defined above.

Applying the Bessel equation, we substitute (3.4.69) into (3.4.43) to obtain

$$\ddot{\eta} + \omega_0^2(1 + \omega\eta^2) = 0, \quad \omega_0 = \frac{x^2}{\sigma}, \quad x = \frac{4H}{3\alpha^2}. \tag{3.4.70}$$

The initial conditions are

$$\eta(0) = 1, \quad \dot{\eta}(0) = 0. \tag{3.4.71}$$

The solution of (3.4.70) that satisfies conditions (3.4.71) is

$$\eta(t) = \mathrm{cn}\,(\sigma_1 t, a) \tag{3.4.72}$$

$$\sigma_1 = \omega_0(1 + x)^{\frac{1}{2}}, \quad a = \left[\frac{x}{2(1 + x)}\right]^{\frac{1}{2}},$$

where the period of the solution $T = 4K(a)/\sigma_1$, $K(a)$ is a complete elliptic integral of the first kind; cn is the Jacobi elliptic cosine function.

The dimensionless circular frequency of the natural vibrations is

$$\omega^* = \frac{2\pi\sigma_1}{T}. \tag{3.4.73}$$

It would be more consistent to use expression (3.4.72) to approximate the time function in finding solutions of DEE-type, but the replacement of the elliptic cosine by the ordinary cosine (the first term of the Fourier-series expansion) makes it possible to simplify the notation considerably in this case without changing the essence of the asymptotic method.

Equation (3.4.70) can be solved approximately by the Bubnov-Galerkin method. The resulting dimensionless frequency agrees with the constant λ determined earlier.

The order of the procedures used to find the DEE can be changed: the solution of (3.4.43) can first be represented in the form (3.4.69), where k, ξ and α are unknown constants; then $\eta(t)$ can be found as a function that depends on the parameters k and α, following the application of the method described above to determine the DEE.

If $u = 0$ and $\theta = 2\pi$, formula (3.4.73) gives asymptotic values of the natural frequencies of the circular plate supported elastically around its contour. For a circular region with hinged boundary, (3.4.43) does not admit of an exact solution (the case $A = 0$ is an exception), in contrast to the case of the similarly restrained rectangular plate.

The frequency calculation can be made more accurate by including terms corresponding to the DEE in the expression for H. Thus, the DEE at a circular boundary is taken into account in the numerical example given below.

Below we present calculated results for the ratio $\omega_N^\star/\omega_L^\star$ for a contour-restrained circular plate; $\omega_N^\star, \omega_L^\star$ are the first natural frequencies of the linear and nonlinear vibrations, respectively, as found by the asymptotic method (the first row) and by the finite element method (the second row) [102]:

$A = 0.2$	$A = 0.4$	$A = 0.6$	$A = 0.8$	$A = 1.0$
1.0074	1.0291	1.0634	1.1083	1.1621
1.0072	1.0284	1.0624	1.1075	1.1619

Tables 3.1 and 3.2 give values of the dimensionless natural frequencies $\omega^\star$ of a contour-restrained circular plate and similarly restrained circular sector, respectively; m_1 is the number of half-waves on the circumference and m_2 is the number along the radius. In either case, the values of the derivative $d\omega^\star/dA$ increase with increasing frequency, and more strongly with the derivative $d\omega^\star/dA$ with m_2 (m_1 fixed). For the sector plates, the analysis indicates that the influence of nonlinearity becomes weaker with decreasing angle θ. Table 3.1 also includes the data from [148] (marked with asterisks), where the natural frequencies were found by an integral-equation method, for plates with the angle $\theta = \pi/2$ and $A = 0$. The comparison indicates satisfactory agreement of the calculated results for the first five frequencies. The largest difference does not exceed 2%.

Table 3.1. Comparison of various approximate solutions

A	$m_1 = 0$ $m_2 = 1$	0 2	0 3	1 1	1 2	1 3	2 1	2 2	2 3
0	10.22	39.77	89,10	21.26	60.83	120.1	34.88	84.58	153.8
0.5	10.67	41.16	91.47	21.73	61.78	121.5	35.57	85.78	155.5
1	11.87	44.98	98.12	23.06	64.52	125.7	37.53	89.22	160.4
1.5	13.56	50.57	108.1	25.02	68.77	132.3	40.48	94.59	168.2
2	15.61	57.39	120.6	27.45	74.20	141.0	44.12	101.5	178.4

Table 3.2. Comparison of various approximate solutions

θ	A	$m_1 = 1$ $m_2 = 2$	1 2	1 3	2 1	2 2	2 3	3 1	3 2	3 3
	0	26.22	67.85	128.6	41.59	94.08	165.4	59.24	122.5	204.3
	0.5	26.65	68.72	129.9	42.29	95.27	167.1	60.18	124.0	206.3
π	1	27.87	71.22	133.8	44.25	98.70	172.0	62.83	128.2	212.1
	1.5	29.69	75.15	140.0	47.20	104.1	179.8	66.85	134.9	221.4
	2	31.95	80.24	148.2	50.86	111.0	190.1	71.85	143.5	233.5
	0	49.18	104.8	178.4	89.57	167.5	262.8	139.2	239.6	356.1
		48.70*	105.1*	–	88.13*	165.3*	–	138.3*	–	–
	0.5	49.82	105.9	180.0	90.66	169.2	265.0	140.7	241.8	359.1
$\frac{\pi}{2}$	1	51.62	109.0	184.4	93.80	173.9	271.2	145.1	248.0	367.0
	1.5	54.34	113.8	191.5	98.61	181.4	281.2	151.8	257.8	379.6
	2	57.72	120.1	200.8	104.7	191.0	294.4	160.4	270.6	396.3
	0	77.95	148.6	236.0	152.7	257.6	378.1	248.0	387.9	541.4
	0.5	78.78	149.8	237.7	154.1	259.6	380.7	250.0	390.6	544.8
$\frac{\pi}{3}$	1	81.14	153.5	242.8	158.3	265.6	388.3	255.8	398.7	554.6
	1.5	84.75	159.4	250.8	164.9	275.1	400.4	265.1	411.4	570.4
	2	89.28	166.9	261.4	173.3	287.5	416.4	277.1	428.3	591.4

3.5 Regular and Singular Asymptotics in the Nonlinear Dynamics of Thin-Walled Structures

3.5.1 Circular Rings and Axisymmetric Cylindrical Shells

Normal modes for nonlinear spatial systems usually cannot be obtained on the basis of full equations. Below we show the use of straightforward asymptotic simplification for constructing simplified nonlinear differential equations (see also [70, 71, 75, 78, 103, 104]).

The equations of free axisymmetric vibrations of an orthotropic cylindrical shell in projections onto the axes of an undeformed coordinate system have the form

$$\frac{\partial}{\partial x}\left(T_1 \cos\overline{\theta} - \frac{\partial M}{\partial x}\sin\overline{\theta}\right) - \rho\frac{\partial^2 u}{\partial t^2} = 0;$$

$$\frac{\partial}{\partial x}\left(T_1 \sin\overline{\theta} - \frac{\partial M}{\partial x}\cos\overline{\theta}\right) + \frac{T_2}{R} + \rho\frac{\partial^2 w}{\partial t^2} + 0. \tag{3.5.1}$$

Here ρ is the mass per unit area.

We will write the geometric and elasticity relations in the form

$$\overline{\varepsilon_1} = \frac{\partial u}{\partial x} + \frac{1}{2}\left(\frac{\partial w}{\partial x}\right)^2; \quad \overline{\varepsilon_2} = \frac{w}{R}; \quad \overline{\theta} = \arcsin\frac{\partial w}{\partial x}; \quad \overline{\kappa} = \frac{\partial\overline{\theta}}{\partial x};$$

$$T_1 = B_1\overline{\varepsilon_1}; \quad T_2 = B_2\overline{\varepsilon_2}; \quad M = D\overline{\kappa}.$$

Here, for the sake of simplicity, it is assumed that the Poisson ratio is equal to zero.

We will introduce the notation $\alpha_1 = D/B_1R^2; \alpha_2 = B_2/B_1; \alpha_3 = H_0/B_1R^2$ (H_0 is the initial energy level) and examine affine transformations of the coordinates leading to various limiting systems.

$$u = \alpha_3 RU; \quad w = \alpha_3 RW; \quad t = \left(\frac{\rho R^4}{D}\right)^1 2\tau.$$

The transformed system has the form

$$\frac{\partial}{\partial \xi}\left(\varepsilon \cos\theta - \alpha_3^{-1}\alpha_1 \frac{\partial \kappa}{\partial \xi}\sin\theta\right) - \alpha_1 \frac{\partial^2 U}{\partial \tau^2} = 0;$$

$$\frac{\partial}{\partial \xi}\left(\varepsilon \sin\theta + \alpha_3^{-1}\alpha_1 \frac{\partial \kappa}{\partial \xi}\cos\theta\right) + \alpha_2 W + \alpha_1 \frac{\partial^2 W}{\partial \tau^2} = 0.$$

Here

$$\varepsilon = \frac{\partial U}{\partial \xi} + \frac{1}{2}\alpha_3\left(\frac{\partial W}{\partial \xi}\right)^2; \quad \theta = \arcsin\left(\alpha_3 \frac{\partial W}{\partial \xi}\right); \quad \kappa = \frac{\partial \theta}{\partial \xi}. \qquad (3.5.2)$$

In the case $\alpha_3 \to 0$, the limiting system describes nonlinear vibrations of a rod. When $\alpha_2 \to 0$, it describes linear axisymmetric vibrations of a cylindrical shell. With $\alpha_1 \to 0, \alpha_2 \to 0$, we arrive at the system

$$\frac{\partial}{\partial \xi}(\varepsilon_0 \cos\theta_0) = 0; \quad \frac{\partial}{\partial \xi}(\varepsilon_0 \sin\theta_0) = 0, \qquad (3.5.3)$$

from which it follows that $\varepsilon_0 = 0$.

We will represent the sought functions with series of the small parameter α_1 and write out the equations of the first approximation:

$$\frac{\partial}{\partial \xi}\left(\varepsilon_0 \cos\theta_0 - \alpha_3^{-1}\frac{\partial \kappa_0}{\partial \xi}\sin\theta_0\right) - \frac{\partial^2 U_0}{\partial \tau^2} = 0;$$

$$\frac{\partial}{\partial \xi}\left(\varepsilon_0 \sin\theta_0 - \alpha_3^{-1}\frac{\partial \kappa_0}{\partial \xi}\cos\theta_0\right) - \frac{\partial^2 W_0}{\partial \tau^2} = 0. \qquad (3.5.4)$$

Excluding the function ε_1 from (3.5.4), we obtain

$$\alpha_3^{-1}\frac{\partial \kappa_0}{\partial \xi}\frac{\partial \theta_0}{\partial \xi} - \frac{\partial^2 U_0}{\partial \tau^2}\cos\theta_0 - \frac{\partial^2 W_0}{\partial \tau^2}\sin\theta_0$$

$$= -\frac{\partial}{\partial \xi}\left[\left(\alpha_3^{-1}\frac{\partial^2 \kappa_0}{\partial \xi^2} + \frac{\partial^2 W_0}{\partial \tau^2}\cos\theta_0 + \frac{\partial^2 U_0}{\partial \tau^2}\sin\theta_0\right)\left(\frac{\partial \theta_0}{\partial \xi}\right)^{-1}\right], \qquad (3.5.5)$$

while from (3.5.3) we find

$$\frac{\partial U_0}{\partial \xi} + \frac{1}{2}\alpha_3\left(\frac{\partial W_0}{\partial \xi}\right)^2 = 0. \qquad (3.5.6)$$

Equations (3.5.5) and (3.5.6) constitute the limiting system corresponding to nonlinear vibrations of a flexible rod with its edges free in the axial direction. This system was obtained by using equations of the first approximation, which is an interesting feature of this case:

$$u = \alpha_1^{1/2}\alpha_3 RU; \quad w = \alpha_3 RW; \quad x = \alpha_1^{1/2}R\xi; \quad t = \left(\frac{\rho R^4}{D}\right)^{\frac{1}{2}}\tau.$$

The transformed system has the form

$$\frac{\partial}{\partial\xi}\left(\varepsilon\cos\theta - \alpha_1^{1/2}\alpha_3^{-1}\frac{\partial\kappa}{\partial\xi}\sin\theta\right) - \alpha_1^{1/2}\frac{\partial^2 U_0}{\partial\tau^2} = 0;$$

$$\frac{\partial}{\partial\xi}\left(\varepsilon\sin\theta + \alpha_1^{1/2}\alpha_3^{-1}\frac{\partial\kappa}{\partial\xi}\cos\theta\right) + \alpha_2 W + \frac{\partial^2 W_0}{\partial\tau^2} = 0. \tag{3.5.7}$$

Here

$$\varepsilon = \frac{\partial U}{\partial\xi} + \frac{\alpha_1^{-1}\alpha_3}{2}\left(\frac{\partial W}{\partial\xi}\right)^2; \quad \kappa = \frac{\partial\theta}{\partial\xi}; \quad \theta = \arcsin\left(\alpha_3\alpha_1^{1/2}\frac{\partial W}{\partial\xi}\right).$$

With $\alpha_1 \to 0, \alpha_3 \sim \alpha 1$, we obtain

$$\frac{\partial\varepsilon_0}{\partial\xi} = 0; \quad \frac{\partial^4 W_0}{\partial\xi^4} + \varepsilon_0\frac{\partial^2 W_0}{\partial\xi^2} + \alpha_2 W_0 + \frac{\partial^2 W_0}{\partial\tau^2} + 0. \tag{3.5.8}$$

The problem is linearized if constant axial forces are applied to the ends, and nonlinear effects can be revealed in the equations of the first approximation.

Let the edge of a shell be fixed so as to prevent displacements in the axial direction. Then, the nonlinear effects in system (3.5.7) are preserved, since it follows from the first equation that

$$\varepsilon_0 = \frac{\alpha_1^{-1}\alpha_3}{2l}\int_0^l\left(\frac{\partial W_0}{\partial\xi}\right)^2 d\xi,$$

where $l = l/R$; l is the shell length.

At the same time, with hinged ends in the second equation, the variables can be separated:

$$w = w_m\sin\frac{m\pi x}{l}\beta_m(\tau), \quad m = 1, 2, \ldots.$$

We obtain for the time function an ordinary differential equation containing a "rigid" nonlinearity which has the following form because of the initial variables

$$\ddot{\beta}_m + \frac{1}{\rho}\left(\frac{B_2}{R^2} + \frac{Dm^4\pi^4}{l^4}\right)\beta_m + \frac{B_1 m^4\pi^4}{4\rho l^4}w_m^2\beta_m^3 = 0. \tag{3.5.9}$$

Each of the functions (3.5.9) is integrated in elliptic functions. For example, with $\beta_m(0) = 1, \beta_m(0) = 0$, we have

$$\beta_m = \text{sp}\left(K_m t, S_m\right), \tag{3.5.10}$$

where

$$K_m = \frac{m^2\pi^2 D^{3/2}}{4l^2\rho^{1/2}}A^{1/2}\left(1 + \frac{B_1 w_m}{A}\right)^2; \quad S_m = \frac{B_1 w_m^2}{2A}\left(1 + \frac{B_1 w_m^2}{A}\right)^{-1};$$

$$A = 4D\left(1 + \frac{B_2 l^4}{DR^2 m^2\pi^4}\right).$$

Thus, the limiting system (3.5.9) has exact solutions corresponding to normal vibration modes. When $\alpha_2 \to 0$, (3.5.10) converts to the familiar equation for a rod with fixed ends:

$$u = \alpha_3 R U; \quad w = \alpha_3 R W; \quad x = R\xi; \quad t = \left(\frac{\rho R^2}{B_1} \right)^{\frac{1}{2}} \tau.$$

The transformed equations of motion have the form

$$\frac{\partial}{\partial \xi} \left(\varepsilon \cos\theta - \alpha_1 \alpha_3^{-1} \frac{\partial \kappa}{\partial \xi} \sin\theta \right) - \frac{\partial^2 U}{\partial \tau^2} = 0;$$

$$\frac{\partial}{\partial \xi} \left(\varepsilon \sin\theta + \alpha_1 \alpha_3^{-1} \frac{\partial \kappa}{\partial \xi} \cos\theta \right) + \alpha_2 W + \frac{\partial^2 W}{\partial \tau^2} = 0,$$

where ε, x, θ are defined by (3.5.2).

With $\alpha_1 \to 0$ we arrive at the limiting system

$$\frac{\partial}{\partial \xi} (\varepsilon \cos\theta) - \frac{\partial^2 U}{\partial \tau^2} = 0; \quad \frac{\partial}{\partial \xi} (\varepsilon \sin\theta) + \alpha_2 W + \frac{\partial^2 W}{\partial \tau^2} = 0, \qquad (3.5.11)$$

corresponding to vibrational tension. However, by virtue of the condition $\varepsilon \ll 1$, system (3.5.11) turns out to be consistent only in the quasilinear case $(\alpha_3 \to 0, \alpha_3 \sim \alpha_1)$.

Let us examine plane nonlinear vibrations of a circular ring (a cylindrical shell of infinite length). We will write the equations of motion in a local coordinate system, the axes of which correspond to the tangential and radial directions at a point on the undeformed axis of the ring:

$$\frac{\partial T}{\partial y} - \frac{1}{R} \frac{\partial \overline{M}}{\partial y} - \rho \frac{\partial^2 v}{\partial t^2} = 0; \quad \frac{\partial^2 \overline{M}}{\partial y^2} + \frac{1}{R} \overline{T} + \frac{\partial^2 w}{\partial t^2} = 0, \qquad (3.5.12)$$

where

$$\overline{T} = T \cos\overline{\theta} - \frac{\partial M}{\partial y} \sin\overline{\theta}; \quad \frac{\partial \overline{M}}{\partial y} = T \sin\overline{\theta} + \frac{\partial M}{\partial y} \cos\overline{\theta};$$

and ρ is the running mass.

We will write the physical and geometric relations in the form

$$T = EF\overline{\varepsilon}; \quad M = E\sqrt{x}; \quad \overline{\kappa} = \frac{\partial \overline{\theta}}{\partial y}; \quad \overline{\varepsilon} = \frac{\partial v}{\partial y} + \frac{w}{R} + \frac{1}{2} \left(\frac{\partial w}{\partial y} - \frac{v}{R} \right)^2;$$

$$\theta = \arcsin \left(\frac{\partial w}{\partial y} - \frac{v}{R} \right).$$

Let $\alpha_1 = J/FR^2$; $\alpha_3 = H_0/EFR$. Subjecting the variables v, w, y, and t to affine transformations, we obtain the following classes of nonequivalent systems:

$$v = \alpha_3 R V; \quad w = \alpha_3 R W; \quad y = R\eta; \quad t = \left(\frac{\rho R^4}{EJ} \right)^{\frac{1}{2}} \tau.$$

The transformed equations have the form

$$\frac{\partial}{\partial \eta}\left(\varepsilon \cos\theta - \alpha_1 \alpha_3^{-1}\frac{\partial \kappa}{\partial \eta}\sin\theta\right) - \varepsilon\sin\theta - \alpha_1\alpha_3^{-1}\frac{\partial \kappa}{\partial \eta}\cos\theta - \alpha_1\frac{\partial^2 V}{\partial \tau^2} = 0;$$

$$\frac{\partial}{\partial \eta}\left(\varepsilon \sin\theta + \alpha_1 \alpha_3^{-1}\frac{\partial \kappa}{\partial \eta}\cos\theta\right) + \varepsilon\cos\theta - \alpha_1\alpha_3^{-1}\frac{\partial \kappa}{\partial \eta}\sin\theta + \alpha_1\frac{\partial^2 W}{\partial \tau^2} = 0.$$

Here

$$\varepsilon = \frac{\partial V}{\partial \tau} + W + \frac{1}{2}\alpha_3\left(\frac{\partial W}{\partial \eta} - V\right)^2; \quad \kappa = \frac{\partial \theta}{\partial \eta};$$

$$\theta = \arcsin\left(\frac{\partial W}{\partial \eta} - V\right). \tag{3.5.13}$$

With $\alpha_3 \to 0$, we obtain the limiting system corresponding to the linear theory. The second limiting system is obtained as $\alpha_1 \to 0$:

$$\frac{\partial}{\partial \eta}(\varepsilon_0 \cos\theta) - \varepsilon_0 \sin\theta = 0; \quad \frac{\partial}{\partial \eta}(\varepsilon_0 \sin\theta) - \varepsilon_0 \cos\theta = 0. \tag{3.5.14}$$

The condition of nontensionability follows from (3.5.14):

$$\frac{\partial V_0}{\partial \eta} + W_0 + \frac{1}{2}\alpha_3\left(\frac{\partial W_0}{\partial \eta} - V_0\right)^2 = 0. \tag{3.5.15}$$

We will write out the equations of the first approximation:

$$\frac{\partial}{\partial \eta}\left(\varepsilon_1 \cos\theta_0 - \alpha_3^{-1}\frac{\partial \kappa_0}{\partial \eta}\sin\theta_0\right) - \varepsilon - 1\sin\theta_0$$

$$-\alpha_3^{-1}\frac{\partial \kappa_0}{\partial \eta}\cos\theta_0 - \frac{\partial^2 V_0}{\partial \tau^2} = 0;$$

$$\frac{\partial}{\partial \eta}\left(\varepsilon_1 \sin\theta_0 + \alpha_3^{-1}\frac{\partial \kappa_0}{\partial \eta}\cos\theta_0\right) + \varepsilon_1 \cos\theta - 0$$

$$-\alpha_3^{-1}\frac{\partial \kappa_0}{\partial \eta}\sin\theta_0 + \frac{\partial^2 W_0}{\partial \tau^2} = 0. \tag{3.5.16}$$

After making the obvious transformations, we may exclude the functions ε_1 from (3.5.16). Considering condition (3.5.15), we obtain the following relations connecting the functions V_0 and W_0:

$$\frac{\partial V_0}{\partial \eta} + W_0 + \frac{1}{2}\alpha_3\left(\frac{\partial W_0}{\partial \eta} - V_0\right)^2 = 0;$$

$$\alpha_3^{-1}\frac{\partial \kappa_0}{\partial \eta}\left(1 + \frac{\partial \theta_0}{\partial \eta}\right) + \frac{\partial^2 V_0}{\partial \tau^2}\cos\theta_0 - \frac{\partial^2 W_0}{\partial \tau^2}\sin\theta_0 \tag{3.5.17}$$

$$= -\frac{\partial}{\partial \eta}\left[\left(\alpha_3^{-1}\frac{\partial^2 \kappa_0}{\partial \eta^2} + \frac{\partial^2 V_0}{\partial \eta^2}\sin\theta_0 + \frac{\partial^2 W_0}{\partial \eta^2}\cos\theta_0\right)\left(1 + \frac{\partial \theta_0}{\partial \eta}\right)^{-1}\right].$$

The nonlinear system (3.5.17) is fairly complex, so it is expedient to take the following for the limiting linearized system ($\alpha_1 \to 0$, $\alpha_3 \to 0$)

$$\frac{\partial V_0}{\partial \eta} + W_0 = 0; \quad \frac{\partial^6 V_0}{\partial \eta^6} + 2\frac{\partial^4 V_0}{\partial \eta^4} + \frac{\partial^2 V_0}{\partial \eta^2} - \frac{\partial^2 V_0}{\partial \tau^2} + \frac{\partial^4 V_0}{\partial \tau^2 \partial \eta^2} = 0, \quad (3.5.18)$$

and to evaluate the nonlinear effects by using the equations of the first approximation ($W = W_0 = \alpha_3 W_1 = ...; V = V_0 = \alpha_3 V_1...$)

$$\frac{\partial V_1}{\partial \eta} + W_1 = -\frac{1}{2}\left(\frac{\partial W_0}{\partial \eta} - V_0\right)^2;$$

$$\frac{\partial^6 V_1}{\partial \eta^6} + 2\frac{\partial^4 V_1}{\partial \eta^4} + \frac{\partial^2 V_1}{\partial \eta^2} - \frac{\partial^2 V_1}{\partial \tau^2} + \frac{\partial^4 V_1}{\partial \tau^2 \partial \eta^2} \qquad (3.5.19)$$

$$= -\frac{\partial^2 W_0}{\partial \tau^2}\theta_0 + \frac{\partial \kappa_0}{\partial \eta}\frac{\partial \theta_0}{\partial \eta} - \frac{\partial}{\partial \eta}\left(\frac{\partial^2 \kappa_0}{\partial \eta^2}\frac{\partial \theta_0}{\partial \eta} + \frac{\partial^2 W_0}{\partial \tau^2}\frac{\partial \theta_0}{\partial \eta} - \frac{\partial^2 V_0}{\partial \tau^2}\theta_0\right).$$

The n-th mode of free vibrations is a particular solution of system (3.5.19):

$$W_{0,n} = w_{0,n}\cos n\eta e^{iw_n\tau}; \quad V_{0,n} = \frac{1}{n}w_{0,n}\sin n\eta e^{iw_n\tau}; \quad w_n = \frac{n^2(n^2-1)^2}{n+1}.$$

Then, the solution of system (3.5.19) satisfying the periodicity conditions has the form

$$V_{1,n} = \frac{B}{2n}w_{0,n}^2\sin 2n\eta e^{2iw_n\tau};$$

$$W_{1,n} = \left[\frac{1}{4}\left(n - \frac{1}{n}\right)^2 - B\right]w_{0,n}^2\cos 2n\eta e^{2iw_n\tau} \qquad (3.5.20)$$

$$- \frac{1}{4}w_{0,n}^2\left(n - \frac{1}{4}\right)e^{2iw_n\tau},$$

where

$$B = \frac{1}{2}\left[\frac{(n^2+1)(4n^2-1)}{(n^2-1)^2} - \frac{4n^2+1}{4n^2-1}\right]^{-1}.$$

Thus, in the case of nonlinear vibrations, the n-th harmonic with respect to the coordinate η is accompanied by a zero harmonic

$$w = \alpha_3 RW; \quad v = \alpha_1^{1/2}\alpha_3 V; \quad y = \alpha_1^{1/2}R\eta; \quad t = \left(\frac{\rho R^4}{EJ}\right)^{\frac{1}{2}}\tau.$$

The limiting ($\alpha_1 \to 0, \alpha_3 \sim \alpha_1$) system has the form

$$\frac{\partial V}{\partial \eta} + W + \frac{\alpha_1^{-1}\alpha_3}{2}\left(\frac{\partial W}{\partial \eta}\right)^2 = 0; \quad \frac{\partial^2 W}{\partial \eta^2} + \frac{\partial^2 W}{\partial \tau^2} = 0. \qquad (3.5.21)$$

The second equation of (3.5.21) is linear, and the nonlinearity of the problem is determined by the first equation of (3.5.21). With certain additive terms, this system was used earlier to investigate nonlinear vibrations of a ring [70, 71, 78]. It turns out that these additive terms play a minor role in investigating the given type of vibrations and are of the same order as the discarded terms.

The particular solutions of system (3.5.21), which satisfy the periodicity conditions, have the form

$$
W_n = \left\{ \begin{array}{c} w_{10,n}\cos n\eta \\ w_{20,n}\sin n\eta \end{array} \right\} e^{iw_n\tau} - \frac{n^2}{4}\alpha_1^{-1}\alpha_3 \left\{ \begin{array}{c} w_{10,n} \\ w_{20,n} \end{array} \right\} e^{2iw_n\tau}
$$

$$
V_n = \frac{1}{n}\left\{ \begin{array}{c} -w_{10,n}\sin n\eta \\ w_{20,n}\cos n\eta \end{array} \right\} e^{iw_n\tau} + \frac{n}{8}\alpha_1^{-1}\alpha_3 \left\{ \begin{array}{c} w_{10,n}^2\sin 2n\eta \\ -w_{20,n}^2\sin 2n\eta \end{array} \right\} e^{2iw_n\tau}.
$$

The axisymmetric component of the radial displacement coincides with the nonlinear correction for W_0 determined from (3.5.21):

$$
w = \alpha_3 RW; \quad v = \alpha_3 RV; \quad y = R\eta; \quad t = \left(\frac{\rho R^2}{EF}\right)^{\frac{1}{2}}\tau.
$$

With $\alpha_1 \to 0$, this transformation makes it possible to obtain the following limiting system:

$$
\frac{\partial(\varepsilon\cos\theta)}{\partial\eta} - \varepsilon\sin\theta - \frac{\partial^2 V}{\partial\tau^2} = 0; \quad \frac{\partial(\varepsilon\sin\theta)}{\partial\eta} - \varepsilon\cos\theta - \frac{\partial^2 W}{\partial\tau^2} = 0,
$$

where ε, κ, and θ are defined by (3.5.13).

Similarly to the case of axisymmetric vibrations, this system is consistent at low amplitudes ($\alpha_3 \to 0$, $\alpha_1 \sim \alpha_3$). Nonlinear effects may be revealed while analysing higher approximations.

3.5.2 Reinforced and Isotropic Cylindrical Shells

Asymptotic methods are extensively used in the theory of thin shells. The asymptotic analysis of the basic equations of the theory of isotropic cylindrical shells has been carried out in, for example [65, 68, 69, 82, 124, 131, 132, 138, 29d]. The results of these studies have been extended to the dynamic case [83, 136], to nonlinear shells [69, 157] and to orthotropic shells [2]. Finally, it is shown in [56d] that in the case of a structurally orthotropic shell the presence of a large number of geometrical rigidity parameters leads to additional possibilities of asymptotic integration even though it complicates the analysis. As a result a number of new approximate equations have been obtained for the main types of reinforced shells, which have no analogy in the isotropic case.

In the present book simplified boundary value problems are formulated for new types of approximate equations of the theory of nonlinear dynamics of eccentrically reinforced cylindrical shells. It is worth noting that ribbed shells were investigated in many papers [49, 162, 1d] etc.

The nonlinear dynamic boundary value problems of the theory of closed circular cylindrical shells eccentrically reinforced in the two principal directions are investigated within the framework of the structurally orthotropic scheme. The middle surface of the shell is chosen as the main one. Detailed discussions of the basic relations of the linear theory of shells can be found in many monographs and papers [51d, 55d] and therefore we discuss only the final results here.

The governing equations of motion are written in the form proposed by Sanders [140] (in comparison with the original equations from [140], some dynamical terms are added):

$$\frac{\partial N_{11}}{\partial x_1} + \frac{\partial N_{12}}{\partial x_2} + \frac{1}{2R}\frac{\partial M_{12}}{\partial x_2} - \frac{1}{2}\frac{\partial}{\partial x_2}[\Phi(N_{11}+N_{22})] - \rho R^2\frac{\partial^2 u_1}{\partial t^2} = 0,$$

$$\frac{\partial N_{12}}{\partial x_1} + \frac{\partial N_{22}}{\partial x_2} - Q_2 - \frac{1}{2R}\frac{\partial M_{12}}{\partial x_1} + (\Phi_1 N_{12} + \Phi_2 N_{22})$$

$$+ \frac{1}{2}\frac{\partial}{\partial x_1}[\Phi(N_{11}+N_{22})] - \rho R^2\frac{\partial^2 u_2}{\partial t^2} = 0,$$

$$\frac{\partial Q_1}{\partial x_1} + \frac{\partial Q_2}{\partial x_2} + N_{22} - \frac{\partial}{\partial x_1}(\Phi_1 N_{11} + \Phi_2 N_{12}) \qquad (3.5.22)$$

$$- \frac{\partial}{\partial x_2}(\Phi_1 N_{12} + \Phi_2 N_{22}) - \rho R^2\frac{\partial^2 w}{\partial t^2} = 0,$$

$$\frac{\partial M_{11}}{\partial x_1} + \frac{\partial M_{12}}{\partial x_2} - RQ_1 = 0, \qquad \frac{\partial M_{12}}{\partial x_1} + \frac{\partial M_{22}}{\partial x_2} - RQ_2 = 0.$$

Here N_{11}, N_{22}, N_{12} are the membrane stresses; M_{11}, M_{22}, M_{12} are the bending and torsion moments; Q_1, Q_2 are shearing forces;

$$\Phi_1 = -\frac{1}{R}\frac{\partial w}{\partial x_1}, \quad \Phi_2 = -\frac{1}{R}\left(\frac{\partial w}{\partial x_2} - u_2\right), \quad \Phi = \frac{1}{2R}\left(\frac{\partial u_2}{\partial x_1} - \frac{\partial u_1}{\partial x_2}\right),$$

u_1, u_2, w are the tangential and normal displacements; $\rho = \rho_0 + \frac{\rho_1}{l_1} + \frac{\rho_2}{l_2}$; ρ_0, $\rho_1(\rho_2)$ are the density of the shell material, and the stringer (ring) material.

It is worth noting that Sanders [140] defined the variant of "moderately small rotation" by setting restrictions on the components of the linearized rotation vector to the effect that the squares of these components can be at most of the order of magnitude of the strains.

The components of the elasticity tensor are defined by [114]

$$N_{11} = B_{11}\varepsilon_{11} + B_{12}\varepsilon_{22} + K_{11}\kappa_{11},$$

$$N_{22} = B_{21}\varepsilon_{11} + B_{22}\varepsilon_{22} + K_{22}\kappa_{22}, \quad N_{12} = B_{33}\varepsilon_{12}, \qquad (3.5.23)$$

$$M_{11} = D_{11}\kappa_{11} + D_{12}\kappa_{22} + K_{11}\varepsilon_{11},$$

$$M_{22} = D_{21}\kappa_{11} + D_{22}\kappa_{22} + K_{22}\varepsilon_{22}, \quad M_{12} = D_{33}\kappa_{12},$$

$$B_{11} = B + \frac{E_s F_s}{l_1}; \quad B_{22} = B + \frac{E_r F_r}{l_2}; \quad B = \frac{Eh}{1-\nu^2}; \quad B_{33} = \frac{Eh}{1-\nu};$$

$$B_{21} = B_{12} = \nu_{21}B_{11} = \nu_{12}B_{22};$$

$$K_{11} = \frac{E_s S_s}{l_1}; \quad K_{22} = \frac{E_r S_r}{l_2}; \quad D_{11} = D + \frac{E_s J_s}{l_1}; \quad D_{22} = D + \frac{E_r J_r}{l_2};$$

$$D = \frac{Eh^3}{12(1-\nu^2)}; \quad D_{33} = \frac{D}{2} + \frac{E_s J_{ks}}{l_1} + \frac{E_r J_{kr}}{l_2};$$

$$D_{21} = D_{12} = \nu_{41}D_{11} = \nu_{14}D_{22} = D;$$

$F_s(F_r)$, $J_s(J_r)$, $J_{ks}(J_{kr})$, $S_s(S_r)$ are the transvese section area, inertia moment, rotation inertia moment, torsion inertia moment and static moment of the stringer (rib); E, $E_s(E_r)$ are the elasticity modulus of the shell material.

The following geometrical relations are accepted here [140]:

$$\varepsilon_{11} = \frac{1}{R}\frac{\partial u_1}{\partial x_1} + \frac{1}{2}\Phi_1^2 + \Phi^2; \quad \varepsilon_{22} = \frac{1}{R}\left(\frac{\partial u_2}{\partial x_2} - w\right) + \frac{1}{2}\Phi_2^2 + \Phi^2;$$

$$\varepsilon_{12} = \frac{1}{2R}\left(\frac{\partial u_2}{\partial x_1} + \frac{\partial u_1}{\partial x_2}\right) + \frac{1}{2}\Phi_1\Phi_2; \quad \kappa_{11} = \frac{1}{R}\frac{\partial \Phi_1}{\partial x_1}; \qquad (3.5.24)$$

$$\kappa_{22} = \frac{1}{R}\frac{\partial \Phi_2}{\partial x_2}; \quad \kappa_{12} = \frac{1}{2R}\left(\frac{\partial \Phi_2}{\partial x_1} + \frac{\partial \Phi_1}{\partial x_2} - \Phi\right).$$

We assume that at the end faces of the shell ($x_1 = 0.1$, $e = L/R$, L is the length of the shell) the following are specified:

(1) $u_2 = 0$; (2) $u_1 = 0$; (3) $w = 0$; (4) $\dfrac{\partial w}{x_1} \equiv w_{x1} = 0$;

(5) $N_{12} = 0$; (6) $N_{11} = 0$; (7) $Q_1 = 0$; (8) $M_{11} = 0$. (3.5.25)

Below we shall denote any variant of the boundary conditions by the symbol G_{ij}^{mn}, where the set of indices corresponds to the numbers of the specified boundary conditions.

Let us introduce the small parameter

$$\varepsilon_1 = \left(\frac{D_1}{B_2 R^2}\right)^{\frac{1}{2}}.$$

This parameter will be used for the estimation of orders of magnitude of various terms and parameters.

Let us also introduce the dimensionless geometrical rigidity parameters

$$\varepsilon_2 = \frac{D_1}{D_2}; \quad \varepsilon_3 = \frac{D_3}{D_1}; \quad \varepsilon_4 = \frac{B_2}{B_1};$$

$$\varepsilon_5 = \frac{B_3}{B_1}; \quad \varepsilon_6 = \frac{K_{11}}{B_1 R}; \quad \varepsilon_7 = \frac{K_{22}}{B_2 R};$$

$$B_1 = B_{11}(1 - \nu_{12}\nu_{21}); \quad B_2 = B_{22}(1 - \nu_{12}\nu_{21});$$

$$\frac{1}{B_3} = \frac{1}{B_{33}} - \frac{B_{21}}{B_{11}B_{22}(1 - \nu_{12}\nu_{21})};$$

$$D_1 = D_{11} - \frac{K_{11}^2}{B_1}; \quad D_2 = D_{22} - \frac{K_{22}^2}{B_2};$$

$$D_3 = D_{12} + D_{33} + \frac{K_{11}K_{22}}{B_1 B_2}B_{12}(1 - \nu_{12}\nu_{21}).$$

Depending on the assumed estimates of the value of these parameters, three types of reinforced shells can be distinguished:

- stringer shells (SS) $\varepsilon_1 \ll 1$, $\varepsilon_2 \sim \varepsilon_1^2$, $\varepsilon_3 \sim \varepsilon_1$, $\varepsilon_4 \sim \varepsilon_5 < 1$, $\varepsilon_6 \sim \varepsilon_1$, $\varepsilon_7 \cong 0$;
- ring stiffened shells (RS) $\varepsilon_1 \ll 1$, $\varepsilon_2 \sim \varepsilon_1^{-1}$, $\varepsilon_3 \sim \varepsilon_1$, $\varepsilon_4 < 1$, $\varepsilon_6 \cong 0$, $\varepsilon_7 \sim \varepsilon_1^{1/2}$;
- "wafer" shells (WS) $\varepsilon_1 \ll 1$, $\varepsilon_2 \sim 1$, $\varepsilon_3 < 1$, $\varepsilon_4 \sim 1$, $\varepsilon_5 < \varepsilon_6 \sim \varepsilon_7 \sim \varepsilon_1$.

Derivatives of various components are estimated in the following manner

$$\frac{\partial}{\partial x_1} \sim \varepsilon_1^{-\alpha_1}; \quad \frac{\partial}{\partial x_2} \sim \varepsilon_1^{-\alpha_2}; \quad \frac{\partial}{\partial t} \sim \varepsilon_1^{-\alpha_3}.$$

We also introduce the parameters α_k $(k = 4 - 6)$ through the relations

$$\frac{w}{R} \sim \varepsilon_1^{\alpha_4}; \quad u_1 \sim \varepsilon_1^{\alpha_5} w; \quad u_2 \sim \varepsilon_1^{\alpha_6} w.$$

It is known that two-dimensional equations of the theory of shells are valid provided that the following estimates are satisfied

$$0 < \alpha_k < 1, \quad k = 1, 2, 3.$$

Now let us briefly describe the asymptotic procedure (for details we refer readers to [83, 29d]).

We pose an expansion for any components of the desired stress-strain shell state U

$$U = U_0 + \varepsilon_1 U_1 + \varepsilon_1 U_1 + \dots \tag{3.5.26}$$

Substituting ansatz (3.5.26) into the governing boundary value problems and comparing the coefficient of ε_1^k, we conclude that the limiting $(\varepsilon_1 \to 0)$ systems are strongly dependent on the value of the parameters α_k. Now we must scan all possible α_k and search all sensible values of these parameters, for which the limiting systems make mathematical (well-posed) and physical sense. It is remarkable that as a result of this routine but very laborious procedure we obtain only a few limiting systems which are analysed below.

Now let us consider the possible simplifications of the general relations of the reinforced shells, which result from the previous assumptions. As a result of the asymptotic procedure, one obtains the following limiting systems

(1) $\alpha_2^* < \frac{1}{2}$, $\alpha_1^* = \alpha_2^*$, $\alpha_3^* = 0$, $\alpha_4^* = 2\alpha_2^*$, $\alpha_5^* = \alpha_6^* = \alpha_2^*$.
Here for WS and SS shells $\alpha_k^* = \alpha_k$ and for RS

$$\alpha_1^* = 2\alpha_1, \quad \alpha_2^* = 2\alpha_2, \quad \alpha_3^* = \alpha_3, \quad \alpha_4^* = 2\alpha_4, \quad \alpha_5^* = \alpha_5, \quad \alpha_6^* = \alpha_6.$$

The corresponding limiting system is

$$\frac{\partial N_{11}}{\partial x_1} + \frac{\partial N_{12}}{\partial x_2} - \left\{ \frac{1}{2} \frac{\partial}{\partial x_2} \left[\Phi(N_{11} + N_{22}) \right] \right\} = 0,$$

$$\frac{\partial N_{12}}{\partial x_1} + \frac{\partial N_{22}}{\partial x_2} + \frac{1}{2} \left\{ \frac{\partial}{\partial x_1} \left[\Phi(N_{11} + N_{22}) \right] + \Phi_1 N_{12} + \Phi_2 N_{22} \right\} = 0, \tag{3.5.27}$$

$$N_{22} - \frac{\partial}{\partial x_1} (\Phi_1 N_{11} + \Phi_2 N_{12}) - \frac{\partial}{\partial x_2} (\Phi_1 N_{12} + \Phi_2 N_{22}) - \rho R^2 \frac{\partial^2 w}{\partial t^2} = 0$$

and

$$N_{11} = B_{11}\varepsilon_{11} + B_{12}\varepsilon_{22}, \quad N_{22} = B_{21}\varepsilon_{11} + B_{22}\varepsilon_{22}, \quad N_{12} = B_{33}\varepsilon_{12};$$

$$\varepsilon_{11} = \frac{1}{R}\frac{\partial u_1}{\partial x_1} + \frac{1}{2}\Phi_1^2; \quad \varepsilon_{22} = \frac{1}{R}\left(\frac{\partial u_2}{\partial x_2} - w\right) + \frac{1}{2}\Phi_2^2;$$

$$\varepsilon_{12} = \frac{1}{2R}\left(\frac{\partial u_2}{\partial x_1} + \frac{\partial u_1}{\partial x_2}\right) + \frac{1}{2}\Phi_1\Phi_2; \tag{3.5.28}$$

$$\Phi_1 = -\frac{1}{R}\frac{\partial w}{\partial x_1}, \quad \Phi_2 = -\frac{1}{R}\frac{\partial w}{\partial x_2} + \left(\frac{u_2}{R}\right), \quad \Phi = \frac{1}{2R}\left(\frac{\partial u_2}{\partial x_1} + \frac{\partial u_1}{\partial x_2}\right).$$

Equations (3.5.27) describe the nonlinear membrane motion.

(2) $\alpha_2^* < \frac{1}{2}$, $\alpha_1^* = -\frac{1}{2} + 2\alpha_2^*$, $\alpha_3^* = -1 + 2\alpha_2^*$, $\alpha_4^* = 2\alpha_2^*$, $\alpha_5^* = \frac{1}{2}$,
$\alpha_6^* = \alpha_2^* (\alpha_k^* = \alpha_k^*$ for WS; $\alpha_1^* = \frac{1}{2} + \alpha_1$, $\alpha_2^* = \alpha_2$, $\alpha_3^* = -1 + \alpha_3$,
$\alpha_4^* = \alpha_4$, $\alpha_5^* = \alpha_5$, $\alpha_6^* = \alpha_6$ for SS; $\alpha_1^* = -\frac{1}{4} + 2\alpha_2 + \alpha_1$,
$\alpha_2^* = 2\alpha_2$, $\alpha_3^* = -\frac{1}{2} + 2\alpha_2 + \alpha_3, \alpha_k^* = 2\alpha_k$, $k = 4, 5, 6$ for RS).

$$\frac{\partial N_{11}}{\partial x_1} + \frac{\partial N_{12}}{\partial x_2} - \left[\frac{1}{2}\frac{\partial}{\partial x_2}(\Phi_1 N_{11})\right] = 0, \tag{3.5.29}$$

$$\frac{\partial N_{12}}{\partial x_1} + \frac{\partial N_{22}}{\partial x_2} - \left[\frac{1}{R}\frac{\partial M_{22}}{\partial x_2} - \frac{1}{2}\frac{\partial}{\partial x_1}(\Phi N_{11}) + \Phi_1 N_{12} + \Phi_2 N_{22}\right] = 0,$$

$$\frac{\partial^2 M_{22}}{\partial x_2^2} + R N_{22} - R\frac{\partial}{\partial x_1}(\Phi_1 N_{11} + \Phi_2 N_{12}) +$$

$$-R\frac{\partial}{\partial x_2}(\Phi_1 N_{12} + \Phi_2 N_{22}) - \rho R^2\frac{\partial^2 w}{\partial t^2} = 0,$$

$$N_{11} = B_{11}\varepsilon_{11} + B_{12}\varepsilon_{22}, \quad 0 = B_{21}\varepsilon_{11} + B_{22}\varepsilon_{22} + K_{22}\kappa_{22},$$

$$N_{12} = B_{33}\varepsilon_{12}, \tag{3.5.30}$$

$$M_{22} = D_{22}\kappa_{22} + K_{22}\varepsilon_{22}, \quad \varepsilon_{11} = \frac{1}{R}\frac{\partial u_1}{\partial x_1} + \frac{1}{2}\Phi_1^2;$$

$$0 = \frac{1}{R}\left(\frac{\partial u_2}{\partial x_2} - w\right) + \frac{1}{2}\Phi_2^2; \quad 0 = \frac{1}{2R}\left(\frac{\partial u_2}{\partial x_1} + \frac{\partial u_1}{\partial x_2}\right) + \frac{1}{2}\Phi_1\Phi_2;$$

$$\Phi_1 = -\frac{1}{R}\frac{\partial w}{\partial x_1}, \quad \Phi_2 = -\frac{1}{R}\frac{\partial w}{\partial x_2} - \frac{u_2}{R}, \quad \Phi = \frac{1}{2R}\left(\frac{\partial u_2}{\partial x_1} + \frac{\partial u_1}{\partial x_2}\right).$$

Equations (3.5.29)–(3.5.30) are the nonlinear quasimembrane shell motion equations (only bending moments in the circumferential direction are considered) without tangential inertia. These equations may be obtained as a result of the reduction of general relations if it is assumed that the following relations are satisfied

$$\varepsilon_{22} = 0, \quad \varepsilon_{12} = 0.$$

These relations denote physical conditions of extension in the circumferential direction and the absence of shear in the middle shell surface.

For $\alpha_k^* > 0$ it is possible to omit the term in brackets in relations (3.5.27)–(3.5.30). The equations of membrane and quasimembrane vibrations are of fourth order with respect to the axial coordinate x_1, and they can be satisfied by two boundary conditions on every shell edge only while integrating the corresponding limiting systems. From the point of view of singular perturbation theory we deal with outer solutions and we must construct inner solutions (boundary layers) [83, 120, 122, 29d].

The boundary layer solution has a large variability index in the x_1 direction, and its variability in the circumferential direction and in time is the same as with the inner solution. Now let us present all the stress–strain state components U as follows

$$U = U^{(0)} + U^{(k)}, \tag{3.5.31}$$

where the indexes $^{(0)}$ and $^{(k)}$ indicate the components of the outer solution and the boundary layer respectively.

It is necessary also to introduce the parameter ν characterizing the order of $w^{(0)}$ with respect to $w^{(k)}$:

$$w^{(k)} \sim \varepsilon_1^\nu w^{(0)}.$$

The value of the parameter ν and the boundary layer variability in the x_1 direction depend on the boundary conditions and are defined in an asymptotic, splitting process. As an illustration of the method used we consider the boundary conditions for the variant G_{45}^{23}. First of all, we write asymptotic orders of the components of the boundary conditions for the inner solutions

$$u_1^{(0)} \sim \varepsilon_1^{1/2} w^{(0)}; \quad \frac{\partial w^{(0)}}{\partial x_1} \sim \varepsilon_1^{1/2 - 2\alpha_2^*} w^{(0)}; \quad N_{12}^{(0)} \sim \varepsilon_1^{3/2 - \alpha_2^*} w^{(0)};$$

$$w^{(0)} \sim \varepsilon_1^{2\alpha_2^*} R; \tag{3.5.32}$$

$$u_1^{(k)} \sim \varepsilon_1^{1/2} w^{(k)} \sim \varepsilon_1^{1/2 + \nu} w^{(0)}; \quad \frac{\partial w^{(k)}}{\partial x_1} \sim \varepsilon_1^{-1/2} w^{(k)} \sim \varepsilon_1^{-1/2 + \nu} w^{(0)};$$

$$N_{12}^{(k)} \sim \varepsilon_1^{1/2 - \alpha_2^*} w^{(k)} \sim \varepsilon_1^{1/2 - \alpha_2^* + \nu} w^{(0)}; \quad w^{(k)} \sim \varepsilon_1^\nu w^{(0)}.$$

We choose the value of the parameter ν from the condition of the absence of a contradiction in the limiting boundary value problems (in other words, the number of the boundary conditions for the limiting system must coincide with the order of the differential equation with respect to x_1). In the case under consideration the unique possible value of ν is

$$\nu = 1 - 2\alpha_2^* > 0. \tag{3.5.33}$$

Let us emphasize that the boundary layer nonlinearity order estimation results immediately from (3.5.33)

$$w^{(k)} \sim \varepsilon_1 R.$$

Estimating (3.5.32) and taking into account (3.5.33), we find

$$w^{(0)} \sim \varepsilon_1^{-(1-2\alpha_2^*)} w^{(k)} \gg w^{(k)},$$

$$u_1^{(0)} \sim \varepsilon_1^{-(1-2\alpha_2^*)} u_1^{(k)} \gg u_1^{(k)},$$

$$\frac{\partial w^{(0)}}{\partial x_1} \sim \frac{\partial w^{(k)}}{\partial x_1}; \quad N_{12}^{(0)} \sim N_{12}^{(k)}.$$

Then, the splitting of the boundary conditions may be represented in the following form

$$w^{(0)}|_{x_1=0,l} = u_1^{(0)}|_{x_1=0,l} = 0; \tag{3.5.34}$$

$$\frac{\partial w^{(k)}}{\partial x_1}\bigg|_{x_1=0,l} = -\frac{\partial w^{(0)}}{\partial x_1}\bigg|_{x_1=0,l}; \quad N_{12}^{(k)}|_{x_1=0,l} = -N_{12}^{(0)}|_{x_1=0,l}. \tag{3.5.35}$$

Therefore, for the outer and inner solutions, boundary conditions (3.5.34) and (3.5.35) must be, respectively, given.

In the same way other boundary conditions are split. The results are presented in Table 3.3.

To obtain the boundary layer equations, let us take (3.5.31) into the initial equations and take into account the outer solution equations (3.5.33)–(3.5.34).

Table 3.3. Splitting of the boundary conditions for the half-membrane state and boundary effect

G	Splitting boundary conditions
G_{34}^{12}	$w^{(0)},\ u^{(1)},\ w_{x_1}^{(k)},$ $\frac{\partial u_2^{(k)}}{\partial x_2} - w^{(k)} + \frac{1}{2R}\left(\frac{\partial w^{(k)}}{\partial x_2}\right)^2 \equiv L_1^{(k)}$
G_{12}^{38}	$w^{(0)},\ w_{x_1}^{(0)},\ L_1^{(k)},\ M_{11}^{(0)} + M_{11}^{(k)}$
G_{46}^{13}	$w^{(0)},\ N_{11}^{(0)},\ L_1^{(k)},\ w_{x_1}^{(0)} + w_{x_1}^{(k)}$
G_{68}^{13}	$w^{(0)},\ N_{11}^{(0)},\ L_1^{(k)},\ M_{11}^{(0)} + M_{11}^{(k)}$
G_{45}^{23}	$w^{(0)},\ u_1^{(0)},\ w_{x_1}^{(k)},\ N_{12}^{(0)} + N_{12}^{(k)}$
G_{56}^{34}	$w^{(0)},\ N_{11}^{(0)},\ w_{x_1}^{(0)} + w_{x_1}^{(k)},\ N_{12}^{(0)} + N_{12}^{(k)}$
G_{58}^{23}	$w^{(0)},\ w_{x_1}^{(0)},\ N_{12}^{(0)} + N_{12}^{(k)},\ M_{11}^{(0)} + M_{11}^{(k)}$
G_{68}^{35}	$w^{(0)},\ N_{11}^{(0)},\ N_{12}^{(0)} + N_{12}^{(k)},\ M_{11}^{(0)} + M_{11}^{(k)}$
G_{47}^{12}	$w^{(0)},\ u_1^{(0)},\ Q_1^{(k)},\ w_{x_1}^{(k)} + w_{x_1}^{(0)}$
G_{67}^{14}	$w^{(0)},\ N_{11}^{(0)},\ w_{x_1}^{(0)} + w_{x_1}^{(k)},\ Q_1^{(k)}$
G_{78}^{12}	$w^{(0)},\ w_{x_1}^{(0)},\ M_{11}^{(0)} + M_{11}^{(k)},\ Q_1^{(k)}$
G_{78}^{16}	$w^{(0)},\ N_{11}^{(0)},\ M_{11}^{(0)} + M_{11}^{(k)},\ Q_1^{(k)}$
G_{57}^{24}	$u_1^{(0)},\ N_{12}^{(0)},\ w_{x_1}^{(0)} + w_{x_1}^{(k)},\ Q_1^{(k)}$
G_{67}^{45}	$N_{11}^{(0)},\ N_{12}^{(0)},\ w_{x_1}^{(0)} + w_{x_1}^{(k)},\ Q_1^{(k)}$
G_{78}^{25}	$u_1^{(0)},\ N_{12}^{(0)},\ M_{11}^{(0)} + M_{11}^{(k)},\ Q_1^{(k)}$
G_{78}^{56}	$N_{11}^{(0)},\ N_{12}^{(0)},\ M_{11}^{(0)} + M_{11}^{(k)},\ Q_1^{(k)}$

As a result, we get the following limiting systems

$$\alpha_2^* < \frac{1}{2}, \ \alpha_1^* = \frac{1}{2}, \ \alpha_3^* = 0, \ \nu^* = 1, \ \alpha_5^* = \frac{1}{2}, \ \alpha_6^* = 1 - \alpha_2^*.$$

Here $\alpha_k^* = \alpha_k$, $(k = 1 - 6)$, $\nu^* = \nu$ for WS and SS; $\alpha_1^* = 2\alpha_1$, $\alpha_5^* = 2\alpha_5$, $\alpha_k^* = \alpha_k$, $(k = 2, 3, 4, 6)$, $\nu^* = \nu - 1/2$ for RS.

The equations of motion have the form

$$\frac{\partial N_{11}^{(k)}}{\partial x_1} + \frac{\partial N_{12}^{(k)}}{\partial x_2} = 0, \qquad \frac{\partial N_{12}^{(k)}}{\partial x_1} + \frac{\partial N_{22}^{(k)}}{\partial x_2} = 0,$$

$$\frac{\partial^2 M_{11}^{(k)}}{\partial x_1^2} + R N_{22}^{(k)} - N_{11}^{(0)} \frac{\partial^2 w^{(k)}}{\partial x_1^2} = 0; \tag{3.5.36}$$

$$0 = B_{11}\varepsilon_{11}^{(k)} + B_{12}\varepsilon_{22}^{(k)} + K_{11}\kappa_{11}^{(k)}, \qquad N_{22}^{(k)} = B_{21}\varepsilon_{11}^{(k)} + B_{22}\varepsilon_{22}^{(k)},$$

$$N_{12}^{(k)} = B_{33}\varepsilon_{12}^{(k)}, \qquad M_{11}^{(k)} = D_{11}\kappa_{11}^{(k)} + K_{11}\varepsilon_{11}^{(k)};$$

$$\varepsilon_{11}^{(k)} = \frac{1}{R}\frac{\partial u_1^{(k)}}{\partial x_1} + \frac{1}{2R}\left(\frac{\partial^2 w^{(k)}}{\partial x_1^2}\right)^2 + \frac{1}{R^2}\frac{\partial w^{(0)}}{\partial x_1}\frac{\partial w^{(k)}}{\partial x_1};$$

$$\varepsilon_{22}^{(k)} = -\frac{w^{(k)}}{R} + \frac{1}{R^2}\frac{\partial w^{(0)}}{\partial x_2}\frac{\partial w^{(k)}}{\partial x_2};$$

$$\varepsilon_{12}^{(k)} = \frac{1}{2R}\left(\frac{\partial u_2^{(k)}}{\partial x_1} + \frac{\partial u_1^{(k)}}{\partial x_2}\right) + \frac{1}{R^2}\frac{\partial w^{(k)}}{\partial x_1}\frac{\partial w^{(k)}}{\partial x_2}$$

$$+ \frac{1}{R^2}\left(\frac{\partial w^{(0)}}{\partial x_1}\frac{\partial w^{(k)}}{\partial x_2} + \frac{\partial w^{(0)}}{\partial x_2}\frac{\partial w^{(k)}}{\partial x_1}\right);$$

$$\kappa_{11}^{(k)} = -\frac{1}{R^2}\frac{\partial^2 w^{(k)}}{\partial x_1^2}.$$

We add equations (3.5.36) to conditions (3.5.35) and get a well-posed approximate boundary value problem to satisfy the boundary conditions.

In (3.5.36) the variable coefficients, which are obtained by the outer solution term, can be "frozen" with respect to x_1 on the shell edges. This is valid because the outer solution variability index in the axial direction is much smaller than the boundary layer index, and in the zone localized near the shell edges, the inner solution may be assumed constant with respect to the x_1 variables.

We can limit the equations for the boundary layers, and thus give the possibility of satisfying the boundary conditions, which differ from (3.5.36) by the presence of the inertial term in the third equation of motion.

The above derived sets of equations are sufficiently accurate to describe the bending state in a shell.

Let us formulate simplified WS boundary value problems:

$$\alpha_1 = \alpha_2 = \frac{1}{2}, \qquad \alpha_3 = 0, \qquad \alpha_4 = 1, \qquad \alpha_5 = \alpha_2 = \frac{1}{2}.$$

In this case the limiting equations are equal to well-known equations for the shallow shell theory. One has to extract the parameter ε_5 for these shells of this class from geometrical–rigidity parameters, which are the tangent rigidity and the extension–compression ratio. If reinforcement is strong enough, these values are small.

Let us introduce the parameters of asymptotic integration $\tilde\alpha_k$ $(k = 1 - 6)$:

$$\frac{\partial}{\partial x_1} \sim \varepsilon_5^{-\tilde\alpha_1}; \quad \frac{\partial}{\partial x_2} \sim \varepsilon_5^{-\tilde\alpha_2}; \quad \frac{\partial}{\partial t} \sim \varepsilon_5^{-\tilde\alpha_3};$$

$$\frac{w}{R} \sim \varepsilon_5^{\tilde\alpha_4}; \quad u_1 \sim \varepsilon_5^{\tilde\alpha_5} w; \quad u_6 \sim \varepsilon_5^{\tilde\alpha_6} w.$$

As a result of the asymptotic procedure we get the following limiting systems, which do not have an analogue in the isotropic case

(a) $\tilde\alpha_1 = \tilde\alpha_2$, $\tilde\alpha_3 = \tilde\alpha_4 = 2\tilde\alpha_2$, $\tilde\alpha_5 = \tilde\alpha_6 = \tilde\alpha_2$;

$$\frac{\partial N_{11}^{(1)}}{\partial x_1} + \frac{\partial N_{12}^{(1)}}{\partial x_2} = 0, \quad \frac{\partial N_{12}^{(1)}}{\partial x_1} + \frac{\partial N_{22}^{(1)}}{\partial x_2} = 0, \tag{3.5.37}$$

$$\frac{\partial^2 M_{11}^{(1)}}{\partial x_1^2} + 2\frac{\partial^2 M_{12}^{(1)}}{\partial x_1 \partial x_2} + \frac{\partial^2 M_{22}^{(1)}}{\partial x_2^2} + RN_{22}^{(1)} - \rho R^2 \frac{\partial^2 w^{(1)}}{\partial t^2} = 0;$$

$$0 = B_{11}\varepsilon_{11}^{(1)} + K_{11}\kappa_{11}^{(1)}, \quad 0 = B_{22}\varepsilon_{22}^{(1)} + K_{22}\kappa_{22}^{(1)}, \quad N_{12}^{(1)} = B_{33}\varepsilon_{12}^{(1)},$$

$$M_{11}^{(1)} = D_{11}\kappa_{11}^{(1)} + K_{11}\varepsilon_{11}^{(1)}; \quad M_{22}^{(1)} = D_{22}\kappa_{22}^{(1)} + K_{22}\varepsilon_{22}^{(1)}; \quad M_{12}^{(1)} = D_{33}\kappa_{12}^{(1)};$$

$$\varepsilon_{11}^{(1)} = \frac{1}{R}\frac{\partial u_1^{(1)}}{\partial x_1} + \frac{1}{2}\left(\Phi_1^{(1)}\right)^2; \quad \varepsilon_{22}^{(1)} = \frac{1}{R}\left(\frac{\partial u_2^{(1)}}{\partial x_2} - w^{(1)}\right) + \frac{1}{2}\left(\Phi_2^{(1)}\right)^2;$$

$$\varepsilon_{12}^{(1)} = \frac{1}{2R}\left(\frac{\partial u_2^{(1)}}{\partial x_1} + \frac{\partial u_1^{(1)}}{\partial x_2}\right) + \frac{1}{2}\Phi_1^{(1)}\Phi_2^{(1)}.$$

The stress-strain state of a WS is approximately described by an equation of three types. Unlike in the case considered above, in this case it is necessary to introduce not one, but two parameters ν_1 and ν_2 characterizing the ratio of the order of magnitudes of the quantities defining each of the three states $w^{(1)}$, $w^{(2)}$, $w^{(3)}$:

$$w^{(2)} \sim \varepsilon_5^{\nu_1} w^{(1)}, \quad w^{(3)} \sim \varepsilon_5^{\nu_2} w^{(1)}. \tag{3.5.38}$$

The ν_1 and ν_2 values are defined in every case by the boundary condition splitting process. Moreover, it is possible to have the following ν_1 and ν_2 values

$$\nu_1 = \frac{3}{2}; \quad \nu_1 = \frac{3}{2}; \quad \nu_1 = 2; \quad \nu_1 = 2;$$

$$\nu_2 = 2; \quad \nu_2 = \frac{5}{2}; \quad \nu_2 = 2; \quad \nu_2 = \frac{5}{2}.$$

The corresponding limiting systems are:

(b) $\tilde\alpha_1 = \tilde\alpha_2 - \frac{1}{2}$, $\tilde\alpha_3 = 2\tilde\alpha_2$, $\nu_1 = \frac{3}{2}(\nu_1 = 2)$, $\tilde\alpha_5 = \tilde\alpha_2 - \frac{3}{2}$, $\tilde\alpha_6 = \tilde\alpha_2$;

$$\frac{\partial N_{11}^{(2)}}{\partial x_1} + \frac{\partial N_{12}^{(2)}}{\partial x_2} = 0, \quad \frac{\partial N_{12}^{(2)}}{\partial x_1} + \frac{\partial N_{22}^{(2)}}{\partial x_2} = 0, \frac{\partial^2 M_{22}^{(2)}}{\partial x_2^2} = 0; \qquad (3.5.39)$$

$$N_{11}^{(2)} = B_{11}\varepsilon_{11}^{(2)}, \quad N_{22}^{(2)} = B_{21}\varepsilon_{11}^{(2)} + B_{22}\varepsilon_{22}^{(2)},$$
$$N_{12}^{(2)} = B_{33}\varepsilon_{12}^{(2)}, \quad M_{22}^{(2)} = D_{22}\kappa_{22}^{(2)};$$
$$\varepsilon_{11}^{(2)} = \frac{1}{R}\frac{\partial u_1^{(2)}}{\partial x_1}; \quad \varepsilon_{22}^{(2)} = \frac{1}{R}\left(\frac{\partial u_2^{(2)}}{\partial x_2} - w^{(2)}\right); \quad \varepsilon_{12}^{(2)} = \frac{1}{2R}\frac{\partial u_1^{(2)}}{\partial x_2};$$
$$\kappa_{22}^{(2)} = -\frac{1}{R^2}\frac{\partial^2 w^{(2)}}{\partial x_2^2};$$

(c) $\tilde{\alpha}_1 = \tilde{\alpha}_2 + \frac{1}{2}, \ \tilde{\alpha}_3 = 2\tilde{\alpha}_2, \ \nu_2 = 2(\nu_2 = \frac{5}{2}), \ \tilde{\alpha}_5 = \tilde{\alpha}_2 - \frac{1}{2}, \ \tilde{\alpha}_6 = \tilde{\alpha}_2 - 2;$

$$\frac{\partial N_{11}^{(3)}}{\partial x_1} + \frac{\partial N_{12}^{(3)}}{\partial x_2} = 0, \quad \frac{\partial N_{12}^{(3)}}{\partial x_1} + \frac{\partial N_{22}^{(3)}}{\partial x_2} = 0, \frac{\partial^2 M_{11}^{(3)}}{\partial x_1^2} = 0; \qquad (3.5.40)$$

$$N_{11}^{(3)} = B_{11}\varepsilon_{11}^{(3)} + B_{12}\varepsilon_{22}^{(3)}, \quad N_{22}^{(3)} = B_{22}\varepsilon_{22}^{(3)},$$
$$N_{12}^{(3)} = B_{33}\varepsilon_{12}^{(3)}, \quad M_{11}^{(3)} = D_{11}\kappa_{11}^{(3)};$$
$$\varepsilon_{11}^{(3)} = \frac{1}{R}\frac{\partial u_1^{(3)}}{\partial x_1}; \quad \varepsilon_{22}^{(3)} = \frac{1}{R}\frac{\partial u_2^{(3)}}{\partial x_2}; \quad \varepsilon_{12}^{(3)} = \frac{1}{2R}\frac{\partial u_2^{(3)}}{\partial x_1};$$
$$\kappa_{11}^{(3)} = -\frac{1}{R^2}\frac{\partial^2 w^{(3)}}{\partial x_1^2};$$

The final results for splitting boundary conditions are given in Table 3.4a.

It is worth noting that (3.5.39), (3.5.40) are everywhere linear, and this is sufficient simplification for the solution of practical problems.

For stringer shells we have the following limiting systems

$$\alpha_2 = \frac{1}{2}, \ \alpha_1 = 0, \ \alpha_3 = -1, \ \alpha_4 = 1, \ \alpha_5 = 1, \ \alpha_6 = \frac{1}{2};$$

$$\frac{\partial N_{11}^{(1)}}{\partial x_1} + \frac{\partial N_{12}^{(1)}}{\partial x_2} = 0, \quad \frac{\partial N_{12}^{(1)}}{\partial x_1} + \frac{\partial N_{22}^{(1)}}{\partial x_2} = 0, \qquad (3.5.41)$$
$$\frac{\partial^2 M_{11}^{(1)}}{\partial x_1^2} + 2\frac{\partial^2 M_{12}^{(1)}}{\partial x_1 \partial x_2} + \frac{\partial^2 M_{22}^{(1)}}{\partial x_2^2} + RN_{22}^{(1)}$$
$$+ \frac{\partial}{\partial x_1}\left(\frac{\partial w^{(1)}}{\partial x_1}N_{11}^{(1)} + \frac{\partial w^{(1)}}{\partial x_2}N_{12}^{(1)}\right)$$
$$+ \frac{\partial}{\partial x_2}\left(\frac{\partial w^{(1)}}{\partial x_1}N_{12}^{(1)} + \frac{\partial w^{(1)}}{\partial x_2}N_{22}^{(1)}\right) - \rho R^2\frac{\partial^2 w^{(1)}}{\partial t^2} = 0;$$

Table 3.4(a). Splitting of boundary conditions

G	"Wafer" shells		
G^{12}_{34}	$w^{(1)}, w^{(1)}_{x_1}, u^{(1)}_1 + u^{(2)}_1, u^{(1)}_2 + u^{(3)}_2$	$\frac{3}{2}$	2
G^{13}_{46}	$w^{(1)}, w^{(1)}_{x_1}, u^{(1)}_2 + u^{(2)}_2, N^{(1)}_{11} + N^{(2)}_{11} + N^{(3)}_{11}$	2	2
G^{12}_{38}	$w^{(1)}, M^{(1)}_{11}, u^{(1)}_1 + u^{(2)}_1, u^{(1)}_2 + u^{(3)}_2$	$\frac{3}{2}$	2
G^{13}_{68}	$w^{(1)}, M^{(1)}_{11}, N^{(1)}_{12} + N^{(1)}_{12} + N^{(3)}_{11}, u^{(1)}_2 + u^{(3)}_2$	2	2
G^{23}_{45}	$w^{(1)}, N^{(1)}_{12} + N^{(2)}_{12} + N^{(3)}_{12}, u^{(1)}_1 + u^{(2)}_1, w^{(1)}_{x_1}$	$\frac{3}{2}$	$\frac{5}{2}$
G^{34}_{56}	$w^{(1)}_{x_1}, w^{(1)}, N^{(1)}_{12} + N^{(3)}_{12}, N^{(1)}_{11} + N^{(2)}_{11}$	2	$\frac{5}{2}$
G^{23}_{58}	$w^{(1)}, M^{(1)}_{11}, N^{(1)}_{12} + N^{(2)}_{12} + N^{(3)}_{12}, u^{(1)}_1 + u^{(2)}_1$	$\frac{3}{2}$	$\frac{5}{2}$
G^{35}_{68}	$w^{(1)}, M^{(1)}_{11}, N^{(1)}_{12} + N^{(3)}_{12}, N^{(1)}_{11} + N^{(2)}_{11}$	2	$\frac{5}{2}$
G^{12}_{47}	$Q^{(1)}_1, w^{(1)}_{x_1}, u^{(1)}_1 + u^{(2)}_1, u^{(1)}_2 + u^{(3)}_2$	$\frac{3}{2}$	2
G^{14}_{67}	$Q^{(1)}_1, w^{(1)}_{x_1}, u^{(1)}_2 + u^{(3)}_2, N^{(1)}_{11} + N^{(2)}_{11} + N^{(3)}_{11}$	2	2
G^{12}_{78}	$Q^{(1)}_1, M^{(1)}_{11}, u^{(1)}_1 + u^{(2)}_1, u^{(1)}_2 + u^{(3)}_2$	$\frac{3}{2}$	2
G^{16}_{78}	$Q^{(1)}_1, M^{(1)}_{11}, N^{(1)}_{11} + N^{(1)}_{11} + N^{(3)}_{11}, u^{(1)}_2 + u^{(3)}_2$	2	2
G^{24}_{57}	$Q^{(1)}_1, N^{(1)}_{12} + N^{(2)}_{12} + N^{(3)}_{12}, u^{(1)}_1 + u^{(2)}_1, w^{(1)}_{x_1}$	$\frac{3}{2}$	$\frac{5}{2}$
G^{45}_{67}	$w^{(1)}_{x_1}, Q^{(1)}_1, N^{(1)}_{12} + N^{(3)}_{12}, N^{(1)}_{11} + N^{(2)}_{11}$	2	$\frac{5}{2}$
G^{25}_{78}	$Q^{(1)}_1, M^{(1)}_{11}, N^{(1)}_{12} + N^{(2)}_{12} + N^{(3)}_{12}, u^{(1)}_1 + u^{(2)}_1$	$\frac{3}{2}$	$\frac{5}{2}$
G^{56}_{78}	$Q^{(1)}_1, M^{(1)}_{11}, N^{(1)}_{12} + N^{(3)}_{12}, N^{(1)}_{11} + N^{(2)}_{11}$	2	$\frac{5}{2}$

$$N^{(1)}_{11} = B_{11}\varepsilon^{(1)}_{11} + B_{12}\varepsilon^{(1)}_{22} + K_{11}\kappa^{(1)}_{11}; \quad 0 = B_{21}\varepsilon^{(1)}_{11} + B_{22}\varepsilon^{(1)}_{22};$$

$$N^{(1)}_{12} = B_{33}\varepsilon^{(1)}_{12}, \quad M^{(1)}_{11} = D_{11}\kappa^{(1)}_{11} + K_{11}\varepsilon^{(1)}_{11};$$

$$M^{(1)}_{22} = D_{21}\kappa^{(1)}_{11} + D_{22}\kappa^{(1)}_{22}; \quad M^{(1)}_{12} = D_{33}\kappa^{(1)}_{12};$$

$$\varepsilon^{(1)}_{11} = \frac{1}{R}\frac{\partial u^{(1)}_1}{\partial x_1} + \frac{1}{2R^2}\left(\frac{\partial w^{(1)}}{\partial x_1}\right)^2;$$

$$0 = \frac{1}{R}\left(\frac{\partial u^{(1)}_2}{\partial x_2} - w^{(1)}\right) + \frac{1}{2R^2}\left(\frac{\partial w^{(1)}}{\partial x_2}\right)^2;$$

$$0 = \frac{1}{2R}\left(\frac{\partial u^{(1)}_2}{\partial x_1} + \frac{\partial u^{(1)}_1}{\partial x_2}\right) + \frac{1}{2R^2}\frac{\partial w^{(1)}}{\partial x_1}\frac{\partial w^{(1)}}{\partial x_2};$$

$$\kappa^{(1)}_{11} = -\frac{1}{R^2}\frac{\partial^2 w^{(1)}}{\partial x_1^2}; \quad \kappa^{(1)}_{22} = -\frac{1}{R^2}\frac{\partial^2 w^{(1)}}{\partial x_2^2}; \quad \kappa^{(1)}_{12} = -\frac{1}{R^2}\frac{\partial^2 w^{(1)}}{\partial x_1 \partial x_2}.$$

Equations (3.5.41) describe the dynamic shell state which varies rapidly in the circumferential direction. The boundary layer varies rapidly both along the guide and the generatrix and it is defined by the system

(1) $\alpha_2 = \frac{1}{2}, \alpha_1 = \frac{1}{2}, \alpha_3 = -1, \nu = 1 \left(\text{or } \nu = \frac{3}{2}\right), \alpha_5 = \alpha_6 = \frac{1}{2};$

$$\frac{\partial N^{(2)}_{11}}{\partial x_1} + \frac{\partial N^{(2)}_{12}}{\partial x_2} = 0, \quad \frac{\partial N^{(2)}_{12}}{\partial x_1} + \frac{\partial N^{(2)}_{22}}{\partial x_2} = 0, \tag{3.5.42}$$

$$\frac{\partial^2 M_{11}^{(2)}}{\partial x_1^2} + RN_{22}^{(2)} + \frac{\partial^2 w^{(1)}}{\partial x_2^2} N_{22}^{(2)} = 0;$$

$$N_{11}^{(2)} = B_{11}\varepsilon_{11}^{(2)} + B_{12}\varepsilon_{22}^{(2)} + K_{11}\kappa_{11}^{(2)}; \quad N_{22}^{(2)} = B_{21}\varepsilon_{11}^{(2)} + B_{22}\varepsilon_{22}^{(2)};$$

$$N_{12}^{(2)} = B_{33}\varepsilon_{12}^{(2)}; \quad M_{11}^{(2)} = D_{11}\kappa_{11}^{(2)} + D_{22}\kappa_{22}^{(2)} + K_{11}\varepsilon_{11}^{(2)};$$

$$\varepsilon_{11}^{(2)} = \frac{1}{R}\frac{\partial u_1^{(2)}}{\partial x_1}; \quad \varepsilon_{22}^{(2)} = \frac{1}{R}\left(\frac{\partial u_2^{(2)}}{\partial x_2} - w^{(2)}\right) + \frac{1}{R^2}\frac{\partial w^{(1)}}{\partial x_2}\frac{\partial w^{(2)}}{\partial x_2};$$

$$\varepsilon_{12}^{(2)} = \frac{1}{2R}\left(\frac{\partial u_2^{(2)}}{\partial x_1} + \frac{\partial u_1^{(2)}}{\partial x_2}\right) + \frac{1}{2R^2}\frac{\partial w^{(2)}}{\partial x_1}\frac{\partial w^{(1)}}{\partial x_2};$$

$$\kappa_{11}^{(2)} = -\frac{1}{R^2}\frac{\partial^2 w^{(2)}}{\partial x_2^2}; \quad \kappa_{22}^{(2)} = -\frac{1}{R^2}\frac{\partial^2 w^{(2)}}{\partial x_2^2}.$$

The final separations of the boundary conditions are shown in Table 3.4b.

Table 3.4(b). Splitting of the boundary conditions

G	Stringer shells			
G_{34}^{12}	$w^{(1)}, L_1^{(2)}, w_{x_1}^{(k)}, \frac{\partial u_1^{(2)}}{\partial x_2} - w_{x_1}^{(k)} \equiv L_2^{(2)}$			1
G_{46}^{13}	$w^{(1)}, w_{x_1}^{(1)}, L_1^{(2)}, N_{11}^{(1)} + N_{11}^{(2)}$			1
G_{38}^{12}	$w^{(1)}, u_1^{(1)}, L_1^{(2)}, M_{11}^{(1)} + M_{11}^{(2)}$			1
G_{68}^{13}	$w^{(1)}, M_{11}^{(1)}, L_1^{(2)}, N_{11}^{(1)} + N_{11}^{(2)}$			1
G_{45}^{23}	$w^{(1)}, L_2^{(2)}, w_{x_1}^{(1)}, N_{12}^{(1)} + N_{12}^{(2)}$			$\frac{3}{2}$
G_{56}^{34}	$w^{(1)}, N_{12}^{(2)}, w_{x_1}^{(1)}, N_{11}^{(1)} + N_{11}^{(2)}$			1
G_{58}^{23}	$w^{(1)}, u_1^{(1)}, N_{12}^{(2)}, M_{11}^{(1)} + M_{11}^{(2)}$			1
G_{68}^{35}	$w^{(1)}, M_{11}^{(1)}, N_{11}^{(1)} + N_{11}^{(2)}, N_{12}^{(2)}$			1
G_{47}^{12}	$w^{(1)}, L_2^{(2)}, w_{x_1}^{(1)}, Q_1^{(1)} + Q_1^{(2)}$			$\frac{3}{2}$
G_{67}^{14}	$w^{(1)}, Q_1^{(2)}, w_{x_1}^{(1)}, N_{11}^{(1)} + N_{11}^{(2)}$			1
G_{78}^{12}	$w^{(1)}, w_{x_1}^{(1)}, M_{11}^{(1)} + M_{11}^{(2)}, Q_1^{(2)}$			1
G_{78}^{16}	$w^{(1)}, M_{11}^{(1)}, N_{11}^{(1)} + N_{11}^{(2)}, Q_1^{(2)}$			1
G_{57}^{24}	$u_1^{(1)}, N_{12}^{(1)} + N_{12}^{(2)}, w_{x_1}^{(1)}, Q_1^{(1)} + Q_1^{(2)}$			1
G_{67}^{45}	$N_{11}^{(1)}, N_{12}^{(2)}, w_{x_1}^{(1)}, Q_1^{(1)} + Q_1^{(2)}$			$\frac{3}{2}$
G_{78}^{25}	$u_1^{(1)}, M_{11}^{(1)}, N_{12}^{(1)} + N_{12}^{(2)}, Q_1^{(1)} + Q_1^{(2)}$			$\frac{3}{2}$
G_{78}^{56}	$M_{11}^{(1)}, N_{11}^{(1)}, N_{12}^{(1)} + N_{12}^{(2)}, Q_1^{(1)} + Q_1^{(2)}$			$\frac{3}{2}$

(2) $\alpha_2 = \frac{1}{2} + \omega$, $\alpha_1 = \omega$, $\alpha_3 = -1 + 2\omega$, $\alpha_4 = 1 + 2\omega$,
$\alpha_5 = 1 + \omega$, $\alpha_6 = \frac{1}{2} + \omega$; $\omega > 0$.

$$\frac{\partial N_{11}}{\partial x_1} + \frac{\partial N_{12}}{\partial x_2} = 0, \quad \frac{\partial N_{12}}{\partial x_1} + \frac{\partial N_{22}}{\partial x_2} = 0, \qquad (3.5.43)$$

$$\frac{\partial^2 M_{11}}{\partial x_1^2} + 2\frac{\partial^2 M_{12}}{\partial x_1 \partial x_2} + \frac{\partial^2 M_{22}}{\partial x_2^2} - \rho R^2 \frac{\partial^2 w}{\partial t^2} = 0;$$

$$N_{11} = K_{11}\kappa_{11}, \quad 0 = B_{21}\varepsilon_{11} + B_{22}\varepsilon_{22},$$
$$N_{12} = B_{33}\varepsilon_{12}, \quad M_{11} = D_{11}\kappa_{11};$$
$$M_{22} = D_{22}\kappa_{22}; \quad M_{12} = D_{33}\kappa_{12}.$$

The stress-strain relations are defined in this case by (3.5.41).
Equations (3.5.43) describe mainly the bending vibrations of the shell.

(3) $\alpha_2 > \frac{1}{2}$, $\alpha_1 = \alpha_2$, $\alpha_3 = -1 + 2\alpha_2$, $\alpha_4 = 1$, $\alpha_5 = \alpha_6 = 1 - \alpha_2$.

$$\frac{\partial N_{11}}{\partial x_1} + \frac{\partial N_{12}}{\partial x_2} = 0, \quad \frac{\partial N_{12}}{\partial x_1} + \frac{\partial N_{22}}{\partial x_2} = 0, \tag{3.5.44}$$

$$\frac{\partial^2 M_{11}}{\partial x_1^2} + \frac{\partial}{\partial x_1}\left(\frac{\partial w}{\partial x_1}N_{11} + \frac{\partial w}{\partial x_2}N_{12}\right) + \frac{\partial}{\partial x_2}\left(\frac{\partial w}{\partial x_1}N_{12} + \frac{\partial w}{\partial x_2}N_{22}\right)$$

$$-\rho R^2 \frac{\partial^2 w}{\partial t^2} = 0;$$

$$\varepsilon_{11} = \frac{1}{R}\frac{\partial u_1}{\partial x_1} + \frac{1}{2R^2}\left(\frac{\partial w}{\partial x_1}\right)^2; \quad \varepsilon_{22} = \frac{1}{R}\frac{\partial u_2}{\partial x_2} + \frac{1}{2R^2}\left(\frac{\partial w}{\partial x_2}\right)^2;$$

$$\varepsilon_{12} = \frac{1}{2R}\left(\frac{\partial u_2}{\partial x_1} + \frac{\partial u_1}{\partial x_2}\right) + \frac{1}{2R^2}\frac{\partial w}{\partial x_1}\frac{\partial w}{\partial x_2}.$$

The formulae for the components of the elasticity tensor do not change in comparison with the governing equations.

Equations (3.5.44) correspond to the SS vibrations with higher frequencies than in the previous case.

Here we have large order variability both in the circumferential and axial directions of the outer solution $(w^{(1)})$ and the boundary layer $(w^{(2)})$. Let us introduce the parameter ν:

$$w^{(2)} \sim \varepsilon_1^\nu w^{(1)}.$$

The asymptotic investigation shows that both the states are dynamic, therefore the splitting of the boundary conditions will not be unique. There are two correct splittings of the boundary conditions represented in Table 3.4c.

For example let us represent the limiting systems for the variant of the boundary conditions G_{46}^{13}. To solve this problem we begin by calculating the outer solution, described by the following equations:

$$\alpha_1 = \alpha_2 = \frac{1}{2}, \ \alpha_3 = 0, \ \alpha_4 = 1, \ \alpha_5 = \alpha_6 = \frac{5}{4}.$$

$$\frac{\partial N_{11}^{(1)}}{\partial x_1} + \frac{\partial N_{12}^{(1)}}{\partial x_2} = 0, \quad \frac{\partial N_{12}^{(1)}}{\partial x_1} + \frac{\partial N_{22}^{(1)}}{\partial x_2} = 0, \tag{3.5.45}$$

$$\frac{\partial^2 M_{22}^{(1)}}{\partial x_2^2} + RN_{22}^{(1)} - \rho R^2 \frac{\partial^2 w^{(1)}}{\partial t^2} = 0;$$

$$N_{11}^{(1)} = B_{11}\varepsilon_{11}^{(1)} + B_{12}\varepsilon_{22}^{(1)}, \quad N_{22}^{(1)} = B_{21}\varepsilon_{11}^{(1)} + B_{22}\varepsilon_{22}^{(1)} + K_{22}\kappa_{22}^{(1)},$$

$$N_{12}^{(1)} = B_{33}\varepsilon_{12}^{(1)}, \quad M_{22}^{(1)} = D_{22}\kappa_{22}^{(1)} + K_{22}\varepsilon_{22}^{(1)};$$

$$\varepsilon_{11}^{(1)} = \frac{1}{R}\frac{\partial u_1^{(1)}}{\partial x_1}; \quad \varepsilon_{22}^{(1)} = \frac{1}{R}\left(\frac{\partial u_2^{(1)}}{\partial x_2} - w^{(1)}\right);$$

$$\varepsilon_{12}^{(1)} = \frac{1}{2R}\left(\frac{\partial u_2^{(1)}}{\partial x_1} + \frac{\partial u_1^{(1)}}{\partial x_2}\right); \quad \kappa_{22}^{(1)} = -\frac{1}{R^2}\frac{\partial^2 w^{(1)}}{\partial x_2^2}.$$

Table 3.4(c). Splitting of the boundary conditions

G	Ring reinforced shells								
G_{34}^{12}	$u_1^{(1)}, u_2^{(1)}, w^{(1)}+w^{(2)}, w_{x_1}^{(2)}$			0	$w^{(2)}, u_2^{(1)}, u_1^{(1)}+u_1^{(2)}, w_{x_1}^{(2)}$				$-\frac{1}{4}$
G_{46}^{13}	$u_2^{(1)}, N_{11}^{(1)}, w^{(1)}+w^{(2)}, w_{x_1}^{(2)}$			0	$w^{(2)}, N_{11}^{(1)}, u_2^{(1)}+u_2^{(2)}, w_{x_1}^{(2)}$				$-\frac{1}{2}$
G_{38}^{12}	$u_1^{(1)}, u_2^{(1)}, w^{(1)}+w^{(2)}, M_{11}^{(2)}$			0	$w^{(2)}, u_2^{(1)}, u_1^{(1)}+u_1^{(2)}, M_{11}^{(2)}$				$-\frac{1}{4}$
G_{68}^{13}	$u_2^{(1)}, N_{11}^{(1)}, w^{(1)}+w^{(2)}, M_{11}^{(2)}$			0	$w^{(2)}, M_{11}^{(2)}, u_2^{(1)}+u_2^{(2)}, N_{11}^{(1)}$				$-\frac{1}{2}$
G_{45}^{23}	$u_1^{(1)}, w_{x_1}^{(2)}, w^{(1)}+w^{(2)}, N_{12}^{(1)}$			0	$w^{(2)}, w_{x_1}^{(2)}, u_1^{(1)}+u_1^{(2)}, N_{12}^{(1)}+N_{12}^{(2)}$				$-\frac{1}{4}$
G_{56}^{34}	$N_{11}^{(1)}, N_{12}^{(1)}, w^{(1)}+w^{(2)}, w_{x_1}^{(2)}$			0	$w^{(2)}, w_{x_1}^{(2)}, N_{12}^{(1)}+N_{12}^{(2)}, N_{11}^{(1)}$				$-\frac{1}{4}$
G_{58}^{23}	$u_1^{(1)}, N_{11}^{(1)}, w^{(1)}+w^{(2)}, M_{11}^{(2)}$			0	$w^{(2)}, M_{11}^{(2)}, u_1^{(1)}+u_1^{(2)}, N_{11}^{(1)}+N_{12}^{(2)}$				$-\frac{1}{4}$
G_{68}^{35}	$N_{11}^{(1)}, N_{12}^{(1)}, w^{(1)}+w^{(2)}, M_{11}^{(2)}$			0	$w^{(2)}, M_{11}^{(2)}, N_{11}^{(1)}+N_{12}^{(2)}, N_{11}^{(1)}$				$-\frac{1}{4}$
G_{47}^{12}	$u_1^{(1)}, u_2^{(1)}, w_{x_1}^{(1)}+w_{x_1}^{(2)}, Q_1^{(2)}$			$\frac{1}{4}$	$Q_1^{(2)}, u_2^{(1)}, u_1^{(1)}+u_1^{(2)}, w_{x_1}^{(2)}$				$-\frac{1}{4}$
G_{67}^{14}	$u_2^{(1)}, N_{11}^{(1)}, w_{x_1}^{(1)}+w_{x_1}^{(2)}, Q_1^{(2)}$			$\frac{1}{4}$	$w_{x_1}^{(2)}, Q_1^{(2)}, u_2^{(1)}+u_2^{(2)}, N_{11}^{(1)}+N_{11}^{(2)}$				$-\frac{1}{2}$
G_{78}^{12}	$u_1^{(1)}, u_2^{(1)}, M_{11}^{(1)}+M_{11}^{(2)}, Q_1^{(2)}$			$\frac{1}{2}$	$Q_1^{(2)}, M_{11}^{(2)}, u_1^{(1)}+u_1^{(2)}, u_2^{(1)}$				$-\frac{1}{4}$
G_{78}^{16}	$u_2^{(1)}, N_{11}^{(1)}, M_{11}^{(1)}+M_{11}^{(2)}, Q_1^{(2)}$			$\frac{1}{2}$	$Q_1^{(2)}, M_{11}^{(2)}, u_2^{(1)}+u_2^{(2)}, N_{11}^{(1)}+N_{11}^{(2)}$				$-\frac{1}{2}$
G_{57}^{24}	$N_{11}^{(1)}, N_{12}^{(1)}, w_{x_1}^{(1)}+w_{x_1}^{(2)}, Q_1^{(2)}$			$\frac{1}{4}$	$Q_1^{(2)}, w_{x_1}^{(2)}, N_{12}^{(1)}+N_{12}^{(2)}, u_1^{(1)}$				$-\frac{1}{4}$
G_{67}^{45}	$u_1^{(1)}, N_{11}^{(1)}, M_{11}^{(1)}+M_{11}^{(2)}, Q_1^{(2)}$			$\frac{1}{4}$	$Q_1^{(2)}, w_{x_1}^{(2)}, N_{11}^{(1)}, N_{11}^{(1)}+N_{12}^{(2)}$				$-\frac{1}{4}$
G_{78}^{25}	$u_1^{(1)}, N_{11}^{(1)}, w_{x_1}^{(1)}+w_{x_1}^{(2)}, Q_1^{(2)}$			$\frac{1}{2}$	$M_{11}^{(2)}, Q_1^{(2)}, u_1^{(1)}+u_1^{(2)}, N_{12}^{(1)}$				$-\frac{1}{4}$
G_{78}^{56}	$N_{11}^{(1)}, N_{12}^{(1)}, M_{11}^{(1)}+M_{11}^{(2)}, Q_1^{(2)}$			$\frac{1}{2}$	$Q_1^{(2)}, M_{11}^{(2)}, N_{11}^{(1)}, N_{12}^{(1)}+N_{12}^{(2)}$				$-\frac{1}{4}$

The boundary layers are defined by following limiting system:

$$\alpha_2 = \frac{1}{4}, \ \alpha_1 = \frac{1}{2}, \ \alpha_3 = 0, \ \nu = 0, \ \alpha_5 = \frac{1}{2}, \ \alpha_6 = \frac{3}{4},$$

$$\frac{\partial N_{11}^{(2)}}{\partial x_1} + \frac{\partial N_{12}^{(2)}}{\partial x_2} = 0, \quad \frac{\partial N_{12}^{(2)}}{\partial x_1} + \frac{\partial N_{22}^{(2)}}{\partial x_2} = 0, \tag{3.5.46}$$

$$\frac{\partial^2 M_{11}^{(2)}}{\partial x_1^2} + \frac{\partial^2 M_{22}^{(2)}}{\partial x_2^2} + RN_{22}^{(2)} + \frac{\partial^2 w^{(2)}}{\partial x_1^2}N_{11}^{(1)} - \rho R^2 \frac{\partial^2 w^{(2)}}{\partial t^2} = 0;$$

$$0 = B_{11}\varepsilon_{11}^{(2)} + B_{12}\varepsilon_{22}^{(2)}, \quad N_{22}^{(2)} = B_{21}\varepsilon_{11}^{(2)} + B_{22}\varepsilon_{22}^{(2)} + K_{22}\kappa_{22}^{(2)},$$

$$N_{12}^{(2)} = B_{33}\varepsilon_{12}^{(2)}, \quad M_{11}^{(2)} = D_{11}\kappa_{11}^{(2)}; \quad M_{22}^{(2)} = D_{22}\kappa_{22}^{(2)} + K_{22}\varepsilon_{22}^{(2)};$$

$$\varepsilon_{11}^{(2)} = \frac{1}{R}\frac{\partial u_1^{(2)}}{\partial x_1} + \frac{1}{2R^2}\left(\frac{\partial w^{(2)}}{\partial x_1}\right)^2; \quad \varepsilon_{22}^{(2)} = -\frac{w^{(2)}}{R};$$

$$\varepsilon_{12}^{(2)} = \frac{1}{2R}\left(\frac{\partial u_2^{(2)}}{\partial x_1} + \frac{\partial u_1^{(2)}}{\partial x_2}\right) + \frac{1}{2R^2}\frac{\partial w^{(2)}}{\partial x_1}\frac{\partial w^{(2)}}{\partial x_2}.$$

If at first one calculates the boundary layer, asymptotic analyses of the governing equations show that $\nu = -0.5$ and the limiting system for the inner state coincides with system (3.5.45), if in the third equation of motion the term $(\partial^2 w^{(2)}/\partial x_1^2)N_{11}^{(1)}$ is introduced.

Let us consider the amplitude–frequency dependencies for nonlinear vibrations of the simply supported stringer shell.

The governing relations may be chosen in the form (3.5.41). Let us pose [70, 71]

$$w(x_1,x_2,t) = f_1(t)\sin s_1 x_1 \cos s_2 x_2 + f_2(t)\sin^2 s_1 x_1. \qquad (3.5.47)$$

Here $s_1 = \pi ml^{-1}$, $s_2 = n$; and m and n are the wave numbers in the axial (circumferential) direction.

It should be pointed out that the time functions f_1 and f_2 are not independent. The connection between them should be taken from the condition of continuity of the displacement u_2 in the circumferential direction [70, 71]:

$$\int\limits_0^{\tilde{\alpha}} \frac{\partial u_2}{\partial x_2}\,dx_2 = 0, \quad \tilde{\alpha} = 2\pi R^{-1}. \qquad (3.5.48)$$

Using geometrical relations and (3.5.47), one obtains from (3.5.48)

$$f_2 = 0.25 R^{-1} s_2^2 f_1^2.$$

The Airy stress function we get from relations (3.5.47) is in the following form:

$$B_1^{-1}\Phi = p^2 s_2^{-2}(1 - \varepsilon_6 s_2^2)\xi \sin s_1 x_1 \cos s_2 x_2 - \frac{5}{16}p^2\xi^2 \cos 2s_2 x_2$$

$$+ 0.5 s_1^2\xi^3 \sin s_1 x_1 \cos 2s_1 x_1 \cos s_2 x_2; \quad p = s_1 s_2^{-1}, \quad \xi = R^{-1}f_1.$$

Now we use the Galerkin procedure for the governing equations

$$\int\limits_0^{\tilde{\alpha}}\int\limits_0^l L_1(w)\sin s_1 x_1 \cos s_2 x_2\,dx_1\,dx_2,$$

$$\int\limits_0^{\tilde{\alpha}}\int\limits_0^l L_1(w)\sin^2 s_1 x_1\,dx_1\,dx_2,$$

where

$$L_1(w) = \nabla_1^4 w - R\left(\frac{\partial^2}{\partial x_1^2} - \nabla_3^4\right)\Phi - L(w,\Phi) + \rho R^2\frac{\partial^2 w}{\partial t^2}.$$

As a result we obtain the following ordinary differential equation with constant coefficients for the time function $\xi(t)$:

$$\frac{d^2\xi}{dt^2} + \alpha\xi\left[\left(\frac{d\xi}{dt_1}\right)^2 + \xi\frac{d^2\xi}{dt_1^2}\right] + A_1\xi + A_2\xi^3 + A_3\xi^5 = 0. \tag{3.5.49}$$

Here $t_1 = \sqrt{B_1(\rho R^2)^{-1}}\, t$; $A_1 = \epsilon_1^2\epsilon_4 + 2\epsilon_1^2\epsilon_3\epsilon_4 p^{-2} + \epsilon_1^2\epsilon_2\epsilon_4 p^{-4} + s_2^{-4}(1 - \epsilon_6^2 s_2^2)^2$; $A_2 = \frac{1}{16} + \frac{1}{2}s_2^4\epsilon_1^2\epsilon_4 - \frac{3}{4}(1 - \epsilon_6^2 s_2^2)$; $A_3 = \frac{1}{4}s_2^4$; $\alpha = \frac{3}{32}s_2^4$.

Let us consider a practically important case of steady-state periodical vibrations and use the method of strained coordinates [119, 120, 122] for solving (3.5.49).

We change the independent variable t_1 to a new one $\tau = \omega t_1$, where ω is an unknown frequency of the periodic solution. Then (3.5.49) must be replaced by

$$\omega^2\ddot{\xi} + \alpha\omega^2\xi(\dot{\xi}^2 + \xi\ddot{\xi}) + A_1\xi + A_2\xi^3 + A_3\xi^5 = 0, \quad (\,\dot{}\,) \equiv \frac{d}{d\tau}. \tag{3.5.50}$$

The initial time point may be chosen in any way because of the periodicity of the solution.. Without loss of generality let

$$\tau = 0, \quad \xi = f, \quad \dot{\xi} = 0. \tag{3.5.51}$$

Let us introduce a formal small parameter and pose

$$\xi(\tau) = \epsilon\xi_1(\tau) + \epsilon^2\xi_2(\tau) + \epsilon^3\xi_3(\tau) + \ldots \tag{3.5.52}$$

$$\omega = \omega_0 + \epsilon\omega_1 + \epsilon^2\omega_2 + \epsilon^3\omega_3 + \ldots \tag{3.5.53}$$

with the constraint that expansion (3.5.53) is uniformly asymptotic [119, 120, 122].

Now substituting (3.5.52), (3.5.53) into (3.5.50) and comparing the coefficients of ϵ^n in the usual way, we find

$$\epsilon^1 : \omega_0^2\ddot{\xi}_1 + A_1\xi_1 = 0, \tag{3.5.54}$$

$$\epsilon^2 : \omega_0^2\ddot{\xi}_2 + A_1\xi_2 = 2\omega_0\omega_1\ddot{\xi}_1, \tag{3.5.55}$$

$$\epsilon^3 : \omega_0^2\ddot{\xi}_2 + A_1\xi_3 = -(\omega_1^2 + 2\omega_0\omega_2)\ddot{\xi}_1 - 2\omega_0\omega_1\ddot{\xi}_2$$
$$-\alpha\omega_0^2(\xi_1\dot{\xi}_1^2 + \ddot{\xi}_1\xi_1^2) - A_2\xi_1^3, \tag{3.5.56}$$

$$\epsilon^4 : \omega_0^2\ddot{\xi}_4 + A_1\xi_4 = -2(\omega_0\omega_3 + \omega_0\omega_2)\ddot{\xi}_1 - (\omega_1^2 + 2\omega_0\omega_2)\ddot{\xi}_2$$
$$-2\omega_0\omega_1\ddot{\xi}_3 - \alpha\left[\omega_0^2(2\xi_1\dot{\xi}_1\dot{\xi}_2 + 2\xi_1\dot{\xi}_1\dot{\xi}_2 + \dot{\xi}_1^2\ddot{\xi}_2 + \xi_1^2\ddot{\xi}_2)\right.$$
$$\left.+2\omega_0\omega_1(\xi_1\dot{\xi}_1^2 + \xi_1^2\ddot{\xi}_1)\right] - 3A_3\xi_1^2\xi_2, \tag{3.5.57}$$

...   .

The initial conditions (3.5.54) give us

$$\xi_1(0) = f, \quad \dot{\xi}_1(0) = 0, \tag{3.5.58}$$

$$\xi_i(0) = 0, \quad \dot{\xi}_i(0) = 0, \quad i = 1, 2, \ldots \tag{3.5.59}$$

The solution of the initial value problem (3.5.57), (3.5.58) is

$$\omega_0 = \sqrt{A_1}; \quad \xi_1 = f \cos \tau. \tag{3.5.60}$$

Let us rewrite (3.5.55), taking into account (3.5.60):

$$\ddot{\xi}_2 + \xi_2 = \frac{2f}{\sqrt{A_1}} \omega_1 \cos \tau. \tag{3.5.61}$$

From the conditions of the absence of secular terms, it follows that $\omega_1 = 0$. Then initial value problem (3.5.59), (3.5.61) has the solution $\xi_2 = 0$.

In the same way one obtains

$$\omega_2 = 0.125\sqrt{A_1}(3C_1 - 2\alpha)f^2; \quad C_1 = A_2 A_1^{-1};$$
$$\xi_3 = -0.03125 C_1(\cos\tau - \cos 3\tau)f^3 + 0.0625\alpha(\cos\tau + \cos 3\tau)f^3;$$
$$\omega_3 = 0; \quad \omega_4 = 0.039062\sqrt{A_1}\left[\gamma_1(\gamma_2 - 6C_1) - 2\gamma_2(\gamma_2 + 8\alpha) + 80C_2\right]f^4;$$
$$C_2 = A_3 A_1^{-1}, \quad \gamma_1 = C_1 - 2\alpha, \quad \gamma_3 = 3C_1 - 2\alpha.$$

As a result, we have approximate expressions for the frequency of the periodic nonlinear vibrations:

$$\omega = \sqrt{A_1}\Omega;$$
$$\Omega = 1 + 0.125\left\{\gamma_2 + 0.03125\left[\gamma_2(\gamma_1 - 2\gamma_2 - 16\alpha)\right.\right.$$
$$\left.\left. - 6C_1\gamma_1 + 80C_2\right]f^2\varepsilon^2\right\}f^2\varepsilon^2.$$

For the clamped edges of the shell the displacement w may be approximated as

$$w(x_1, x_2, t) = f_1(t)\sin^2 s_1 x_1 \cos s_2 x_2 + f_2(t)\sin^2 s_1 x_1,$$

where $f_2 = \frac{3}{16} s_2^2 R^{-1} f_1^2$.

The coefficients of time equation (3.5.49) in this case are

$$A_1 = \frac{8}{3}\left[2\varepsilon_1^2\varepsilon_4 + \varepsilon_1^2\varepsilon_3\varepsilon_4 p^{-2} + \frac{3}{8}\varepsilon_1^2\varepsilon_2 p^{-4} + 2s_2^{-4}(1 - \varepsilon_6 s_2^2)^2\right];$$

$$A_2 = \frac{1}{12} + \frac{3}{2}\left[(\varepsilon_6 s_2^2 - 1) + 0.125\varepsilon_1^2\varepsilon_4 s_2^4\right];$$

$$A_3 = \frac{21}{128}s_2^4; \quad \alpha = \frac{9}{64}s_2^4.$$

The numerical results for the frequency Ω are obtained for the following values of the parameters

$$\varepsilon_1 = 0.015; \quad \varepsilon_2 = 0.00016; \quad \varepsilon_3 = 0.012; \quad \varepsilon_4 = 0.6; \quad \varepsilon_5 = 0.75;$$
$$\varepsilon_6 = 0.005; \quad \varepsilon_7 = 0; \quad LR^{-1} = 2; \quad Rh^{-1} = 500; \quad \overline{f} = fRh^{-1}; \quad \varepsilon = 1;$$
$$m = 1, n = 8.$$

The typical amplitude–frequency dependencies are represented in Fig. 3.7.

From these results it follows that the vibration frequency Ω decreases with increasing amplitude. Consequently, for the stringer shell we have the weak nonlinearity of the soft type.

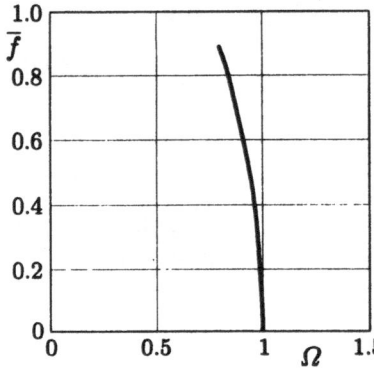

Fig. 3.7. Frequency–amplitude dependence for the stringer shell

As can be seen from the above simplified boundary value problem, after asymptotic decomposition the computation of the shell can be done in several stages; at each stage all equations not higher than the fourth order in x_1 must be considered with the corresponding boundary conditions. Simplified boundary value problems, which are novel in the literature concerning shells, may serve as a basis for calculations of a very wide class of problems for elastic shells. They can also be used as a starting point when seeking further reduced equations under additional simplifying assumptions. We note that the results obtained with the use of this separation are in good agreement with the results of numerical computations carried out without separation of the boundary conditions.

3.5.3 Nonlinear Oscillations of a Cylindrical Panel

The equations of the nonlinear vibrations of an elastic rectangular cylindrical panel [151] together with the initial and boundary conditions can be written in the dimensionless form

$$\Delta_1^2 w + \delta^4 \partial_t^2 w - k\delta^2 \partial_y^2 \Phi = \delta^4 q(x,t) + \delta^2 L(w,\Phi), \tag{3.5.62}$$

$$\Delta_1^2 \Phi + \frac{1}{2}\alpha\delta^2 L(w,w) + \alpha k\delta^2 \partial_y^2 w = 0,$$

$$[w, \partial_t w]_{t=0} = 0, \tag{3.5.63}$$

$$[\partial_x^2 \Phi, \partial_x \partial_y \Phi \partial_y^2 w + \nu\delta^2 \partial_x^2 w, \partial_y^3 w + (2-\nu)\delta^2 \partial_x^2 \partial_y w]_{y=\pm 1} = 0, \tag{3.5.64}$$

$$[w, \partial_x^2 w, \delta^2 \partial_x^2 \Phi - \nu\partial_y^2 \Phi + (2+\nu)\partial_x \partial_y^2 \Phi]_{y=\pm 1} = 0, \tag{3.5.65}$$

$$\Delta_1 = \partial_y^2 + \delta^2 \partial_x^2, \delta = \frac{a_2}{a_1}.$$

Here

$$x_1 = a_1 x, \quad x_2 = a_2 x, \quad W = a_1 w, \quad F = D\Phi, \quad \tau = ct,$$
$$c^2 = \rho h a_1^4 D^{-1}, \quad \alpha = Eha_1^2 D^{-1}, \quad k = a - 1R^{-1}, \quad Q = qDa_1^{-3}. \tag{3.5.66}$$

It is assumed that the transverse load Q is a function of the longitudinal coordinate x_1 and the time τ. The panel platform occupies the rectangle $|x_\beta| \le a_\beta$, $\beta = 1, 2$. The boundary conditions (3.5.64) correspond to a free edge, and (3.5.65) to a fixed hinge support.

Besides problem (3.5.62)–(3.5.65), the nonlinear integro-differential equation of the vibrations of a circular arch, written below in dimensionless form

$$(1 - \nu^2)\partial_x^4 w + \partial_t w - \frac{\alpha}{2}(k + \partial_x^2 w) \int_{-1}^{1} \left[\frac{1}{2}(\partial_x w)^2 - kw \right] dx = q \qquad (3.5.67)$$

$$[w, \partial_t w]_{t=0} = 0, \qquad [w, \partial_x^2 w]_{x\pm 1} = 0$$

is considered.

A natural small parameter δ occurs in the system of (3.5.62)–(3.5.65). Therefore, there is a problem of constructing an asymptotic form as $\delta \to 0$.

Asymptotic expansions are constructed in the form

$$w = \sum_{m=0}^{\infty} \delta^m \left[w_m(x, y, t) + u_m(\frac{1+x}{\delta}, y, t) + v_m(\frac{1+x}{\delta}, y, t) \right], \qquad (3.5.68)$$

$$\Phi = \sum_{m=0}^{\infty} \delta^m \left[\Phi_m(x, y, t) + \varphi_m(\frac{1+x}{\delta}, y, t) + \psi_m(\frac{1+x}{\delta}, y, t) \right].$$

The functions w_m, Φ_m are found by using a first iteration process. For this the solution is sought in the form

$$\{w, \Phi\} = \sum_{m=0}^{\infty} \delta^m \{w_m, \Phi_m\}. \qquad (3.5.69)$$

We substitute (3.5.69) into (3.5.62)–(3.5.65) and collect coefficients of identical powers of δ. Equating the coefficients of δ^0 and δ^1 to zero, to determine w_0, Φ_0 and w_1, Φ_1 we obtain

$$\partial_y^4 w_m = 0; \quad [\partial_y^2 w_m, \partial_y^3 w_m]_{y=\pm 1} = 0; \quad [w_m, \partial_x^2 w_m]_{x=\pm 1} = 0; \qquad (3.5.70)$$
$$m = 0, 1$$
$$\partial_y^4 \Phi_m = 0; \quad [\partial_x^2 \Phi_m, \partial_x \partial_y \Phi_m]_{y=\pm 1} = 0; \quad [\partial_y^2 \Phi_m, \partial_x \partial_y^2 \Phi_m]_{x=\pm 1} = 0;$$

Seeking w_m, Φ_m in the form

$$\{w_m, \Phi_m\} = \sum_{j=0}^{3} y^j \{w_{m,j}, \Phi_{m,j}\}$$

we have from (3.5.70)

$$w_m = w_{m,0}(x, t) + y w_{m,1}(x, t), \quad [w_{m,0}, \partial_x^2 w_{m,0}]_{x=\pm 1} = 0, \quad \Phi_m = 0. (3.5.71)$$

The function $w_{0,0}$ is still uknown and will be determined below. The function Φ_0 is taken to be equal to zero since it follows from the formulation of problem (3.5.62)–(3.5.65) that the function Φ is determined to the accuracy of the linear components in x and y. Continuing the iteration process it is found that the functions $w_{m,j}$, $\Phi_{m,j}$ vanish for odd values of m and j. Consequently, henceforth in this chapter we speak only about evaluating the function $w_{m,j}$, $\Phi_{m,j}$ for even m and j.

Equating the expression for δ^2 to zero and taking (3.5.71) into account, we deduce

$$w_2 = w_{2,0}(x,t) + y^2 w_{2,2}(x,t); \quad 2w_{2,2} = -\nu\partial_x^2 w_{0,0}; \quad (3.5.72)$$
$$\Phi = C_2(t)y^2.$$

The function $w_{2,0}$ is still unknown and will be determined below. At this stage of the first iteration process, the conditions on the boundary $x = \pm 1$ are not satisfied. The discrepancies occurring here are later compensated by using boundary layer functions.

To determine $C_2(t)$ we will use the well-known identity connecting the functions Φ and w for a fixed reinforcement of the boundary $x = \pm 1$ in the longitudinal direction

$$\int_{-1}^{1} (\partial_y^2\Phi - \nu\delta^2\partial_x^2\Phi)\,dx = \alpha\delta^2\int_{-1}^{1}\left[\frac{1}{2}(\partial_x w)^2 - kw\right]dx. \quad (3.5.73)$$

Using (3.5.69) and (3.5.72), we deduce from (3.5.73) that

$$C_2(t) = \frac{\alpha}{4}\int_{-1}^{1}\left[\frac{1}{2}(\partial_x w_{0,0})^2 - kw_{0,0}\right]dx. \quad (3.5.74)$$

Equating the expression for δ^4 to zero, we obtain the system of equations

$$\partial_y^4 w_4 + 2\partial_x^2\partial_y^2 w_2 + \partial_x^4 w_0 - (k + \partial_x^2 w_{0,0})\partial_y^2\Phi_2 = q, \quad (3.5.75)$$
$$[\partial_y^4 w_4 + \nu\partial_x^2 w_2, \partial_y^3 w_4 + (2-\nu)\partial_x^2\partial_y w_2]_{y=\pm 1} = 0,$$
$$\partial_y^4\Phi_4 + \alpha(k + \partial_x^2 w_{0,0})\partial_y^2 w_2 = 0, \quad [\partial_x^2\Phi_4, \partial_x\partial_y\Phi_4]_{y=\pm 1} = 0.$$

We find from (3.5.75)

$$w_4 = \sum_{m=0}^{2} y^{2m} w_{4,2m}(x,t), \quad w_{4,4} = \frac{\nu - 2}{12}\partial_x^2 w_{2,2} \quad (3.5.76)$$

$$w_{4,2} = (1-\nu)\partial_x^2 w_{2,2} - \frac{\nu}{2}\partial_x^2 w_{2,0}.$$

Here $w_{4,0}$ is also an uknown function. Taking account of (3.5.72) and (3.5.74) to determine the principal term of the expansion (3.5.69) from (3.5.75) and (3.5.76), we obtain the integro-differrential equation (3.5.67) for which the zeroth initial and boundary conditions are derived from (3.5.63) and (3.5.65) by using (3.5.71).

Changing to dimensional variables in (3.5.67) by means of (3.5.66), we arrive at the well-known equation of arch vibrations. Furthermore, we find

Φ_2, the principal term of the expansion (3.5.69) for the function Φ from (3.5.72) and (3.5.74).

Let us now construct the next terms of the asymptotic form. It can be shown that w_{2m}, Φ_{2m} are determined in the form

$$\{w_{2m}, \Phi_{2m}\} = \sum_{j=0}^{m} y^{2j}\{w_{2m,2j}, \Phi_{2m,2j}(x,t)\}.$$

In particular, we have from (3.5.72), (3.5.73) and (3.5.75)

$$\Phi_{4,0} = \Phi_{4,4} = \frac{\alpha \nu f}{24}\partial_x^2 w_{0,0}; \quad \Phi_{4,2} = -2\Phi_{4,4}\int_{-1}^{1}\Phi_{4,4}\,dx + \frac{g}{2}.$$

To determine $w_{2,0}$ we derive

$$(1 - \nu^2)\partial_x^4 w_{2,0} + \partial_t^2 w_{2,0} - 2\Phi_{2,2}\partial_x^2 w_{2,0} - fg = 2f\int_{-1}^{1}\Phi_{4,4}\,dx$$

$$+\frac{\nu}{6}\partial_x^2 q + \frac{2}{3}\nu^2(1 - \nu)\partial_x^6 w_{0,0}, \quad w_{2,0}|_{x=\pm 1} = 0 \qquad (3.5.77)$$

$$\left(f = k + \partial_x^2 q w_{0,0}; \quad g = -\frac{\alpha}{2}\int_{-1}^{1}w_{2,0}f\,dx\right).$$

We note that, unlike (3.5.67), equation (3.5.77) is linear. Equating the expressions for δ^{2m+2}, δ^{2m+4} ($m = 2, 3, ...$) to determine the functions w_{2m} and Φ_{2m}, we obtain

$$(1 - \nu^2)\partial_x^4 w_{2m,0} + \partial_t^2 w_{2m,0} - 2f\Phi_{2m+2} - 2\Phi_{2,2}\partial_x^2 w_{2m,0} = l_{2m,0} \quad (3.5.78)$$

$$\Phi_{2m+2,2} = C_{2m+2}(t) - \sum_{j=2}^{m+1} j\Phi_{2m+2,2j}; \quad \Phi_{2m+2,0} = \sum_{j=2}^{m+1}(j-1)\Phi_{2m+2,2j}.$$

The functions $l_{2m,0}$, $\Phi_{2m+2,2j}$ ($j = 2, ..., m+1$) are found in the previous stages of the first iteration process, while the functions $w_{2m,2j}$ are calculated in terms of derivatives of the functions $w_{0,0}$, $w_{2,0}$, ..., $w_{2m-4,0}$. The functions $C_{2m+2}(t)$ ($m > 1$) are determined from identity (3.5.73) on substituting expansion (3.5.68).

The boundary layer functions u_m, $\varphi_m(\nu_m, \psi_m)$, concentrated in the neighbourhood of $x = -1$ ($x = 1$), compensate for the discrepancies in satisfying boundary conditions (3.5.65). They are determined by using the second iteration process. The boundary values for $\partial_x^2 w_{2,0}$, $w_{2m,0}$, $\partial_x^2 w_{2m,0}$ ($m \geq 2$), needed to close (3.5.77) and (3.5.78) are obtained here simultaneously. We substitute (3.5.68) into (3.5.62)–(3.5.65), we take account of the results of the first iteration process, we make a change of the variables $x = -1 + \delta\xi$ ($x = 1 + \delta\zeta$) and we collect coefficients of identical powers of δ. Equating the coefficients for δ^0 to zero, we find a system of nonlinear equations with zero right-hand side for u_0, φ_0 from which we obtain $u_0 + \varphi_0 = 0$. Equating the coefficients for $\delta^1, \delta^2, \delta^3$ to zero we deduce

$$u_1 = \varphi_1 = u_2 = u_3 = 0, \quad \Delta_2^2\varphi_2 = 0, \quad [\partial_\xi^2\varphi_2, \partial_\xi\partial_y\varphi_2]_{y+\pm1} = 0 \quad (3.5.79)$$

$$A\varphi_2|_{\xi=0} = 2\nu C_2(t), \quad B\varphi_2|_{\xi=0} = 0, \quad [A\varphi_2, B\varphi_2]_{\xi=l\to\infty} \to 0$$

$$\Delta_2^2 u_4 = k\partial_y^2\varphi_2, \quad [u_4, \partial_\xi^2 u_4]_{\xi=l\to\infty} \to 0 \quad (3.5.80)$$

$$[\partial_y^2 u_4 + \nu\partial_\xi^2 u_4, \partial_y^3 u_4 + (2-\nu)\partial_\xi^2\partial_y u_4]_{y=\pm1} = 0$$

$$u_4|_{\xi=0} = -w_4|_{x=-1}, \quad \partial_\xi^2 u_4|_{\xi=0} = -\partial_x^2 w_2|_{x=-1}$$

$$(\Delta_2 = \partial_\xi^2 + \partial_y^2, \quad A = \partial_\xi^2 - \nu\partial_y^2, \quad B = \partial_\xi^3 + (2+\nu)\partial_\xi\partial_y^2, \quad l = \frac{2}{\delta}.$$

We note that the boundary value problems for u_m, φ_m are linear for $m \geq 1$. The functions ν_m, ψ_m are found analogously.

We will illustrate the calculation of the boundary layer function u_4 for the case of a rectangular plate ($k = 0$). We construct the solution in the form

$$u_4 = a_0 e^{s_0\xi} F_0(y) + 2\mathrm{Re}\sum_{m=1}^{\infty} a_m e^{-s_m\xi} F_m(y).$$

The Papkovich functions $F_m(y)$ are determined from the boundary value problem (the prime denotes the derivative with respect to y):

$$F_m^{IV} + 2s_m^2 F_m'' + s_m^4 F_m = 0 \quad (3.5.81)$$

$$[F_m'' + \nu s_m^2 F_m, F_m''' + (2-\nu)s_m^2 F_m']_{y=\pm1} = 0$$

(s_0, s_m are, respectively, the real and complex roots of the equation $\Psi(s) = (3 + \nu)\sin 2s -- (1 - \nu)2s = 0$).

To calculate a_m from boundary conditions (3.5.80), the problem is posed of representing the two real functions $f_1 = -w_4(-1, y, t)$ and $f_2 = -\partial_x^2 w_2(-1, y, t)$ in the form of the series

$$\{f_1, f_2\} = \sum_{m=1}^{\infty}\{1, s_m^2\}a_m F_m(y). \quad (3.5.82)$$

Here the time t plays the role of a parameter.

To obtain the initial conditions for $t = 0$ for the function $w_{2m,0}$, we substitute (3.5.68) into (3.5.63), and we collect the coefficients of identical powers of δ and equate them to zero. In particular, the coefficient for δ^0 yields the initial conditions written in (3.5.67) for $w_{0,0}$. The consistency conditions

$$q(\pm1, 0) = \partial_x^2 q(\pm1, 0) = \partial_t q(\pm1, 0) = 0$$

should be satisfied here.

The coefficients of δ^2 and δ^4 are reduced, respectively, to the zero–initial conditions for the functions $w_{2,0}, \partial_t w_{2,0}$ and $w_{4,0}, \partial_t w_{4,0}$. Analogous consistency conditions on the higher derivatives of q are added to construct the next terms of the expansion.

After evaluating the principal terms of expansion (3.5.68), the process of constructing the next terms of the asymptotic form is continued analogously: functions of the first and second iteration processes are determined alternately. The boundary values of the functions of the first iteration process $w_{2m,0}$, $\partial_x^2 w_{2m,0}$ are determined simultaneously in the solution of the boundary layer problems.

In the case of rigid clamping of the panel edges $x_1 = a_1$ ($[w, \partial_x w]_{x=\pm 1} = 0$) the principal term of the expansion is also determined from the equation of arch vibrations, but with the boundary conditions $[w, \partial_x w]_{x=\pm 1} = 0$.

In the case of the hinge of supports or rigid clamping of the boundaries $x_2 = \pm a_2$ there is no passage to the limit from the equations of the vibrations of a cylindrical panel to the equations of the vibrations of an arch.

3.5.4 Stability of Thin Spherical Shells Under Dynamic Loading[1]

As shown below, the theory of two-point Padé approximants gives the possibility of obtaining a solution of very complicated problems. The very interesting example of using this technique is matching by TPPA coefficients of limiting equations and constructing on this basis the constituting equation, which may be exploited for any values of the parameters.

In the course of solving the problem of the stability of shells under dynamic loads, it becomes necessary to describe the motion of its mid-plane with deflections that are large when compared to the thickness. Characteristic forms from the linear theory are usually chosen as the approximating functions when approximate analytical methods are used to reduce the initial system of partial differential equations to a Cauchy problem for ordinary diferential equations. However, the effectiveness of such approaches is limited to the region of small values of the deflection amplitude. Here an asymptotic method is used to obtain the corresponding differential equation describing the motion of a shell with significant deflections. Since one may use as a small parameter a quantity which is proportional to the ratio of the thickness of the shell to the amplitude of its deflection w_0, the resulting equation will be more accurate, the greater the deflection of the shell. To describe the motion of the structure throughout the entire range of displacements, we obtain an ordinary differential equation whose coefficients are determined by combining the corresponding expansions for large and small deflections. The thus-formulated Cauchy problem is solved numerically by the Runge–Kutta method. The efficiency of the proposed approach is evaluated by comparing the results of calculations with known experimental data.

As the initial equations, we will examine the equations of motion of an orthotropic spherical shell written in terms of the stress function Φ and w for the case of the axisymmetric deformation

---

[1] By courtesy of A.Yu. Evkin

$$D_{11}\left(w^{IV} + 2\frac{w^{III}}{r}\right) - D_{22}\left(\frac{w^I}{r^2} - \frac{w^I}{r^3}\right) = \frac{h}{r}\frac{\partial}{\partial r}\left(\frac{\partial w}{\partial r}\frac{\partial \Phi}{\partial r}\right)$$

$$+\frac{h}{rR}\frac{\partial}{\partial r}\left(r\frac{\partial \Phi}{\partial r}\right) + q + -\rho h\frac{\partial w^2}{\partial t^2}; \tag{3.5.83}$$

$$\frac{B_{11}\left(\Phi^{IV} + 2\frac{\Phi^{III}}{r}\right) - B_{22}\left(\frac{\Phi^I}{r^2} - \frac{\Phi^I}{r^3}\right)}{B_{11}B_{22} - B_{12}^2} + \frac{1}{r}\frac{\partial^2 w}{\partial r^2}\frac{\partial w}{\partial r} + \frac{1}{rR}\frac{\partial}{\partial r}\left(r\frac{\partial w}{\partial r}\right) = 0.$$

After the substitution of variables

$$z = \frac{r^2}{w_0 R}, \quad F = \frac{h\Phi}{w_0}\sqrt{\frac{B_{11}}{D_{11}(B_{11}B_{22} - B_{12}^2)}}, \quad W = \frac{w}{w_0},$$

$$\bar{q} = \frac{qR^2}{4}\sqrt{\frac{B_{11}}{D_{11}(B_{11}B_{22} - B_{12}^2)}}, \quad \varepsilon^2 = \frac{4}{w_0}\sqrt{\frac{B_{11}D_{11}}{(B_{11}B_{22} - B_{12}^2)}},$$

we obtain

$$\varepsilon^2[z^2 W^{IV} + 4z W^{III} + W^{II}(9 - \lambda_0)/4]$$

$$= 2(zw^I F^I)^I + (zF^I)^I\bar{q} - c\frac{\partial^2 w}{\partial \tau^2}, \tag{3.5.84}$$

$$\varepsilon^2[z^2 F^{IV} + 4z F^{III} + F^{II}(9 - a)/4]$$

$$= W^I(2zW^I + W^I) + (zW^I)^I, \tag{3.5.85}$$

where

$$\lambda_0 = \frac{D_{22}}{D_{11}}, \quad a = \frac{B_{22}}{B_{11}}, \quad C = \frac{w_0^2\rho h^2 R^2}{4}\sqrt{\frac{B_{11}}{D_{11}(B_{11}B_{22} - B_{12}^2)}},$$

$$w_0 = \frac{w}{h},$$

$\tau = w_0 t$, and w_0 is the natural frequency of linear vibrations of the shell. In the case of an isotropic sphere

$$\varepsilon^2 = \frac{2h}{w_0\sqrt{3(1 - \nu^2)}}, \quad \lambda_0 = 1, \quad a = 1,$$

$$c = \frac{\rho R^2 w_0^2\sqrt{3(1 - \nu^2)}}{2E}.$$

When the amplitudes of the deflection w_0 are sufficiently great when compared to the thickness of the shell, the parameter ε^2 becomes small and can be used in an asymptotic integration of system (3.5.84)–(3.5.85). In [36d], the corresponding procedure was performed for the case of a static load on an isotropic spherical shell. It was established that the main approximation of the asymptote yields good results when $w_0/h \geq 4$. In the case of an orthotropic sphere, the parameter ε^2 can also be regarded as small for the corresponding deflection if we exclude from consideration shells in which there is a substantial increase in flexural rigidity in the meridional direction D_{11}.

When $\varepsilon = 0$, (3.5.85) has two solutions. The first, $W^I = 0$, corresponds to the momentless state of the shell. The second solution, $W^I = -1$, corresponds to the mirror reflection of the part of the shell relative to the plane whose intersection with the sphere gives a circle of radius $r_1 = \sqrt{w_0 R}$. Thus, in the case of large deflections, the form of the shell becomes determinated and in the initial variables it is described by the function

$$w = \begin{cases} 0 & r \geq r_1, \\ w_0 \left(1 - \frac{r^2}{w_0 R}\right) & r \leq r_1. \end{cases}$$

In the neighbourhood of $r = r_1$ ($z = 1$), discontinuities in the derivatives are compensated for by rapidly changing functions of the internal boundary layer. We obtain the following equations for them in the main approximation of the asymptote:

$$v^{IV} + 2(v^I u^I)^I - u^{II} = 0, \quad u^{IV} + v^I(1 - 2v^I) = 0, \tag{3.5.86}$$

these equations coinciding with the corresponding equations of the static problem [36d]. Here, the functions u and v are differentiated with respect to the variable $x = (1 - z)/\varepsilon$, $W = \varepsilon(\tau)v(x)$, $F = \varepsilon(\tau)u(x)$.

The boundary conditions for the sphere which is rigidly fixed at its boundary (at $r = r_0$) take the form

$$v^I = 0, \quad u^{II} = 0, \quad x = -x_0; \tag{3.5.87}$$
$$v^I = 1, \quad u^I = 0, \quad x \to +\infty, \tag{3.5.88}$$

where

$$x_0 = \frac{z_0 - 1}{\varepsilon}, \quad z_0 = \frac{r_0^2}{w_0 R}.$$

The relations of boundary value problem (3.5.86)–(3.5.87) were obtained with the assumption that

$$\bar{q} \ll 1, \quad c\frac{\partial^2 w}{\partial \tau^2} \sim 1. \tag{3.5.89}$$

In this case, neither the load nor the inertial term enters into the equations or the boundary conditions for the functions u and v describing the stress state of the internal boundary layer. It should be noted that the satisfaction of relations (3.5.89) is necessary only for large deflections. Thus, it is satisfied in all cases of practical importance. In particular, it is possible to study the action of shock loads in which the parameter q can rearch large values. However, due to the instantaneous nature of these loads, the given parameter remains small even when the deflections are substantial.

The equation needed to determine the function $w_0(t)$ or $\varepsilon(\tau)$ can be obtained by the variational method. As in [36d], we obtain the following equation for the total potential energy of the system when a uniformly distributed radial pressure acts on the surface:

$$U = \frac{D_1}{\varepsilon^4}(J_0\varepsilon - \bar{q}),$$

where

$$D_1 = \frac{32\pi D_{11}h\sqrt{b}}{R}, \quad b = \frac{D_{11}B_{11}}{h^2(B_{11}B_{22} - B_{12}^2)},$$

$$J_0 = \int_{-x_0}^{\infty} \left[(u^{II})^2 + (v^{II})^2 \right] dx = 0.56 + \frac{0.2}{x_0^{3/2}}.$$

An expression for $J_0(x_0)$ was obtained after the numerical solution of boundary value problem (3.5.86)–(3.5.87) for different values of x_0 and a subsequent approximation of the corresponding function in the interval $0,5 \leq x_0 \leq \infty$. The last term is connected with the effect of the edge on the deformation of the shell.

One obtains the following expression in the main approximation for the kinetic energy of the system:

$$K = \frac{\pi h^4 R\rho \dot{f} f^2 \omega_0^2}{2},$$

where $f = w_0/h$.

The corresponding equation of motion has the form

$$\ddot{f}f + \dot{f}^2 + w_*^2\sqrt{f} = \frac{4f\sqrt{b}\,(\bar{q}(\tau) + \bar{q}_0)}{\lambda}. \tag{3.5.90}$$

Here, one has isolated the dynamic $q(\tau)$ and static q_0 components of the load:

$$\lambda = \frac{\omega_0^2 \rho R^2 B_{11}}{B_{11}B_{22} - B_{12}^2}, \quad w_*^2 = \frac{6\bar{J}_0 b^{3/2}}{\lambda},$$

$$\bar{J}_0 = 0.56 + \frac{0.1\sqrt{\bar{f}}b^{3/4}}{\left(\sqrt{2H/h} - \sqrt{f}\right)^{5a}} \quad (f < 2H),$$

where H is the camber of the shell.

One may describe the motion of a shell with small deflections by the Ritz method. Here, we make use of the following approximation of the deflection function:

$$w(r,t) = \begin{cases} f(t)h\left[1 - \left(\frac{r}{r_*}\right)^2\right]^2 & 0 \leq r \leq r_* \\ 0 & r_* \leq r \leq r_0, \end{cases} \tag{3.5.91}$$

where r_0 is the radius of the circumference of the shell in the plane.

Such a function was used in [96] for the case when $r_* = r_0$, in the study of the stability of an isotropic medium under dynamic loads. Here, let us examine an orthotropic shell. It is also assumed that the quantity r_* is arbitrary within the interval $0 \leq r_* \leq r_0$. The stress function for which the strain-compatibility equation (3.5.83) is satisfied and continuity of the displacements at $r_* = r_0$ is assured has the form

$$\frac{\partial \Phi}{\partial r} = B_{22}\frac{4r_* f(1 - \nu_{12}\nu_{21})}{R}\left(B_1 d - \frac{d^3}{9-a} + \frac{d^5}{25-a}\right)$$
$$-B_{22}\frac{8f^2(1 - \nu_{11}\nu_{21})}{r_*}\left(B_2 d + \frac{d^3}{9-a} - \frac{d^5}{25-a} + \frac{d^7}{49-a}\right),$$

where

$$\nu_{12} = \frac{B_{12}}{B_{11}}, \quad \nu_{21} = \frac{B_{12}}{B_{22}}, \quad B_1 = \frac{1}{1-\nu_{12}}\left(\frac{\nu_{12}-5}{25-a} - \frac{\nu_{12}-3}{9-a}\right);$$
$$B_2 = \frac{1}{1-\nu_{12}}\left(\frac{\nu_{12}-3}{9-a} - \frac{2\nu_{12}-10}{25-a} + \frac{\nu_{12}-7}{49-a}\right),$$
$$d = \frac{r}{r_*}.$$

One obtains the equation of motion in the form

$$\ddot{f} + f + \beta f^2 + \eta f^3 = \frac{4\gamma\sqrt{b}\bar{q}(\tau)}{\lambda}, \tag{3.5.92}$$

where

$$\lambda = \frac{40}{k^2}\left[b(9 - \lambda_0) + k^2\left(B_1 - \frac{1}{2(9-a)} + \frac{3}{10(25-a)}\right) - \bar{q}_0 k\sqrt{b}\right];$$
$$\beta = -\frac{80}{2k\lambda}\left[B_1 - B_2 - \frac{9}{10(9-a)} + \frac{4}{5(25-a)} - \frac{1}{5(49-a)}\right];$$
$$k = \frac{r_*^2}{Rh}, \quad \gamma = \frac{5}{3},$$
$$\eta = -\frac{160}{3\lambda k^2}\left[B_2 + \frac{2}{5(9-a)} - \frac{2}{5(25-a)} + \frac{4}{35(49-a)}\right].$$

An equation describing the motion of the shell with both large and small deflections can be written in the form

$$A_0\ddot{f} + A_1\dot{f}^2 + A_2 f = \frac{4A_3\sqrt{b}\bar{q}(\tau)}{\lambda} + \frac{4\alpha f\sqrt{b}\bar{q}_0}{\lambda}, \tag{3.5.93}$$

where the coefficients A_i are obtained by combining the corresponding asymptotic representations of the coefficients of (3.5.93) in the form (3.5.90) and (3.5.92). Using the Padé approximation we find

$$A_0 = 1 + af, \quad A_1 = \frac{af}{1+f}, \quad A_2 = \frac{1 + a\beta w_*^2\sqrt{f}}{1 + a\beta w_*^2\sqrt{f} + \beta f},$$
$$A_3 = \gamma + af, \quad a = \frac{\pi Rh^3 f^2}{\iint\limits_F w^2 dF},$$

where $w(r)$ is the deflection function describing the form of the shell with small deflections. In accordance with (3.5.91), $\alpha = 5/k$.

It is easily shown that as $f \to 0$ we obtain (3.5.92), since

$$A_0 \to 1, \quad A_1 \to 0, \quad A_2 = 1 + \beta f + O(f^2), \quad A_3 \to \gamma.$$

Similarly, if we pass to the limit with $1/f \to 0$, we obtain

$$A_0 \sim af, \quad A_1 \to a \quad A_2 \sim \frac{a\omega_*^2}{\sqrt{f}}, \quad A_3 \sim af,$$

which, to within the constant factor α, corresponds to (3.5.90).

In accordance with the above mathematical model describing the motion of a shell with large deflections, Fig. 3.8 shows three qualitatively different states of shells corresponding to different levels of deflection. In the case of relatively small deflections ($f \sim 1$), the shell undergoes bending in the region of the vertex with radius r_*. With an increase in deflection ($f \gg 1$), the form of the shell becomes close to a mirror reflection of its surface perpendicular to the axis of rotation. The reflection region (in which the shell undergoes bending) increases in size with an increase in deflection. The radius of its circumference is $r_1 = \sqrt{fhR}$. Such a shape is energetically advantageous for the shell, since the membrane strains are concentrated within a narrow zone of the internal boundary layer (at $r \approx r_1$). The third state of the shell in the case of snap-through is characterized by the effect of the fastened edge. The latter prevents a further increase in deflection without membrane strains, which in turn leads to a sharp increase in the stiffness of the structure.

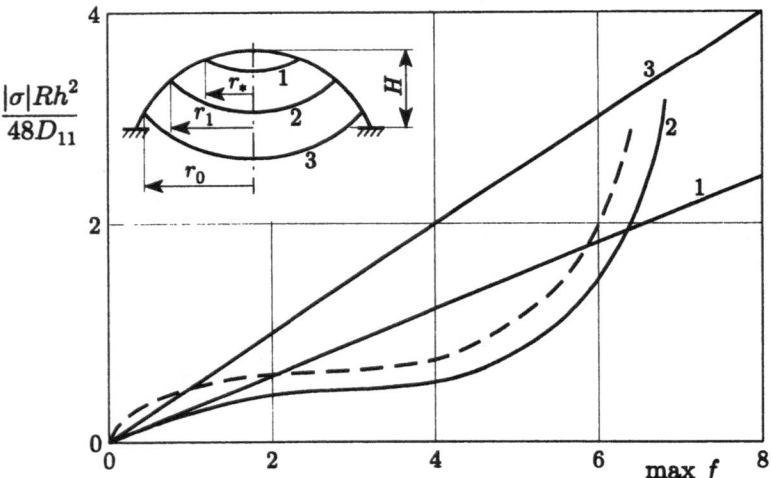

Fig. 3.8. Comparison of the various theoretical approaches

Formulas to determine the largest bending stresses due to the change in the curvature of the shell in the meridional direction were obtained for the corresponding ranges of deflections. With small f, we have

$$\max |\sigma| = \frac{48 D_{11} f}{k R h^2} \quad (r = r_*).\tag{3.5.94}$$

With large f, we obtain the asymptotic formula

$$\max |\sigma| = \frac{5.7 D_{11} \sqrt{f}}{R h^2 b^{1/4}} \left(1 + \frac{0.33 b^{1/4}}{\sqrt{2H/h - \sqrt{f}}} \right) \quad \text{at } r = r_1.\tag{3.5.95}$$

After combining the given expansions, we arrive at the relation

$$\max |\sigma| = \frac{48 D_{11}}{k R h^2} \frac{f}{1 + \frac{48 b^{1/4} \sqrt{f}}{5.7 k} \left(1 + \frac{0.33 b^{1/4}}{\sqrt{2H/h - \sqrt{f}}} \right)^{-1}},\tag{3.5.96}$$

$$f < \frac{2H}{h},$$

which corresponds to curve 2 ($k = 4$) for the isotropic case in Fig. 3.8. For comparison, for a shell with the characteristic parameter $k = 6.87$, relations 1 and 3 obtained using (3.5.94) with $k = 6.87$ and $k = 4$ are shown, respectively. The dashed line in Fig. 3.8 corresponds to (3.5.95) ($k = 4$).

Equation (3.5.93) was solved in specific cases by the Runge–Kutta method in order to determine the maximum response of the system to a specified uniform external pressure. Here, the value of k – determining the form of the shell on the initial section of motion – was chosen from values within the range from 0 to $k = r_0^2/(Rh)$. It was found that if $r_0^2/(Rh) \leq 4$, then the maximum deflection will be $k = r_0^2/(Rh)$ for various shell parameters and different types of loads with a characteristic duration comparable to the period of natural vibrations of the system. In the opposite case, $k = 4$. We henceforth fixed k in our calculations.

To check the validity of the proposed mathematical model, we compared the results of the calculation with known empirical data. In particular, we examined data for spherical shells of steel 3 tested in shock loading with a uniform external pressure after having been fixed at their boundaries [150]. The test data (shells of the series 1USD) are shown in Table 3.5. The results calculated in accordance with the loading conditions achieved in the experiment are also shown for comparison. Since the experimental criterion for attainment of the critical value of q_{0n} was the appearance of appreciable residual deflections, as the critical value in the calculations we took the pressure q_p at which, in accordance with (3.5.96), the maximum bending stresses reached the yield point. The experimental and theoretical data shown in the table agree well with one another. The maximum relative deflections max f due to dynamic loading are also shown.

It should be noted that the formation of several different indentations was observed experimentally, the number of such indentations generally ranging from two to four for the given series. Thus, the mode of buckling was not symmetrical if the axis of rotation of the shell was regarded as the symmetry axis. However, the circular form of the dents seen in the experiment and their

Table 3.5. Comparison of the theoretical and experimental data

No.	h [mm]	H [mm]	R/h	$q_{n0} \cdot 10^{-5}$ [N/m^2]	$Q_p \cdot 10^{-5}$ [N/m^2]	max f
1	0.27	27.0	720	2.32	2.8	1.9
2	0.27	25.3	768	2.36	2.6	2.1
3	0.27	20.7	941	2.06	1.9	2.7
4	0.27	16.2	1200	1.56	1.6	4.2

closeness to the isometric transformation of the sphere obtained by mirror reflection of the part of it relative to a plane make it possible to suggest that in the absence of a substantial effect from the fastening of the edge of the shell and constrained growth of the dents due to their interaction, each of them can be examined within the framework of the axisymmetric theory. Here, the symmetry axis for each dent will be its "own" axis. Its position will be determined by the point of the middle surface of the shell with the maximum deflection. Thus, we can expect that the relations obtained above might also be effectively used in other cases of nonaxisymmetric buckling of shells with large deflections.

Figure 3.9 compares the results of calculations (curve 1) and tests [97] performed for an isotropic shell with the parameters $2H/h = 6.87$ and $R/h = 365$. The shell deformed in the elastic stage during buckling. The conditions under which the shell was loaded by external pressure can be represented in the form

$$\bar{q}(\tau) = \begin{cases} 10\tau_0 \bar{q}_* & 0 \leq \tau_0 \leq 0.1 \\ \dfrac{\bar{q}_*(5.9 - \tau_0)}{5.8} & 0.1 \leq \tau_0 \leq 3 \\ \dfrac{\bar{q}_*(178 - \tau_0)}{350} & 3 \leq \tau_0 \leq 178, \end{cases}$$

where $\tau_0 = \tau/\lambda$.

The results from theory and experiment agree well with one another. The dashed line in Fig. 3.9 shows the results calculated by the method in [49d], which corresponds to the solution of (3.5.92) with $k = r_0^2/(Rh) = 6.87$. The agreement with the test data is somewhat poorer in this case. Calculations performed with both mathematical models showed that the difference between the results increases with an increase in the parameter $2H/h$. Also the difference in the description of the stress state by the different theoretical methods is substantial, as illustrated by curves 1 and 2 (relation 2 was obtained by the method proposed in the present study).

Curves 2–4 in Fig. 3.9 show the way in which the dynamic stability of the shell is affected by the additional static component of the external pressure with $q_0 = 0.1, 0.2$, and 0.3, respectively. In the ranges of q_0 of practical importance, there is a slight decrease in the critical dynamic pressure.

As an example, we also examined the well-studied theoretical case of dynamic loading by an instantaneous applied external pressure which then

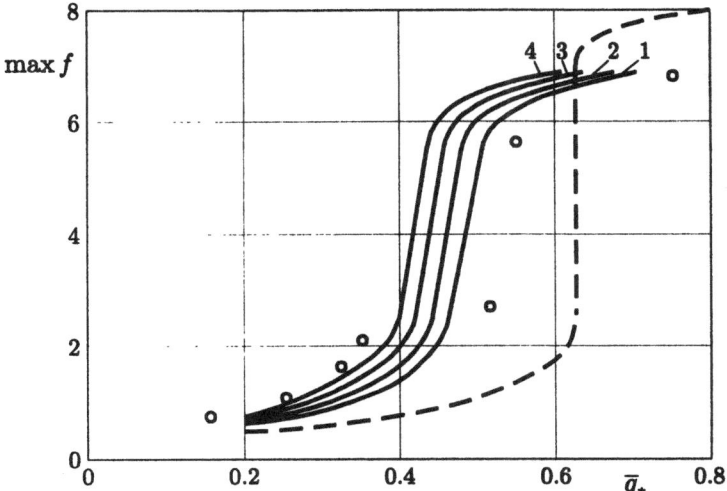

Fig. 3.9. Comparison of the theoretical and numerical results

remains constant over time. We took the load at which the corresponding unstable intermediate equilibrium position is attained as the critical value. For this load, there was also a sharp increase in the amplitude of the deflection. Figure 3.10 shows the results of the calculation performed by the proposed method (curve 1) along with the results obtained in [149, 44d] by powerful numerical methods (curves 2-3). The experimental data from [112] are represented by clear circles. The dark circles represent the critical values of pressure calculated in accordance with the asymmetric theory [149].

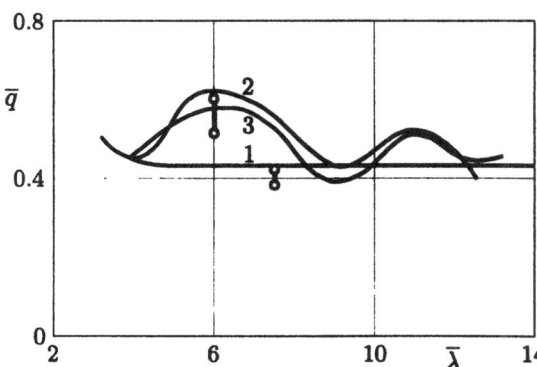

Fig. 3.10.
Comparison of numerical results with the two-point Padé approximants formula

Here, $\lambda = 2[3(1 - \nu^2)]^{1/4}(H/h)^{1/3}$. It is evident from Fig. 3.10 that the numerical solutions give relations which oscillate and decay with a decrease in the thickness of the shell and approach the above-found asymptotic value $q \approx$

0.43. This phenomenon can be exploited to estimate the critical load, with consideration of the experimental data and the potential for the asymmetric buckling of spherical shells.

3.5.5 Asymptotic Investigation of the Nonlinear Dynamic Boundary Value Problem for a Rod

As is well known, asymptotic approaches for nonlinear dynamics of continuous systems are well developed for infinite spatial variables. For systems of finite size we have an infinite number of resonances, and the Poincaré–Lighthill method does not work. The use of an averaging procedure [59d] or the method of multiple scales [107] leads to infinite sytems of nonlinear algebraic or ordinary differential equations, and a subsequent truncation method does not provide the possibility of obtaining all the important properties of the solutions. In this chapter we use an asymptotic procedure which is based on the introduction of an artificial small parameter.

Let us assume a governing boundary value problem in the following form

$$\frac{\partial^2 U}{\partial x^2} - \frac{\partial^2 U}{\partial \tau^2} = -\varepsilon U^3, \tag{3.5.97}$$

where all variables are nondimensional, and ε is a nondimensional small parameter ($\varepsilon \ll 1$). From the physical point of view we have longitudinal vibrations of a rod with nonlinear drag. Let us introduce a change of the variable

$$t = \omega \tau. \tag{3.5.98}$$

We will now search for solutions using the ansatzes

$$U = U_0 + \varepsilon U_1 + \varepsilon^2 U_2 + \ldots,$$
$$\omega = 1 + \varepsilon \omega_1 + \varepsilon^2 \omega_2 + \ldots. \tag{3.5.99}$$

After substituting expresions (3.5.98) and (3.5.99) into the governinig boundary value problem (3.5.97) and splitting it with respect to ε, one obtains

$$\frac{\partial^2 U_0}{\partial x^2} - \frac{\partial^2 U_0}{\partial t^2} = 0, \tag{3.5.100}$$

$$\frac{\partial^2 U_1}{\partial x^2} - \frac{\partial^2 U_1}{\partial t^2} = 2\omega_i \frac{\partial^2 U_0}{\partial t^2} - U_0^3. \tag{3.5.101}$$

The solution (3.5.100) may be written in the form

$$U_0 = C_1 \sin x \sin t + C_2 \sin 2x \sin 2t + \ldots = \sum_{i=1}^{\infty} C_i \sin ix \sin it.$$

Here C_1 is the amplitude of the fundamental oscillation, while the constants C_i for $i > 1$ provide the next approximations. After some routine but cumbersome transformations, we arrive at the following infinite nonlinear algebraic equations

$$-32\omega_1 C_1 \sin x \sin t = \sum_{a=1}^{\infty} C_a C_b (3C_{a+b-1} + 3C_{a+b+1}) + 6C_1 \sum_{a=1}^{\infty} C_a^2,$$

$$-32\omega_1 i^2 C_i \sin ix \sin it = \sum_{a=1}^{\infty} C_a C_b (3C_{a+b-i} + 3C_{a+b+i} + C_{i-a-b})$$

$$+6C_i \sum_{a=1}^{\infty} C_a^2. \tag{3.5.102}$$

Systems like (3.5.102) may be obtained in various ways and the main problem in this approach consists in its solution. The truncation of the infinite system (3.5.102) cannot give any information about resonances of a higher order. We propose to introduce an artificial small parameter μ, writing it near all nondiagonal members of system (3.5.102), and representing the unknown coeficients as an expansion:

$$C_n = C_n^{(0)} + C_n^{(1)}\mu + C_n^{(2)}\mu^2 + \ldots, \quad n = 2, 3, \ldots$$
$$\omega_1 = \omega_1^{(0)} + \omega_1^{(1)}\mu + \omega_1^{(2)}\mu^2 + \ldots.$$

After splitting with respect to μ, solutions may be obtained routinely. It may be easily shown that for even n

$$C_n^{(k)} = 0$$

and

$$\omega_1^{(0)} = -0.281250 C_1^2, \quad \omega_1^{(0)} = -0.001438 C_1^2,$$
$$C_3^{(0)} = 0.0144927 C_1, \quad C_5^{(0)} = 0.0002071 C_1, \quad C_7^{(0)} = 0.0000030 C_1.$$

Numerical results (the dependencies of the fundamental nondimensional frequency ω upon the nondimensional amplitude C_1) are displayed in Fig. 3.11 for various values of the small parameter ε.

3.6 One-Point Padé Approximants Using the Method of Boundary Condition Perturbation

An analysis of plates under mixed boundary conditions represents a significant practical value: a lot of problems, arising in machine design, civil engineering, etc., are reduced to similar ones. These problems are usually solved using numerical methods [42d]. Nevertheless, the numerical approach does not adequately fit the requirements of optimal structural design or any other kind of optimal structural design ideology. The approximate analytical expansion, accurate enough, will be of great practical advantage for these needs.

The basic idea of the present method may be described as follows. The parameter ε is introduced in the boundary conditions in such a way that $\varepsilon = 0$ corresponds to the common problem under consideration [1]. Then

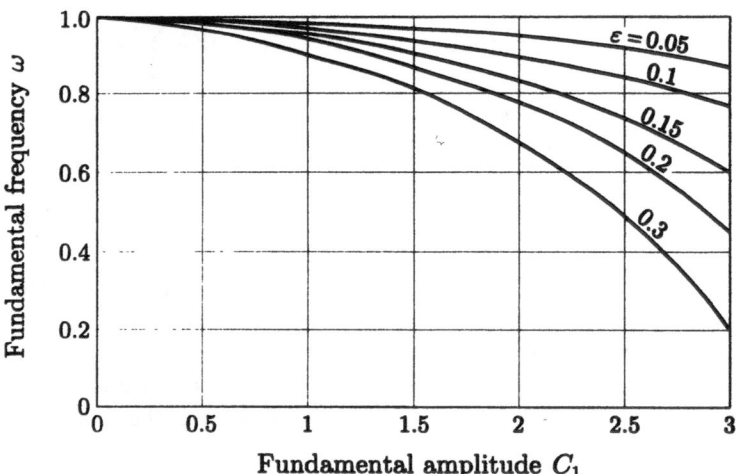

Fig. 3.11. Amplitude–frequency dependencies for fundamental oscillations for various values of the small parameter ε

the ε-expansion of the solution is obtained. As a rule, the expansion fails to converge at the point $\varepsilon = 1$. The PA may be used to eliminate that drawback. Let us produce the PA-definition.

For an expansion given by (3.6.1)

$$F(\varepsilon) = \sum_{i=0}^{\infty} C_i \varepsilon^i \qquad (3.6.1)$$

the fraction-rational function is

$$F(\varepsilon)[m/n] = \left(\sum_{i=0}^{m} a_i \varepsilon^i\right)\left(\sum_{i=0}^{n} b_i \varepsilon^i\right)^{-1}. \qquad (3.6.2)$$

$F(\varepsilon)[m/n]$ represents the PA of expansion (3.6.1) if the Maclaurin series of $F(\varepsilon)$ shows the coincidence of its coefficients with the corresponding coefficients of (3.6.1) up to terms of $(m + n + 1)$-th order. The features of the PA are as follows: it possesses uniqueness while m and n are chosen; it performes a meromorphic continuation of the function; for its definition from the source expansion (3.6.1) the linear algebraic problem arises [42, 43].

It is quite often convenient to perform an asymptotic expansion in terms of a small parameter. Namely, the parameter ε is introduced into the initial equations or the boundary conditions in such a way that for $\varepsilon = 1$ we have the initial boundary problem, and for $\varepsilon = 0$, a simplified problem admitting a simple solution. Then, an expansion in terms of the parameter ε is constructed. Unfortunately, the obtained series is usually not convergent for $\varepsilon = 1$. To eliminate this disadvantage, we can apply the PA. Here is an example. Dorodnitsyn A.A. has proposed a method of perturbation of the form of the boundary conditions based on the introduction of a formal parameter ε. In order to remove the divergence for $\varepsilon = 1$, analytic continuation has been used but the efficiency of that technique is not high. The PA, on the contrary, gives good results.

Let us consider the flexural vibration of a rectangular plate ($-0.5k \leq x \leq 0.5k$, $-0.5 \leq y \leq 0.5$), simply supported at $x = \pm0.5k$, and having mixed boundary conditions of the "clamped–simple supported" type, symmetrical to the y axis on the sides $y = \pm0.5$. The initial equation is

$$\nabla^4 w - \lambda w = 0.$$

The boundary conditions have the form

$$w = 0, \quad w_{,xx} = 0 \quad \text{for} \quad x = \pm0.5k;$$
$$w = 0, \quad w_{,yy} = \bar{H}(x)\varepsilon(w_{,yy} \mp w_{,y}) \quad \text{for} \quad y = \pm0.5,$$

where $\bar{H}(x) = H(x - \mu) + H(-x - \mu)$; $H(x)$ is Heaviside's function.

Substituting w and λ into the form of the ε-series

$$w = w_0 + \varepsilon w_1 + \ldots, \tag{3.6.3}$$
$$\lambda = \lambda_0 + \varepsilon \lambda_1 + \ldots, \tag{3.6.4}$$

after applying the usual procedure of the perturbation method, we have

$$\lambda_0 = \pi^4 \psi^2, \qquad \lambda_1 = 4\pi^2 n^2 \gamma_{mn},$$

$$\lambda_2 = 4\pi^2 n^2 \gamma_{mn} \left\{ 1 - \frac{\gamma_{mn}}{\pi^2 \psi} \left[\frac{\pi \alpha}{2} \mathrm{cth}^{(-1)^m} \frac{\pi \alpha}{2} + \frac{n^2}{\psi} - \frac{3}{2} \right] \right\}$$

$$- \frac{2n^2}{\psi} \sum_{\substack{\{ i = 1, 3, 5, \ldots \\ i = 2, 4, 6, \ldots \} \\ \{ i^2 > m^2 + n^2 k \\ i^2 < m^2 + n^2 k \}}}^{\infty,} \gamma_{im} \left[\alpha_i \mathrm{cth}^{(-1)^i} \frac{\alpha_i}{2} + \left\{ \begin{matrix} \phi_i \mathrm{cth}^{(-1)^i} \phi_i/2 \\ \beta_i \mathrm{cth}^{(-1)^i} \beta_i/2 \end{matrix} \right\} \right],$$

Here $\psi = n^2 + \frac{m^2}{k^2}$, $\alpha = \sqrt{2\frac{m^2}{k^2} + n^2}$, $\alpha_i = \sqrt{\frac{i^2 + m^2}{k^2} + n^2}$, $\beta_i = \pi\sqrt{\frac{m^2 - i^2}{k^2} + n^2}$, $\phi_i = \sqrt{\frac{i^2 - m^2}{k^2} - n^2}$,

$$\gamma_{im} = \left\{ \begin{matrix} 2(0.5 - \mu) + \frac{(-1)^m}{\pi m} \sin 2\pi\mu m, & \text{for } i = m \\[2mm] \frac{4}{\pi} \frac{1}{(m^2 - i^2)} \left[\left\{ \begin{matrix} i \\ m \end{matrix} \right\} \sin \pi\mu i \cos \pi\mu m + \right. \\[2mm] \left. - \left\{ \begin{matrix} m \\ i \end{matrix} \right\} \sin \pi\mu m \cos \pi\mu i \right] & \text{for } i \neq m, \end{matrix} \right.$$

and $\sum'$ denotes the summation without the component $i = m$.

Let us compare the frequencies given by this method with the exact values for the limit case ($\mu = 0$) when both sides $y = \pm0.5$ are completely clamped. For the square plate $\lambda = (1.4783\pi)^4$. The PA of a segment of series (3.6.4) is

$$\lambda(\varepsilon) = \frac{a_0 + a_1\varepsilon}{b_0 + b_1\varepsilon},$$

where $a_0 = \lambda_0$, $b_0 = 1$, $a_1 = \lambda_1 + b_1\lambda_0$, $b_1 = -\lambda_2/\lambda_1$.

For $\varepsilon = 1$ one obtains $\lambda = (1.7081\pi)^4$; the numerical solution is $\lambda = (1.7050\pi)^4$.

Figure 3.12 presents the diagram of the relation of λ to μ for the cases of the symmetrical and the nonsymmetrical restraint layout.

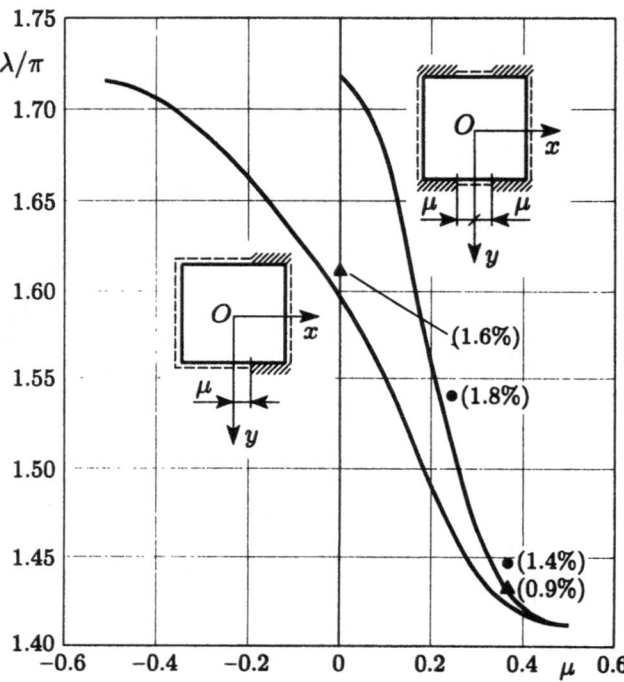

Fig. 3.12. The relationship between the vibration frequency of the plate partially clamped on two opposite sides and the clamped segment length

Let us consider the application of this approach to the static analysis of the rectangular plate ($-0.5a < \bar{x} < 0.5a$; $-0.5b < \bar{y} < 0.5b$), subjected to a uniform lateral load $\bar{q}$. The plate is simply supported along $\bar{x} = \pm a/2$ and subjected to mixed boundary conditions ("clamped-hinged"), symmetrical with respect to $\bar{y}$.

The governing differential equation may be written as

$$D\nabla^4\bar{W} = \bar{q}. \tag{3.6.5}$$

Let us denote

$$W = \frac{\bar{W}}{b}, \quad y = \frac{\bar{y}}{b}, \quad x = \frac{\bar{x}}{b},$$

$$k = \frac{a}{b}, \quad q = \frac{\bar{q}b^4}{D}. \tag{3.6.6}$$

Taking (3.6.6) into account, the governing equation (3.6.5) may be rewritten as

$$\nabla^4 W = q. \tag{3.6.7}$$

The boundary conditions may be formed as

$$W = 0, \quad W_{xx} = 0, \quad \text{when } x = \pm\frac{k}{2}; \tag{3.6.8}$$

$$W = 0, \quad (1 - \bar{H}(x))W_{yy} \pm \bar{H}(x)W_y = 0, \quad \text{when } y = \pm\frac{1}{2}, \tag{3.6.9}$$

where $\bar{H}(x) = H(x - \mu k) + H(-x - \mu k)$.

Introducing the parameter ε into the boundary condition according to the procedure, one obtains

$$W = 0, \quad W_{yy} = \bar{H}(x)\varepsilon(W_{yy} \pm W_y) \quad \text{when } y = \pm\frac{1}{2}. \tag{3.6.10}$$

The case $\varepsilon = 0$ gives us a plate which is simply supported along the boundary; the case $\varepsilon = 1$ corresponds to the problem under consideration (3.6.8)–(3.6.9).

The intermediate values of ε are related to mixed conditions of a "simply supported-elastic clamping" kind with the elastic support coefficient $u = \varepsilon/(1 - \varepsilon)$.

In order to solve the problem, let us represent the deflection of the plate as

$$W = W_1 + W_2, \tag{3.6.11}$$

$$W_1 = \frac{q}{8k} \sum_{m=1,3,5,\dots}^{\infty} \frac{(-1)^{(m-1)/2}}{\alpha_m} \left(1 - \frac{\alpha_m \text{th}\,\alpha_m + 2}{2\text{ch}\,\alpha_m} \text{ch}\, 2\alpha_m y \right.$$

$$\left. + \frac{\alpha_m}{\text{ch}\,\alpha_m} y\text{sh}\, 2\alpha_m y \right) \cos 2\alpha_m x; \tag{3.6.12}$$

$$W_2 = \frac{1}{8} \sum_{m=1,3,5,\dots}^{\infty} \frac{(-1)^{(m-1)/2}}{\alpha_m^2} \frac{A_m}{\text{ch}\,\alpha_m} (\alpha_m \text{th}\,\alpha_m \text{ch}\,\alpha_m y$$

$$- 2\alpha_m y\text{sh}\, 2\alpha_m y) \cos 2\alpha_m x; \tag{3.6.13}$$

$$\alpha_m = \frac{\pi m}{2k}. \tag{3.6.14}$$

Expression (3.6.12) describes the deflection of the simply supported plate subjected to the uniform lateral load q. Expression (3.6.13) describes the deflection of the simply supported plate, caused by the edge bending moments, distributed along $y = \pm0.5$:

$$M_y|_{y=\pm0.5} = \sum_{m=1,3,5,\dots}^{\infty} A_m(-1)^{(m-1)/2} \cos 2\alpha_m x. \tag{3.6.15}$$

Satisfying boundary condition (3.6.10), one obtains the infinite linear algebraic system for the coefficients A_i as the unknowns:

$$A_i(-1)^{(i-1)/2} = \varepsilon \sum_{m=1,3,5,\ldots}^{\infty} \gamma_{im}(-1)^{(m-1)/2} A_m \left[1 - \frac{1}{4\alpha_m}\left(\frac{\alpha_m}{\text{ch}\,^2\alpha_m}\right.\right.$$

$$\left.\left. + \text{th}\,\alpha_m\right)\right] + +\varepsilon\frac{q}{8k}\sum_{m=1,3,5,\ldots}^{\infty}\gamma_{im}\frac{(-1)^{(m-1)/2}}{\alpha_m^4}$$

$$\cdot\left(\frac{\alpha_m}{\text{ch}\,^2\alpha_m} - \text{th}\,\alpha_m\right), \quad i = 1, 3, 5, \ldots, \tag{3.6.16}$$

$$\gamma_{im} = \begin{cases} 2\left(0.5 - \mu - \frac{1}{2\pi\mu}\sin 2\pi m\mu\right), & i = m \\ \frac{4}{\pi}\cdot\frac{1}{(m^2-i^2)}\left(i\sin\pi\mu i\cos\pi\mu m - m\sin\pi\mu m\cos\pi\mu m\right), & i \neq m. \end{cases}$$

Let us apply the perturbation technique to system (3.6.16), representing A_i as the ε-expansion

$$A_i = \sum_{j=0}^{\infty} A_{i(j)}\varepsilon^j. \tag{3.6.17}$$

Substituting (3.6.17) into system (3.6.16) and splitting it into the powers of ε, one obtains the reccurent formulas for A_i:

$$A_{i(0)} = 0; \tag{3.6.18}$$

$$A_{i(1)} = (-1)^{(i-1)/2}\sum_{m=1,3,5,\ldots}^{\infty}\gamma_{mi}\frac{q}{8k}\frac{(-1)^{(m-1)/2}}{\alpha_m^4}(\alpha_m$$

$$- \text{th}\,\alpha_m(1 + \alpha_m\text{th}\,\alpha_m)); \tag{3.6.19}$$

$$A_{i(n)} = (-1)^{(i-1)/2}\sum_{m=1,3,5,\ldots}^{\infty}\gamma_{mi}(-1)^{(m-1)/2}A_{m(n-1)}$$

$$\cdot\left\{1 + \frac{1}{4\alpha_m}\left(\text{th}\,\alpha_m(\alpha_m\text{th}\,\alpha_m - 1) - \alpha_m\right)\right\}. \tag{3.6.20}$$

The truncated perturbation expansion (with holding three initial non-zero terms) may be PA-transformed:

$$A_i[1/1](\varepsilon) = \varepsilon(a_0 + a_1\varepsilon)(b_0 + b_1\varepsilon)^{-1}, \tag{3.6.21}$$

$a_0 = A_i(1)$, $b_0 = 1$, $a_1 = A_i(2) + b_1 A_i(1)$, $b_1 = -A_i(3)/A_i(2)$.

Let us consider the limit cases for (3.6.21). Firstly, $\mu = 0.5$ corresponds to the simply supported edge $y = \pm 0.5$. The $W\gamma_{mi} = 0$, therefore $A_{i(j)} = 0$, $W_2 = 0$, $W = W_1$, i.e., the exact solution for the plate, simply supported along the boundary. Secondly, $\mu = 0.0$ corresponds to the fully clamped edge $y = 0.5$. Here

$$\gamma_{mi} = \begin{cases} 1, & \text{when } i = m \\ 0, & \text{when } i \neq m \end{cases}$$

and the recurrent relation yields

$$A_{i(0)} = 0;$$ (3.6.22)

$$A_{i(1)} = \frac{q}{8k} \frac{1}{\alpha_i^4} \left(\frac{\alpha_i}{\text{ch}^2 \alpha_i} - \text{th} \alpha_i \right);$$ (3.6.23)

$$A_{i(n)} = \frac{q}{8k} \frac{1}{\alpha_i^4} \left(\frac{\alpha_i}{\text{ch}^2 \alpha_i} - \text{th} \alpha_i \right) \left[1 - \frac{1}{4\alpha_i^4} \left(\frac{\alpha_i}{\text{ch}^2 \alpha_i} + \text{th} \alpha_i \right) \right]^{n-1}$$ (3.6.24)

For $\varepsilon = 1$, the PA for the truncated expansion (3.6.22)–(3.6.24) is

$$A_i[1/1](\varepsilon = 1) = \frac{q}{2\alpha_i^3} \cdot \frac{\alpha_i - \text{th} \alpha_i (\alpha_i \text{th} \alpha_i + 1)}{\alpha_i - \text{th} \alpha_i (\alpha_i \text{th} \alpha_i - 1)}.$$ (3.6.25)

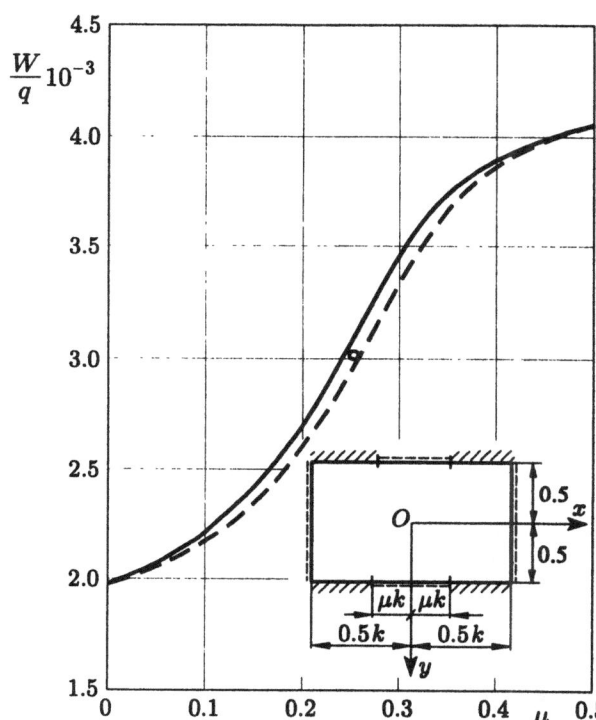

Fig. 3.13.
The relationship between the normal displacement of the plate partially clamped on two opposite sides and the clamped segment length

The formula (3.6.11), taking (3.6.25) into account, describes the plate deformation when the $x = \pm 0.5$ edge is simply supported and the $y = \pm 0.5$ egde is clamped.

The analysis, listed below, was carried out for the square plate. Expansion (3.6.21) for A_i was truncated to ten (initial) terms for $\varepsilon = 1$. The deflection and bending moments in the centre of the plate are calculated for several given values of the parameter μ (see Fig. 3.13–3.14).

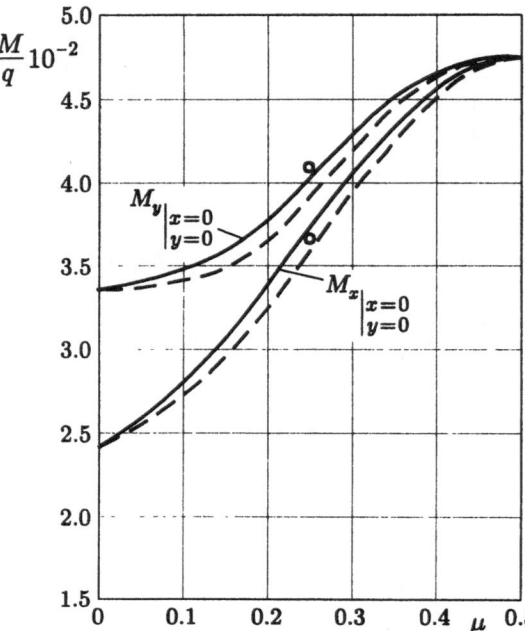

Fig. 3.14. The relationship between the moment M_y in the plate partially clamped on two opposite sides and the clamped segment length

The results, obtained by the coupled series method [70d], are shown as dots. The dotted curves display the data computed by the finite element method.

The discrepancy of the deflection, as well as the bending moments, does not exceed 5%, which confirms the acceptable accuracy of this method.

Figure 3.15 shows the values of the edge moments M_y along $y = \pm 0.5$.

Various problems of statics and dynamics of plates and shells, subjected to mixed boundary conditions, may be solved effectively on the basis of the approach presented here. Nonlinear problems can very often be solved by means of the perturbative approach.

3.7 Two-Point Padé Approximants: A Plate on Nonlinear Support

The present section deals with the problem of oscillations of a plate (a beam in the limit case) on a nonlinear elastic support. This problem may be solved by numerical or asymptotic methods. In the latter case quasilinear asymptotics are usually used [66, 67]. Then, one cannot obtain solutions for large amplitudes. Here a new nonquasilinear asymptotic is proposed. Heuristically it may be described as follows. In the long-wave approximation plate bending rigidity may be neglected, and we may investigate oscillations of a rigid body

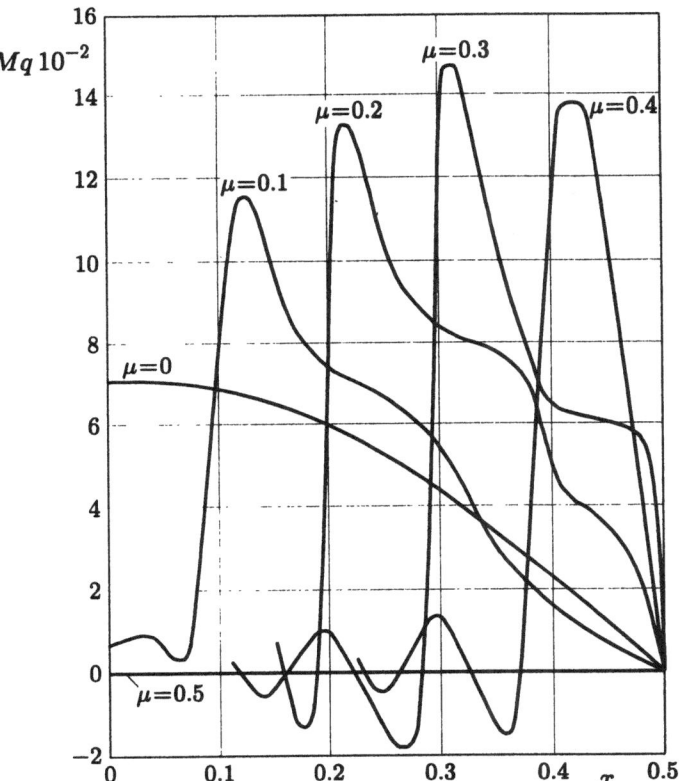

Fig. 3.15. The values of the moment M_y alone the line $y = \pm 0.5$ for various μ

on an elastic nonlinear spring. In the short-wave approximation nonlinearity of the foundation may be neglected.

It is typical for asymptotic methods that approximate solutions may be formulated for some limiting values of the parameters. For the intermediate case the analysis is very difficult. It is possible to overcome these difficulties using two-point Padé approximants.

Let us consider free vibrations of a beam on a nonlinear elastic support. The governing equation may be written in the form

$$EJw_{xxxx} + k(w + \beta_1 h^2 w^3) + \mu_{tt}w = 0. \tag{3.7.1}$$

Here $h_1 = J^{1/4}$; $w = \tilde{w}/h_1$; $x = (\pi x)/L$; $\beta = k_2/k_1$; $k = k_1 L^4/\pi^4$; $\mu = \overline{\mu}L^4/\pi^4$: k_1, k_2 are foundation coefficients.

Let (without loss of the general character of the solution) the beam be simply supported:

$$w\left(\frac{\pm\pi}{2}, t\right) = w_{xx}\left(\frac{\pm\pi}{2}, t\right) = 0. \tag{3.7.2}$$

The initial conditions we assumed are as follows:

$$w(x,0) = A \cos nx; \quad w_t(x,0) = 0. \tag{3.7.3}$$

Taking into consideration the new independent variables ξ, τ, (3.7.1) may be rewritten as

$$w_{\xi\xi\xi\xi} + \varepsilon(w + \alpha w^3) + \omega^2 w_{\tau\tau} = 0. \tag{3.7.4}$$

Here $\xi = nx$; $\tau = n^2(EJ/\mu)^{1/2}tw$; $\varepsilon = k/(EJn^4)$; $\alpha = \beta h_1^2$.

We assume $\alpha \cong 1$ (this is the case of the essentially nonlinear foundation) and investigate two limit cases. For $n \cong 1$ (the long-wave case) $\varepsilon \gg 1$, for $n \gg 1$ (the short-wave case) $\varepsilon \ll 1$.

Let $\varepsilon \ll 1$ (the short-wave case). The displacement w and the "frequency" ω may be expressed as the ε-expansions

$$w = w_0 + \varepsilon w_1 + \varepsilon^2 w_2 + \dots,$$
$$\omega = \omega_0 + \varepsilon \omega_1 + \varepsilon^2 \omega_2 + \dots \tag{3.7.5}$$

After substituting (3.7.5) into (3.7.4) and splitting it by ε, the following recurrent sequence of equations may be obtained:

$$w_0^{(4)} + \omega_0^2 w_{0\tau\tau} = 0,$$
$$w_1^{(4)} + \omega_0^2 w_{1\tau\tau} = -2\omega_0\omega_1 w_{0\tau\tau} - w_0 - \alpha_0 w^3,$$
$$\dots\dots\dots$$

Satisfying the boundary and initial conditions (3.7.2), (3.7.3), we obtain

$$w = A \cos \zeta \cos \tau. \tag{3.7.6}$$

The conditions of the absence of the secular term lead to the expressions

$$\omega_0 = 1; \quad \omega_1 = 0.5 + \frac{9}{32}\alpha A^2; \tag{3.7.7}$$
$$\omega_2 = -0.125 - \frac{9}{32}\alpha A^2 - \frac{459}{2048}\alpha^2 A^4.$$

Now we are going to investigate the long-wave case ($\varepsilon \gg 1$). The beam displacement and frequency square "ansatzes" are

$$w = w_0 + \varepsilon^{-1}w_1 + \varepsilon^{-2}w_2 + \dots,$$
$$\omega = \varepsilon^{0.5}\left[\omega^{(0)} + \varepsilon^{-1}\omega^{(1)} + \varepsilon^{-2}\omega^{(2)} + \dots\right]. \tag{3.7.8}$$

Substituting expressions (3.7.8) into (3.7.4) and performing the ε-splitting, one obtains the system of equations which permits us to determine the unknown expansion coefficient

$$\omega^{(0)} = 1 + \alpha A^2 \cos \zeta. \tag{3.7.9}$$

Using two-point Padé approximants, one may obtain an analytical solution for any value of the parementer ε.

In our case we have

$$\omega = \frac{\omega_0 + (\omega_1 + \overline{\omega})\varepsilon^{0.5} + \omega^{(0)}\overline{\omega}\varepsilon^{1.5}}{1 + \overline{\omega}\varepsilon^{0.5}}, \tag{3.7.10}$$

where $\overline{\omega} = \omega_2/(\omega^{(0)} - \omega_1)$.

We can obtain the solution for the linear case from formulae (3.7.7), (3.7.9), (3.7.10), assuming $\beta = 0$.

There is an exact solution in the linear case, and we can compare it with the approach presented above.

The numerical results are plotted in Fig. 3.16, where the curves correspond to: 1 - the exact solution; 2 - the matched spectrum expression (3.7.10) ($\beta = 0$). The results are consistent with the physics of the problem and confirm the reliability of the approximate solution. Two-point Padé approximants overcome the locality of the asymptotic expansions. Curve 2 coincides satisfactorily with the exact solution everywhere in the interval considered.

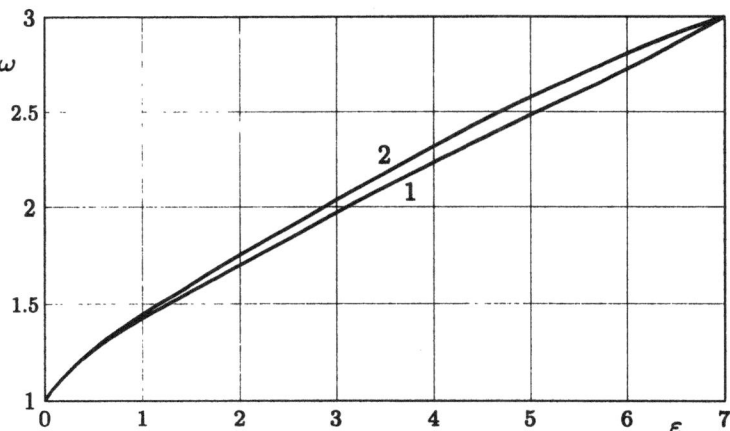

Fig. 3.16. Comparison of the exact solution and the two-point Padé approximants formula

Now we investigate the case of a plate. Only final results of the asymptotic analyses are displayed here. The initial partial differential equation may be written as

$$\overline{w}_{\zeta\zeta\zeta\zeta} + 2l^2\overline{w}_{\zeta\zeta\eta\eta} + l^4\overline{w}_{\eta\eta\eta\eta} + \varepsilon(\overline{w} + \overline{\alpha}\overline{w}^3) + \omega^2\overline{w}_{\tau_1\tau_1} = 0.$$

Here $\overline{w} = \tilde{w}/h$; $\eta = my$; $y = (\pi y_1)/L_1$; $l = L/L_1$; $\overline{\alpha} = \beta_1 h^2$; $\varepsilon_1 = k_0(mn)^4$; $\tau_1 = p^2(mn)^4\omega$; $p^2 = L^4/\mu(D\pi^4)$; $k_0 = k_1/(D\pi^4)$.

Let the plate be simply supported:

$$\overline{w} = \overline{w}_{\zeta\zeta} = 0 \quad \text{for } \zeta = \pm\frac{\pi n}{2}; \quad n = 1, 3, 5, \ldots$$

$$\overline{w} = \overline{w}_{\eta\eta} = 0 \quad \text{for } \zeta = \pm\frac{\pi m}{2}; \quad m = 2, 4, 6, \ldots$$

and the initial conditions be

$$\overline{w} = A \cos \zeta \cos \eta; \quad \overline{w}_\tau = 0 \quad \text{for } \tau = 0.$$

As a result of this description of the asymptotic procedure, one obtains

$$\varepsilon \gg 1, \quad \omega^2 = \tilde{\omega}^{(0)} + \varepsilon \tilde{\omega}^{(1)} + \ldots = 1 + A^2 \cos^2 \zeta \cos^2 \eta,$$
$$\varepsilon \ll 1, \quad \omega^2 = \tilde{\omega}_0 + \varepsilon^{-1} \tilde{\omega}_1 + \varepsilon^{-2} \tilde{\omega}_2 + \ldots,$$

where:

$$\tilde{\omega}_0 = (n^{-2} + l^2 m^{-2})^2;$$

$$\tilde{\omega}_1 = 1 + \frac{27\alpha A^2}{64};$$

$$\tilde{\omega}_2 = -\frac{9\alpha A^2}{512\tilde{\omega}_0} + \frac{9\alpha A^2}{2048} \left(\frac{6}{\tilde{\omega}_0 - \omega_{01}} + 6(\tilde{\omega}_0 - \omega_{02}) \right.$$
$$\left. + \frac{1}{9\tilde{\omega}_0 - \omega_{01}} + \frac{1}{9\tilde{\omega}_0 - \omega_{02}} - \frac{0.284}{\tilde{\omega}_0} \right);$$

$$\omega_{01} = 9n^{-2} + m^{-2}l^2;$$

$$\omega_{02} = n^{-2} + 9m^{-2}l^2.$$

Using the two-point Padé approximant we have

$$\omega^2 = \frac{\tilde{\omega}_0 + (\tilde{\omega}_0 + \tilde{\omega}_0\overline{\overline{w}})\varepsilon + \tilde{\omega}^{(0)}\overline{\overline{w}}}{(1 + \overline{\overline{w}}\varepsilon}$$

where $\overline{\overline{w}} = \tilde{\omega}_2/(\tilde{\omega}^{(0)} - \tilde{\omega}_1)$.

This formula approximately describes the plate frequency spectrum for any ε.

3.8 Solitons and Soliton-Like Approaches in the Case of Strong Nonlinearity

The investigation of the dynamics of a number of structures, particularly the heat exchangers of electric stations, results in the barely studied computational schemes with many impact pairs [133, 134] that do not allow direct application of traditional methods of the theory of vibration-impact systems [165]. Utilization of the analytical approximation of shock interaction permits reduction of the initial problem, under specific conditions, to the computation of a mechanical system (beam, string) with nonlinear elastic supports. Consequently, certain general regularities of the dynamical behaviour of such systems, associated with the existence of regimes of soliton type therein, are successfully clarified. This section is devoted to a detailed numerical investigation of such regimes in the simplest system of the class under consideration, a one-dimensional chain of masses connected by means of a weightless string and interacting with strongly nonlinear elastic supports (unlike the cubic nonlinearity in Sect. 3.1).

Let us note that chains with longitudinal exponential interaction have been well studied at this time [153].

The equations of motion of the system under consideration have the form

$$m\ddot{\bar{w}}_j + c(2\bar{w}_j - \bar{w}_{j-1} - \bar{w}_{j+1}) + 2n\dot{\bar{w}}_j + F(\bar{w}_j) = 0, \tag{3.8.1}$$

where $\bar{w}_j$ is a vector with the components $w_j^{(1)}$, $w_j^{(2)}$; m is the magnitude of each of the concentrated masses; c is the stiffness of a linear spring connecting two successive masses. If S is the string tension and l the spacing between the masses, then the stiffness is $c = S/l$. For infinite strings $j = \ldots, -2, -1, 0, 1, 2, \ldots$; in the case of a finite length $j = 1, 2, \ldots, N$, where $w_0 = w_{N+1}$ is in conformity with the conditions for fastening the string in the transverse direction. The nonlinear function $F(\bar{w}_j)$ that describes the interaction of impact type (an abrupt rise in the reaction for definite magnitudes of the displacements) is given as

$$F(\bar{w}_j) = a\frac{\bar{w}_j}{|w_j|}\operatorname{sh} b|\bar{w}_j|, \quad a > 0, \ b > 0. \tag{3.8.2}$$

Certain typical characteristics of impact pairs in vibrational impact systems cited in [165], say, can be approximated by (3.8.2) (for an appropriate selection of a and b).

The equations of motion of an infinite chain allow exact periodic solutions in the form of standing waves:

$$\omega_j^{(1)} = (-1)^j\varphi(t), \quad \omega_j^{(2)} = \alpha\omega_j^{(1)},$$

where the function $\varphi(t)$ satisfies the ordinary differential equation (α is any real number)

$$\ddot{q} + 4\frac{c}{m}\varphi + 2\frac{n}{m}\dot{\varphi} + \frac{a}{m\sqrt{1 + \alpha^2}}\operatorname{sh} b\sqrt{1 + \alpha^2}\varphi = 0.$$

A soliton can be represented in the form of a modulated wave

$$\omega_j^{(1)} = (-1)^j v_j^{(1)}(t), \quad \omega_j^{(2)} = \alpha\omega_j^{(1)},$$

where the functions $v_j^{(1)}(t)$ are not identical for different subscripts. If the set of functions $v_j^{(1)}(t)$ describing the modulation depends smoothly on the subscript j, it can be replaced approximately by a function of two variables $v(x,t)$ for which, after the change of variables

$$u = \sqrt{1 + \alpha^2}bv, \quad \tau = \sqrt{\frac{ab}{m}}t, \quad \xi = \sqrt{\frac{ab}{cl^2}}x,$$

the following partial differential equation can be written

$$\frac{\partial^2 u}{\partial\tau^2} + \frac{\partial^2 u}{\partial\xi^2} + r^2 u + 2\delta\dot{u} + \operatorname{sh} u = 0, \tag{3.8.3}$$

where $r^2 = 4c/ab$ and $\delta = n/\sqrt{abm}$. For $\delta = 0$ and $r^2 \ll 1$, (3.8.3) has a solution in the form of a localized standing wave, an envelope soliton

$$u(\xi, \tau) = 4\,\mathrm{arth}\left[\frac{\sqrt{\omega^2 - 1}\cos\omega\tau}{\omega\mathrm{ch}\left(\sqrt{\omega^2 - 1}\xi\right)}\right],\tag{3.8.4}$$

where

$$\omega_j^{(1)}(\tau) = (-1)^j \frac{1}{b\sqrt{1 + \alpha^2}} u(\xi_j, \tau), \quad \omega_j^{(2)}(\tau) = \alpha\omega_j^{(1)}(\tau).\tag{3.8.5}$$

The parameter ω characterizing the frequency of vibrations in the variable τ, determine wave amplitude and its degree of spatial localization at the same time. As ω increases the soliton becomes so narrow that the conditions for applicability of the long-wave approximation are disturbed.

In order to reduce the constraints on the soliton profile, the localized standing waves were investigated numerically on the basis of the initial system of equations (3.8.1), which was converted by the change of variables

$$u_j = bw_j, \quad \tau = \sqrt{\frac{ab}{m}}t$$

to the form

$$\frac{\mathrm{d}^2 u_j}{\mathrm{d}\tau^2} + \beta(2u_j - u_{j+1} - u_{j-1}) + 2\delta\dot{u}_j + \frac{u_j}{|u_j|}\mathrm{sh}\,|u_j| = 0,\tag{3.8.6}$$

where $\beta = r^2/4$.

The number of masses for numerical integration was taken as 40 or 41. The length of the chain was assumed to be equal to one, the numbering of the masses was from the left edge of the chain, and the edges were assumed fixed. The system of the first-order equations corresponding to system (3.8.6) was integrated numerically by the Runge–Kutta method with automatic sampling of the integration spacing. Assignment of the initial conditions was by a separate subprogram. During the computation the values of u_j were displayed (in graphical form) as were also those for $\mathrm{d}u_j/\mathrm{d}\tau$ and $F(u_j)$. At the end of the computation, the value of the total system energy was printed out (this quantity was conserved to 1% accuracy during the computation process).

Integration of system (3.8.6) under conditions of applicability of the long-wave approximation was performed first. As should have been expected, the computed data here are practically in agreement with the analytic solution, demonstrating the characteristic features of an envelope soliton under appropriate initial conditions. However, from the viewpoint of realizing intensive impact regimes, the cases of greatest interest are when the conditions mentioned are not satisfactory. As before, the initial conditions were given in the form corresponding to the solution of (3.8.5), (3.8.4), for $\tau = 0$ for the numerical investigation of system (3.8.6) in these cases. Evolution of the initial perturbation in the case of its comparatively small amplitude is shown in Figs. 3.17a and 3.17b.

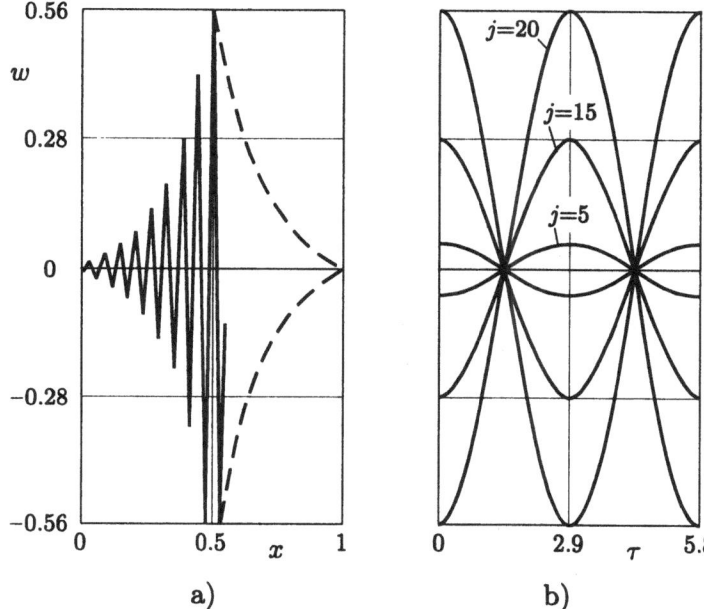

Fig. 3.17. Space distribution of displacements (a); time evolution of displacements of selected masses (b)

The significant magnitude of its coupling parameter ($\beta = 1$) results in the fact that the time development of the process does not follow (3.8.4). Nevertheless, the typical behaviour of the soliton of an envelope is present; and the synchronized motion of all the masses with the spatial configuration attenuating from the centre of the soliton is conserved. Therefore, the initial perturbation of the form

$$u_j^{(1)}(0) = u(\xi_j, 0) = \frac{4}{\sqrt{1 + \alpha^2}}(-1)^j \operatorname{arth} \frac{\sqrt{\omega^2 - 1}}{\omega \operatorname{ch}\left[\sqrt{\frac{\omega^2 - 1}{\beta}} M\left(\xi_j - \frac{1}{2}\right)\right]}, (3.8.7)$$

$$\frac{\mathrm{d}u_j(0)}{\mathrm{d}\tau} = \frac{\mathrm{d}u}{\mathrm{d}\tau}(\xi_j, 0) = 0, \quad u_j^{(2)}(0) = \alpha u_j^{(1)}(0), \quad \frac{\mathrm{d}u_j^{(2)}(0)}{\mathrm{d}\tau} = 0,$$

where M is the number of masses, practically corresponds, in this case, to the exact solution for the soliton of an envelope, although its time evolution cannot possibly be predicted quantitatively from (3.8.5) and (3.8.4).

A systematic investigation of the influence of the coupling parameter β on the nature of the localized waves under intensive excitations of impact type is reflected in Figs. 3.18 and 3.19. The initial conditions were given in the form of (3.8.7). In the weak coupling case, a quite definitive spatial localization of the process is observed (Fig. 3.18a), which is completely conserved in time. Attention is turned to the synchronization of the behaviour of the

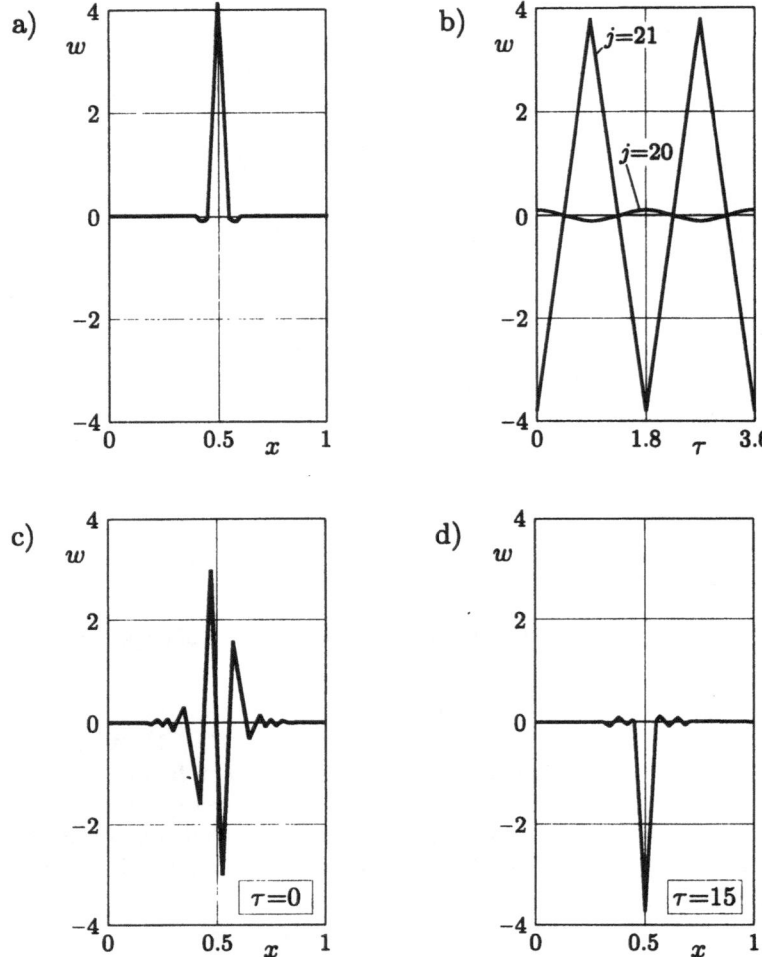

Fig. 3.18. Space distribution of displacements in the case of weak coupling (**a**); time evolution of displacements of the most excited masses (**b**); initial space distribution of displacements, corresponding to strongly excited central masses (**c**); final space distribution of displacements, corresponding to one strongly excited mass (**d**)

strongly and weakly excited masses. The time dependence of the displacement (Fig. 3.18b) is characteristic for a regular impact regime. Let us note that, in this case, the analytic solution (3.8.5) and (3.8.4) is inapplicable despite the low value of β, since the degree of localization is too great and the passage to the long-wave approximation (3.8.3) is not yet justified.

Figures 3.18c and 3.18d reflect the tendency to localization in the case when the initial excitation differs somewhat from (3.8.7). The initial mode with just one mass oscillates in the impact regime. The deviation from the "exact" initial conditions for the envelope soliton results in the origination of

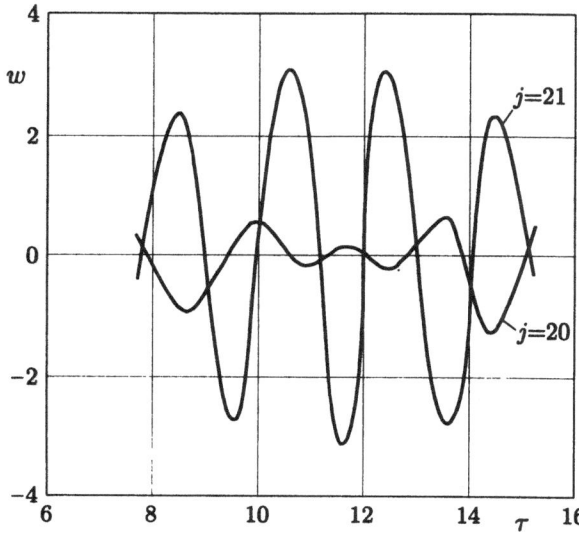

Fig. 3.19.
Time evaluation of
displacements of the
most excited masses

an indefinite "background" that interacts with the soliton. This interaction
is manifested in the quasi-periodicity of the process, the recurrence that is
seen well on the time trajectories of the masses. An analogous tendency to
localization is also observed as the coupling increases further. The appropriate
time graphs (Fig. 3.19) also reflect the influence of the "background" and
demonstrate a periodic return to almost the initial relationship between the
amplitudes of the different masses.

The existence of a spatially localized stationary solution makes the possibility of two, three soliton, etc., regimes evident. This possibility is illustrated
in Fig. 3.20, where results are presented for a numerical integration of system
(3.8.6) under the initial conditions

$$u_j^{(1)}(0) = \frac{4}{\sqrt{1+\alpha^2}}(-1)^j \left\{ \text{arth} \frac{\sqrt{\omega^2-1}}{\omega\text{ch}\left[\sqrt{\frac{\omega^2-1}{\beta}}M\left(\xi_j - \frac{1}{4}\right)\right]} \right.$$

$$\left. + \text{arth} \frac{\sqrt{\omega^2-1}}{\omega\text{ch}\left[\sqrt{\frac{\omega^2-1}{\beta}}M\left(\xi_j - \frac{3}{4}\right)\right]} \right\},$$

$$u_j^{(2)}(0) = \alpha u_j^{(1)}(0), \quad \frac{du_j^{(1)}(0)}{d\tau} = \frac{du_j^{(2)}(0)}{d\tau} = 0.$$

Conservation of the initial spatial distribution of the amplitudes in time indicates the weak interaction between the solitons.

An analysis of the numerical results shows that in the case of strong coupling and weak spatial localization, when the impact interaction is not manifested, the first terms play a fundamental role in the power-law expansion

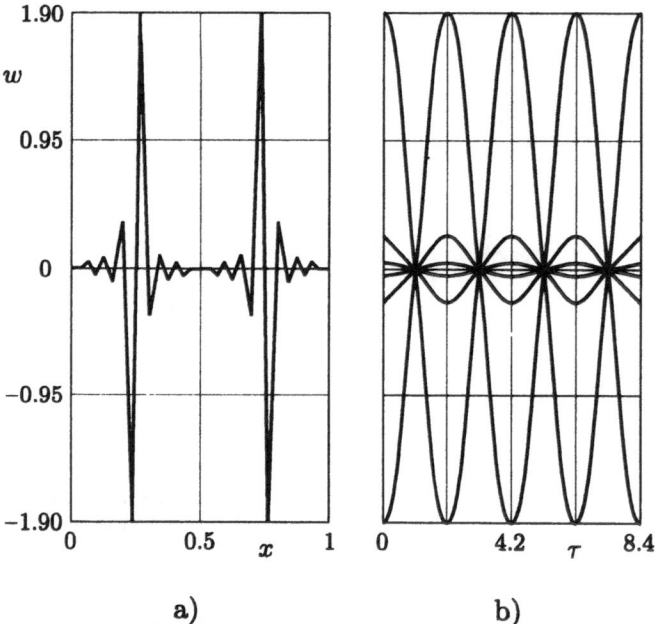

Fig. 3.20. Space distribution of the displacements in two-soliton regime (a); time evolution of displacements of selected masses (b)

of the nonlinear characteristic. The long-wave approximation (3.8.3) can be used here, but without neglecting the term r^2 as compared with one. If the coupling along the chain is weak, while the localization is quite definite (the impact interaction case), then the total time period is determined with good accuracy by the equation of motion of the fundamental mass performing a sawtooth oscillation.

We now discuss briefly the influence of damping. In the case of a "pure" soliton, comparatively little viscous friction will result in smooth damping of the process with the fundamental features conserved in a specific time. For a certain deviation from the initial conditions corresponding to a "pure" soliton, energy pumping is observed from the main mass to the adjacent mass such that the amplitudes of the latter can even grow substantially in the initial stage of motion.

The features described above for an envelope soliton are conserved completely even in the case when impact interaction is concentrated in a number of masses along the chain. The initial conditions were given here also in conformity with the analytic solution for an envelope soliton. The spatial and time dependences of the displacements, as well as the impact interaction forces, reflect the synchronized motion of the masses and the quite definite localization of the process.

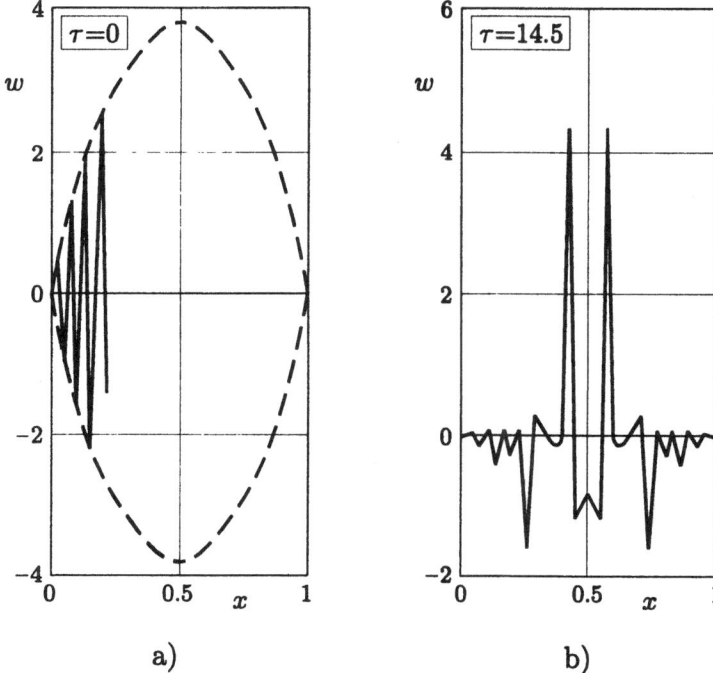

Fig. 3.21. Initial distribution of displacements (**a**); signal distribution of displacements (**b**)

Until now we have spoken about the evolution of the initial perturbation which is similar in shape to the soliton or the multi-soliton solution. Another extreme case is the excitation over nonlocalized modes, for instance over the spatial harmonic with a comparatively small number of half-waves. Investigation of the development of the initial perturbations by means of the first, third, and tenth harmonics indicates the tendency of the system to almost "sawtooth" configurations. This latter transformation of the vibrational process (the case of the first harmonic) can be assessed from Fig. 3.21 in which the tendency to destruction of the "teeth" with subsequent localization is seen.

3.9 Nonlinear Analysis of Spatial Structures

3.9.1 Introduction

In a large number of mechanical and hydromechanical systems there are stationary patterns which possess, in many situations, a coherent structure. For example, in Rayleigh–Bénard convection of simple liquids both isotropic and anisotropic quasi-two-dimensional structures and normal and oblique rolls can be detected. As shown by experimental results, convective instabilities in nematic liquid crystals are followed by transitions to normal rolls, wavy roll structures, as well as rectangular and chaotic roll structures. Although there exist some theoretical attempts to explain the observed phenomena, many problems are still open (in fact, only the normal-oblique transition has been even roughly estimated by either linear or nonlinear analysis). Additionally, this problem is a significant one in the theory of generation of vortices in superconductors, and plasmas or in liquid crystals [39, 53, 92, 129, 143, 47d, 61d]. From the mathematical point of view, the problem reduces to the non-linear analysis of the so-called envelope (or amplitude) equation. The idea of introducing dynamics into the theory of coherent structures (governed by the envelope equation) was presented by Infeld *et al.* [47d]. The crucial point of their approach included a time-dependent, supercritical structure and multi-ple time and space expansions. Two variants of the Ginzburg–Pitaevski uni-versal model equations have supported their general ideas. Motivated by their observation that temporal development is more crucial for the theory of spa-tial structures in media such as superconductors, liquid crystals, simple fluids or plasmas than for static situations, we have outlined time-dependent terms in our mechanical analogy model of the above mentioned pattern-forming structures.

In [129], the mechanical model of the buckling instability of an anisotropic elastic plate embedded in an elastic medium has been used for the interpre-tation of the periodic structures of anisotropic systems. Our idea is the ex-tension of such a mechanical analogy by including real time-dependent forces (such as inertia and damping) and thus to examine their infuence on the stability threshold location of the possible stationary patterns.

3.9.2 Modified Envelope Equation

We consider the small lateral displacement of a thin elastic plate extended in the X_1- and Y_1-directions, which is embedded in an elastic medium with a Duffing-type stiffness.

The governing equation has the form

$$m\frac{\partial^2 u}{\partial t^2} + h\frac{\partial u}{\partial t} + \kappa u + \gamma u^3 = -\left(\lambda_1 \frac{\partial^4 u}{\partial x_1^4} + \lambda_2 \frac{\partial^4 u}{\partial y_1^4} + 2\lambda_3 \frac{\partial^2 u}{\partial x_1^2}\frac{\partial^2 u}{\partial y_1^2}\right.$$
$$\left. +\mu_1\frac{\partial^2 u}{\partial x_1^2} + \mu_2\frac{\partial^2 u}{\partial y_1^2}\right), \qquad (3.9.1)$$

where $u(x_1, y_1, t)$ is a neutral surface of the plate, λ_i $(i = 1, 2, 3)$ are the bending coefficients, the loading is defined by μ_1 and μ_2, m and h correspond to the mass and damping, and κ and γ are stiffness coefficients. In what follows, after rescaling of time, length and surface, we obtain

$$\frac{\partial^2 v}{\partial \tau^2} + c\frac{\partial v}{\partial \tau} + v_0^3 = R\left\{-(1+\nabla^2)^2 v - d\frac{\partial^4 v}{\partial y^4} - 2\eta\frac{\partial^2 v}{\partial x^2}\frac{\partial^2 v}{\partial y^2} + rv\right\}, (3.9.2)$$

where the related connections between the parameters are

$$x_1 = \left(\frac{2\lambda_1}{\mu_1}\right)^{1/2} x, \quad y_1 = \left(\frac{2\lambda_1\mu_2}{\mu_1}\right)^{1/2} y, \quad d = \frac{\lambda_2\mu_1^2}{\lambda_1\mu_2^2} - 1,$$

$$\eta = \frac{\lambda_3\mu_1^2}{\lambda_1\mu_2^2} - 1, \quad r = 1 - \frac{4\lambda_4\kappa}{\mu_1^2}, \quad c = \frac{h}{m}\left(\frac{\kappa}{\gamma}\right)^{1/2}, \tag{3.9.3}$$

$$\tau = \left(\frac{\kappa}{m}\right)^{1/2} t, \quad u = \left(\frac{\kappa}{\gamma}\right)^{1/2} v, \quad R = \frac{\mu_1^2}{4\lambda_1\kappa}.$$

When the inertial term in (3.9.2) disappears, the problem reduces to the so-called "Swift–Hohenberg" model. In order to analyse the unstructured state locally ($u = 0$), we put

$$v = \exp\{\sigma t + \mathrm{i}(qx + py)\}, \tag{3.9.4}$$

into (3.9.2) and we find that the eigenvalue σ can be obtained from the characteristic equation

$$\sigma^2 + c\sigma + R^{-1}\{(1 - q^2 - p^2)^2 + dp^4 + 2\eta q^2 p^2 - r\} = 0, \tag{3.9.5}$$

in which p and q are the wave numbers of the mode considered. Furthermore, we consider the case in which the unstructured state $u = 0$ has one zero eigenvalue (the other cases will be consider elsewhere). Both eigenvalues are real and the second is equal to $-c$. The necessary condition which satisfies our requirements leads to the instability threshold defined by

$$q_c^2 = 1, \quad p_c^2 = 0, \quad r_c = 0, \tag{3.9.6}$$

which is, for $c = \eta = 0$, in agreement with reference [129]. Furthermore, we take a small (positive) parameter, defined as $r = \varepsilon$, and apply the second appropriate rescaling together with the introduction a blown-up version of the damping parameter c i.e.,

$$X = \frac{1}{2}\varepsilon^{1/2}x, \quad Y = \frac{1}{2}\varepsilon^{1/2}y, \quad T = \varepsilon^{1/2}t,$$

$$c = \varepsilon^{1/2}S, \quad \left(S = \frac{\mu_1^2 h^2 \kappa}{m^2\gamma(\mu_1^2 - 4\lambda_1\kappa)}\right). \tag{3.9.7}$$

We search for a solution of the form

$$v = \varepsilon^{1/2}\left(v_0 + \varepsilon^{1/2}v_1 + \varepsilon v_2 + \ldots\right), \tag{3.9.8}$$

and the dependence on X, Y and T reflects the disturbed structure. Applying well-known perturbation techniques, we obtain a recurrent set of linear partial differential equations.

For $\varepsilon^{1/2}$ we have

$$
R\left\{\frac{\partial^2 v_0}{\partial T^2} + S\frac{\partial v_0}{\partial T}\right\} = \left(-(1+\nabla^2)^2 v_2 - d\frac{\partial^4 v_2}{\partial y^4} - 2\eta\frac{\partial^2 u_2}{\partial x^2}\frac{\partial^2 v_2}{\partial y^2}\right)
$$
$$
- 2\left\{\frac{\partial v_1}{\partial x}\left(1 + \frac{\partial v_1}{\partial x^2}\right)\frac{\partial v_1}{\partial X} + \frac{\partial v_1}{\partial y}\frac{\partial v_1}{\partial Y} + (1+d)\frac{\partial^3 v_1}{\partial y^3}\frac{\partial v_1}{\partial Y}\right.
$$
$$
\left. + (1+\eta)\left(\frac{\partial v_1}{\partial y}\frac{\partial v_1}{\partial X} + \frac{\partial v_1}{\partial x}\frac{\partial v_1}{\partial Y}\right)\frac{\partial v_1}{\partial x}\frac{\partial v_1}{\partial y}\right\}
$$
$$
- \left\{\frac{3}{2}\frac{\partial^2 v_0}{\partial x^2}\frac{\partial^2 v_0}{\partial X^2} + \frac{1}{2}\frac{\partial^2 v_0}{\partial X^2} + \frac{3}{2}(1+d)\frac{\partial^2 v_0}{\partial Y^2}\frac{\partial^2 v_0}{\partial Y^2} + \frac{1}{2}\frac{\partial^2 v_0}{\partial Y^2}\right.
$$
$$
\left. + \frac{1}{2}(1+\eta)\left(\frac{\partial^2 v_0}{\partial y^2}\frac{\partial^2 v_0}{\partial X^2} + \frac{\partial^2 v_0}{\partial x^2}\frac{\partial^2 v_0}{\partial Y^2} + 4\frac{\partial v_0}{\partial x}\frac{\partial v_0}{\partial y}\frac{\partial v_0}{\partial X}\frac{\partial v_0}{\partial Y}\right)\right.
$$
$$
\left. - 1 + Rv_0^2\right\}. \tag{3.9.9}
$$

We look for a solution of the form

$$
v_0 = (A(X,Y,T)e^{i(x+y)} + \bar{A}(X,Y,T)e^{-i(x+y)})/\sqrt{3}, \tag{3.9.10}
$$

where A and $\bar{A}$ are complex conjugates, and from (3.9.9) we find

$$
\frac{\partial^2 A}{\partial T^2} + S\frac{\partial S}{\partial T} = \frac{1}{R}\left(\frac{\partial^2 A}{\partial X^2} + \frac{\partial^2 A}{\partial Y^2} + A\right) - |A|^2 A. \tag{3.9.11}
$$

The stationary solutions of (3.9.11) are sought in the form

$$
A = F\exp i(QX + PY), \tag{3.9.12}
$$

and we obtain $F = 0$, and $F^2 = R^{-1}(1 - P^2 - Q^2)$. We examine both of them locally by introducing a small perturbation v (complex),

$$
A = (F + v)\exp(iZ), \qquad Z = QX + PY, \tag{3.9.13}
$$

which results in the equation

$$
\frac{\partial^2 v}{\partial T^2} + \frac{\partial v}{\partial T} = R^{-1}\left\{\frac{\partial^2 v}{\partial X^2} + \frac{\partial^2 v}{\partial Y^2} + 2iQ\frac{\partial v}{\partial X}\right.
$$
$$
\left. + 2iP\frac{\partial v}{\partial Y} - F^2 v - F^2\bar{v}\right\}, \tag{3.9.14}
$$

where v and $\bar{v}$ are complex conjugates. We put

$$
v = \left(v_1 e^{i(KX+LY)} + v_2 e^{-i(KX+LY)}\right), \tag{3.9.15}
$$

into (3.9.15) and obtain the characteristic equation

$$
a_0\sigma^4 + a_1\sigma^3 + a_2\sigma^2 + a_3\sigma + a_4 = 0, \tag{3.9.16}
$$

where

$$a_0 = 1, \quad a_1 = 25, \quad a_2 = \frac{2}{R}\left(K^2 + L^2\right) + 2F^2 + S^2,$$

$$a_4 = \frac{1}{R^2}\left\{\left(K^2 + L^2\right)^2 - 4\left(KQ + LP\right)^2\right\} + \frac{2F^2}{R}\left(K^2 + L^2\right). \qquad (3.9.17)$$

Applying the Routh–Hurwitz criterion to equation (3.9.11), we first analyse the threshold for instability of the unstructured state $u = 0$, which corresponds to $F = 0$. The problem can be reduced to the consideration of the equation

$$E = \frac{1}{R^2}\left\{K^2\left(K^2 - 4Q^2\right) + L^2\left(L^2 - 4P^2\right) + 2KL\left(KL - 4QP\right)\right\}. (3.9.18)$$

It is clearly seen that the relation $K/Q = L/P = N$ governs the occurrence of the oblique rolls. For $N < 2$ all roots of (3.9.16) have negative real parts, which proves the stability of the unstructured state. The threshold of stability is found for the critical value $N = 2$. Investigating the second stationary solution, we find that the stability limit is defined by

$$Q^2 + P^2 + Q - 1 = 0, \qquad (3.9.19)$$

which corresponds to the usual Eckhaus criterion: i.e., in the case of symmetry if $P = Q > 1/\sqrt{3}$, the unstructured state becomes unstable.

Because the characteristic equation is of the fourth order only, one pair of conjugate roots can occur. The examination of this possibility leads to the equation

$$F^4 + S^2\left\{\frac{1}{R}\left(K^2 + L^2\right)^2 + F^2\right\} + 4\left(LQ + LP\right)^2$$

$$+ \frac{1}{R}2F^2\left(K^2 + L^2\right) = 0, \qquad (3.9.20)$$

which is never satisfied. Thus, in the case of the modified envelope equation only Eckhaus-type instability can occur.

4. Discrete–Continuous Systems

4.1 Periodic Oscillations
of Discrete–Continuous Systems
with a Time Delay

4.1.1 The KBM Method

One of the important problems of mechanical and automatic control engineering is active control of the oscillations of mechanical objects by means of control units, which can be treated as inertial systems with concentrated parameters and a time delay [137]. The objects to be controlled can be nonlinear mechanical systems with concentrated (futher referred to as discrete mechanical systems) or distributed parameters. The mixed situation, referred to as discrete–continuous systems, are dealt with in this chapter.

In real control systems of this type, the control unit influences the object subjected to control and the state of the controlled object is monitored only in certain isolated points. It is usually possible to find controlled objects which are governed by partial differential nonlinear equations, as well as control units, which can be modelled by ordinary nonlinear differential equations.

Let us consider a discrete–continuous system governed by the following equations:

$$\frac{\partial^2 u(t,x)}{\partial t^2} = L_x^{(2m)}\{u(t,x)\} + \varepsilon f_1\{x, u(t,x), y(t-\mu)\},$$

$$\frac{dy(t)}{dt} = \sum_{p=0}^{P} A_p y(t-\tau_p) + \varepsilon F_1\{y(t-\mu), u(t-\mu), \xi\} \tag{4.1.1}$$

subjected to the following non-homogeneous boundary conditions

$$L_x^{(h,j)}\{u(t,x)\}|_{x\in S} = \varepsilon g_{hj}\{y(t-\mu)\}, \qquad h = 1,\ldots,m. \tag{4.1.2}$$

The coordinate t denotes time and $t \in R$; x is the vector of the coordinates and $x \in (G \cup S)$, while S is the limiting set of G; $u(t,x)$ is a certain scalar function determined in the set $R \times G$, and $L_x^{(h,j)}$ is a linear operator of order $2m$ on x; $L_x^{(2m)}$ is the linear differential operator of $j \le 2m - 1$; y and F_1 are vectores of an m-dimensional space; A_p are constant matrices of order

$(m \times m)$; F_1, f_1 and g_{hj} are functions of $y(t - \mu)$, $u(t - \mu, \xi)$, $\xi \in (G \cup S)$, while τ_p and μ are time delays. Finally, we assume that ε and μ are small positive parameters.

Thanks to this mathematical formulation of the problem, the presented analytical approach can be further used for many different discrete–continuous mechanical systems governed by (4.1.1). Thus we will continue our consideration first in a general form, and then, in order to demonstrate the physical insight of the problem, we will illustrate the method with an example from the area of mechanics [31].

The problem, including the non-homogeneous boundary conditions (4.1.2), can be reduced [31, 137] to one of homogeneous boundary conditions. Thus, we analyse the following system

$$\frac{\partial^2 \nu(t, x)}{\partial t^2} = L_x^{(2m)} \{\nu(t, x)\} + \varepsilon f_1 \{x, \nu(t, x), y(t - \mu)\},$$

$$\frac{dy(t)}{dt} = \sum_{p=0}^{P} A_p y(t - \tau_p) + \varepsilon F_1 \{y(t - \mu), \nu(t - \mu), \xi\}, \qquad (4.1.3)$$

where $\nu(t, x)$ fulfils the homogeneous boundary conditions

$$L_x^{(h,j)} \{\nu(t, x)\}|_{x \in S} = 0, \qquad h = 1, \ldots, m. \qquad (4.1.4)$$

From the first equation system (4.1.3), and for $\varepsilon = 0$, we obtain

$$L_x^{(2m)} \{X(x)\} + \sigma X(x) = 0,$$
$$L_x^{(h,j)}|_{x \in S} = 0, \qquad h = 1, \ldots, m, \qquad (4.1.5)$$

while from the other one, we obtain the characteristic equation

$$D(\rho) = \det \left\{ \sum_{p=0}^{P} A_p e^{-\tau_p \rho} - E\rho \right\}. \qquad (4.1.6)$$

In this dynamical system, oscillations will appear if $\sigma_s = \omega_{\nu_s}^2$ and/or if the characteristic equation (4.1.6) has the imaginary eigenvalues $\rho_k = \pm i\omega_{yk}$.

Here we shall consider the case where $\sigma_1 = \omega_1^2$ and the other eigenvalues of the first equation of system (4.1.5) amount to $\sigma_s \neq \{(p/q)\omega_1\}^2$, where p and q are integers. Moreover, it is assumed that the characteristic equation (4.1.6) does not possess imaginary eigenvalues. We seek a one-frequency solution of the dynamic system (4.1.1) with the frequency approaching ω_1 for $\varepsilon \to 0$ and $\mu \to 0$. The approach suggested by Krylov, Bogolubov and Mitropolski will be used. We look for a solution in the form

$$\nu(t, x) = a(t) X_1(x) \cos \psi t + \sum_{k=1}^{K} \sum_{l=0}^{L} \varepsilon^k \mu^l V_{kl} \{x, a(t), \psi(t)\},$$

$$y(t) = \sum_{k=1}^{K} \sum_{l=0}^{L} y_{kl} \{a(t), \psi(t)\}, \qquad (4.1.7)$$

where

$$\frac{\mathrm{d}a}{\mathrm{d}t} = \sum_{k=1}^{K}\sum_{l=0}^{L} \varepsilon^k \mu^l A_{kl}\{a(t)\}, \tag{4.1.8}$$

$$\frac{\mathrm{d}\psi}{\mathrm{d}t} = \omega_1 + \sum_{k=1}^{K}\sum_{l=0}^{L} \varepsilon^k \mu^l B_{kl}\{a(t)\}, \tag{4.1.9}$$

and X_1 is the solution of boundary problem (4.1.5). From the first equation of (4.1.7), we obtain

$$\frac{\partial \nu}{\partial t} = \left\{\frac{\mathrm{d}a}{\mathrm{d}t}\cos\psi - a\frac{\mathrm{d}\psi}{\mathrm{d}t}\sin\psi\right\} X_1(x)$$

$$+ \sum_{k=1}^{K}\sum_{l=0}^{L} \varepsilon^k \mu^l \left\{\frac{\partial V_{kl}}{\partial a}\frac{\mathrm{d}a}{\mathrm{d}t} + \frac{\partial V_{kl}}{\partial \psi}\frac{\mathrm{d}\psi}{\mathrm{d}t}\right\},$$

$$\frac{\partial^2 \nu}{\partial t^2} = \left\{\frac{\mathrm{d}^2 a}{\mathrm{d}t^2}\cos\psi - 2\frac{\mathrm{d}a}{\mathrm{d}t}\frac{\mathrm{d}\psi}{\mathrm{d}t}\sin\psi - a\frac{\mathrm{d}^2\psi}{\mathrm{d}t^2}\sin\psi \right. \tag{4.1.10}$$

$$\left. - a\left(\frac{\mathrm{d}\psi}{\mathrm{d}t}\right)^2\cos\psi\right\} X_1(x) + \sum_{k=1}^{K}\sum_{l=0}^{L} \varepsilon^k \mu^l \left\{\frac{\partial^2 V_{kl}}{\partial a^2}\left(\frac{\mathrm{d}a}{\mathrm{d}t}\right)^2\right.$$

$$\left. + 2\frac{\partial^2 V_{kl}}{\partial a \partial \psi}\frac{\mathrm{d}\psi}{\mathrm{d}t}\frac{\mathrm{d}a}{\mathrm{d}t} + \frac{\partial V_{kl}}{\partial a}\frac{\mathrm{d}^2 a}{\mathrm{d}t^2} + \frac{\partial^2 V_{kl}}{\partial \psi^2}\left(\frac{\mathrm{d}\psi}{\mathrm{d}t}\right)^2 + \frac{\partial V_{kl}}{\partial \psi}\frac{\mathrm{d}^2\psi}{\mathrm{d}t^2}\right\}.$$

From the first equation of (4.1.3), we calculate

$$\frac{\partial^2 \nu}{\partial t^2} - L_x^{(2m)}\{\nu(t,x)\} = \left\{\frac{\mathrm{d}^2 a}{\mathrm{d}t^2}X_1(x) - \left\{\left(\frac{\mathrm{d}\psi}{\mathrm{d}t}\right)^2 X_1(x)\right.\right.$$

$$\left. + L_x^{(2m)}\{X_1(x)\}\right\}a\right\}\cos\psi - \left\{2\frac{\mathrm{d}a}{\mathrm{d}t}\frac{\mathrm{d}\psi}{\mathrm{d}t} + a\frac{\mathrm{d}^2\psi}{\mathrm{d}t^2}\right\} X_1(x)\sin\psi$$

$$+ \sum_{k=1}^{K}\sum_{l=0}^{L} \varepsilon^k \mu^l \left\{\frac{\partial^2 V_{kl}}{\partial a^2}\left(\frac{\mathrm{d}a}{\mathrm{d}t}\right)^2 + 2\frac{\partial^2 V_{kl}}{\partial a \partial \psi}\frac{\mathrm{d}\psi}{\mathrm{d}t}\frac{\mathrm{d}a}{\mathrm{d}t} + \frac{\partial V_{kl}}{\partial a}\frac{\mathrm{d}^2 a}{\mathrm{d}t^2}\right.$$

$$\left. + \frac{\partial^2 V_{kl}}{\partial \psi^2}\left(\frac{\mathrm{d}\psi}{\mathrm{d}t}\right)^2 + \frac{\partial V_{kl}}{\partial \psi}\frac{\mathrm{d}^2\psi}{\mathrm{d}t^2} - L_x^{(2m)}\{V_{kl}\}\right\}. \tag{4.1.11}$$

From the second equation of (4.1.7), we obtain

$$\frac{\mathrm{d}y}{\mathrm{d}t} = \sum_{k=1}^{K}\sum_{l=0}^{L} \varepsilon^k \mu^l \left\{\frac{\partial y_{kl}}{\partial a}\frac{\mathrm{d}a}{\mathrm{d}t} - \frac{\partial y_{kl}}{\partial \psi}\frac{\mathrm{d}\psi}{\mathrm{d}t}\right\}. \tag{4.1.12}$$

Moreover, taking (4.1.9) into account, we calculate

$$\frac{\mathrm{d}^2 a}{\mathrm{d}t^2} = \left\{\sum_{k=1}^{K}\sum_{l=0}^{L} \varepsilon^k \mu^l \frac{\mathrm{d}A_{kl}}{\mathrm{d}a}\right\}\left\{\sum_{k=1}^{K}\sum_{l=0}^{L} \varepsilon^k \mu^l A_{kl}\right\}$$

$$= \varepsilon^2 A_{10}\frac{dA_{10}}{da} + \varepsilon^2\mu\left\{\frac{dA_{10}}{da}A_{11} + \frac{dA_{11}}{da}A_{10}\right\} \qquad (4.1.13)$$

$$+\varepsilon^3\left\{\frac{dA_{20}}{da}A_{10} + \frac{dA_{10}}{da}A_{20}\right\} + O(\varepsilon^k\mu^l\,;k+l=4),$$

$$\left(\frac{d\psi}{dt}\right)^2 X_1(x) + L_x^{(2m)}\{X_1(x)\} = \left\{\sum_{k=1}^{K}\sum_{l=0}^{L}\varepsilon^k\mu^l B_{kl}(a)\right\}^2 X_1$$

$$+2\omega_1\sum_{k=1}^{K}\sum_{l=0}^{L}\varepsilon^k\mu^l B_{kl}(a)X_1 = 2\varepsilon\omega_1 B_{10}X_1 + \varepsilon^2\{2\omega_1 B_{20} + B_{10}^2\}X_1$$

$$+2\varepsilon\mu\omega_1 B_{11}X_1 + 2\varepsilon^2\mu\{B_{10}B_{11} + \omega_1 B_{21}\}X_1 \qquad (4.1.14)$$

$$+2\varepsilon\mu^2\omega_1 B_{12}X_1 + 2\varepsilon^2\{\omega_1 B_{30} + B_{20}B_{10}\} + O(\varepsilon^k\mu^l\,;k+l=4),$$

because in accordance with the first equation of (4.1.5), we have $X_1(x)\omega_1^2 + L_x^{(2m)}\{X(x)\} = 0$ and

$$\frac{da}{dt}\frac{d\psi}{dt} = \sum_{k=1}^{K}\sum_{l=0}^{L}\varepsilon^k\mu^l A_{kl}\left\{\omega_1 + \sum_{k=1}^{K}\sum_{l=0}^{L}\varepsilon^k\mu^l B_{kl}\right\}$$

$$= \varepsilon\omega_1 A_{10} + \varepsilon^2\{\omega_1 A_{20} + A_{10}B_{10}\} + \varepsilon\mu\omega_1 A_{11} + \varepsilon^2\mu\{\omega_1 A_{21}$$

$$+A_{11}B_{10} + A_{10}B_{11}\} + \varepsilon\mu^2 A_{12}\omega_1 + \varepsilon^3\{A_{30}\omega_1 \qquad (4.1.15)$$

$$+A_{20}B_{10} + A_{10}B_{20}\} + O(\varepsilon^k\mu^l\,;k+l=4),$$

$$\frac{d^2\psi}{dt^2} = \left\{\sum_{k=1}^{K}\sum_{l=0}^{L}\varepsilon^k\mu^l\frac{dB_{kl}}{da}\right\}\left\{\sum_{k=1}^{K}\sum_{l=0}^{L}\varepsilon^k\mu^l A_{kl}\right\}$$

$$= \varepsilon^2\frac{dB_{10}}{da}A_{10} + \varepsilon^2\mu\left\{\frac{dB_{10}}{da}A_{11} + \frac{dB_{11}}{da}A_{10}\right\} \qquad (4.1.16)$$

$$+\varepsilon^3\left\{\frac{dB_{10}}{da}A_{20} + \frac{dB_{20}}{da}A_{10}\right\} + O(\varepsilon^k\mu^l\,;k+l=4),$$

$$\left(\frac{da}{dt}\right)^2 = \left\{\sum_{k=1}^{K}\sum_{l=0}^{L}\varepsilon^k\mu^l A_{kl}\right\}^2 = \varepsilon^2 A_{10}^2 + 2\varepsilon^2\mu A_{10}A_{11} \qquad (4.1.17)$$

$$+2\varepsilon^2 A_{20}A_{10} + O(\varepsilon^k\mu^l\,;k+l=4),$$

$$\left(\frac{d\psi}{dt}\right)^2 = \left\{\omega_1 + \sum_{k=1}^{K}\sum_{l=0}^{L}\varepsilon^k\mu^l B_{kl}\right\}^2 = \omega_1^2 + 2\varepsilon\omega_1 B_{10} + 2\varepsilon\mu B_{11}\omega_1$$

$$+\varepsilon^2(2\omega_1 B_{20} + B_{10}^2) + 2\varepsilon\mu^2 B_{12}\omega_1 + 2\varepsilon^2\mu(\omega_1 B_{21} + B_{11}B_{10})$$

$$+2\varepsilon^3(\omega_1 B_{30} + B_{20}B_{10}) + O(\varepsilon^k\mu^l\,;k+l=4). \qquad (4.1.18)$$

Since y and ν can be expressed as the power series

$$y(t - \mu) = \sum_{n=0}^{N} \frac{1}{n!} \frac{d^n y(t)}{dt^n} (-\mu)^n,$$

$$\nu(t - \mu, \xi) = \sum_{n=0}^{N} \frac{1}{n!} \frac{d^n \nu(t, \xi)}{dt^n} (-\mu)^n, \tag{4.1.19}$$

and after limiting the calculation to $n = 1$, the functions εf, εF, y and ν can be expanded in a power series of the small parameters μ and ε.

Comparing coefficients of the same powers of $\varepsilon^k \mu^l$, we determine a sequence of reccurent linear differential equations:

$$\varepsilon : \quad \omega_1^2 \frac{\partial^2 V_{10}(x, a, \psi)}{\partial \psi^2} = L_x^{(2m)}\{V_{10}\} + 2\omega_1 B_{10} X_1 a \cos \psi$$
$$+ 2\omega_1 A_{10} X_1 \sin \psi + f_\varepsilon(x, a, \psi),$$

$$\omega_1 \frac{\partial y_{10}(a, \psi)}{\partial \psi} = \sum_{p=0}^{P} A_p y_{10}(a, \psi - \tau_p \omega_1) + F_\varepsilon(a, \psi);$$

$$\varepsilon^2 : \quad \omega_1^2 \frac{\partial^2 V_{20}(x, a, \psi)}{\partial \psi^2} = L_x^{(2m)}\{V_{20}\} + 2\omega_1 B_{20} X_1 a \cos \psi$$
$$+ 2\omega_1 A_{20} X_1 \sin \psi + f_{\varepsilon^2}(x, a, \psi),$$

$$\omega_1 \frac{\partial y_{20}(a, \psi)}{\partial \psi} = \sum_{p=0}^{P} A_p y_{20}(a, \psi - \tau_p \omega_1) + F_{\varepsilon^2}(a, \psi);$$

$$\varepsilon\mu : \quad \omega_1^2 \frac{\partial^2 V_{11}(x, a, \psi)}{\partial \psi^2} = L_x^{(2m)}\{V_{11}\} + 2\omega_1 B_{11} X_1 a \cos \psi$$
$$+ 2\omega_1 A_{11} X_1 \sin \psi + f_{\varepsilon\mu}(x, a, \psi),$$

$$\omega_1 \frac{\partial y_{11}(a, \psi)}{\partial \psi} = \sum_{p=0}^{P} A_p y_{11}(a, \psi - \tau_p \omega_1) + F_{\varepsilon\mu}(a, \psi);$$

$$\varepsilon^3 : \quad \omega_1^2 \frac{\partial^2 V_{30}(x, a, \psi)}{\partial \psi^2} = L_x^{(2m)}\{V_{30}\} + 2\omega_1 B_{30} X_1 a \cos \psi$$
$$+ 2\omega_1 A_{30} X_1 \sin \psi + f_{\varepsilon^3}(x, a, \psi), \tag{4.1.20}$$

$$\omega_1 \frac{\partial y_{30}(a, \psi)}{\partial \psi} = \sum_{p=0}^{P} A_p y_{30}(a, \psi - \tau_p \omega_1) + F_{\varepsilon^3}(a, \psi);$$

$$\varepsilon\mu^2 : \quad \omega_1^2 \frac{\partial^2 V_{12}(x, a, \psi)}{\partial \psi^2} = L_x^{(2m)}\{V_{12}\} + 2\omega_1 B_{12} X_1 a \cos \psi$$
$$+ 2\omega_1 A_{12} X_1 \sin \psi + f_{\varepsilon\mu^2}(x, a, \psi),$$

$$\omega_1 \frac{\partial y_{12}(a, \psi)}{\partial \psi} = \sum_{p=0}^{P} A_p y_{12}(a, \psi - \tau_p \omega_1) + F_{\varepsilon\mu^2}(a, \psi);$$

$$\varepsilon^2 \mu \, : \quad \omega_1^2 \frac{\partial^2 V_{21}(x, a, \psi)}{\partial \psi^2} = L_x^{(2m)}\{V_{21}\} + 2\omega_1 B_{21} X_1 a \cos \psi$$

$$+ 2\omega_1 A_{21} X_1 \sin \psi + f_{\varepsilon^2 \mu}(x, a, \psi),$$

$$\omega_1 \frac{\partial y_{21}(a, \psi)}{\partial \psi} = \sum_{p=0}^{P} A_p y_{21}(a, \psi - \tau_p \omega_1) + F_{\varepsilon^2 \mu}(a, \psi);$$

where $\varepsilon < \mu$ and the functions $f_{(\cdot)}$, $F_{(\cdot)}$ can be successively defined, as will be shown by an example.

After expanding the function $f_{(*)}$ into a Fourier series, one obtains

$$f_{(*)} = \sum_{n=1}^{\infty} \{b_{(*)n}(a) \cos n\psi + c_{(*)n}(a) \sin n\psi\}, \qquad (4.1.21)$$

where

$$b_{(*)n}(a) = \frac{1}{2\pi l} \int_0^l dx \int_0^{2\pi} f_{(*)}(x, a, \psi) X_1(x) \cos n\psi \, d\psi,$$

$$c_{(*)n}(a) = \frac{1}{2\pi l} \int_0^l dx \int_0^{2\pi} f_{(*)}(x, a, \psi) X_1(x) \sin n\psi \, d\psi. \qquad (4.1.22)$$

If we equate the coefficients of $X_1(x) \sin \psi$ and $X_1(x) \cos \psi$ to zero, we obtain A_{kl} and B_{kl}, which are given below:

$$A_{10}(a) = -\frac{c_{(\varepsilon)1}(a)}{2\omega_1}, \qquad B_{10}(a) = -\frac{b_{(\varepsilon)1}(a)}{2\omega_1 a},$$

$$A_{20}(a) = -\frac{c_{(\varepsilon^2)1}(a)}{2\omega_1}, \qquad B_{20}(a) = -\frac{b_{(\varepsilon^2)1}(a)}{2\omega_1 a},$$

$$A_{30}(a) = -\frac{c_{(\varepsilon^3)1}(a)}{2\omega_1}, \qquad B_{30}(a) = -\frac{b_{(\varepsilon^3)1}(a)}{2\omega_1 a},$$

$$A_{11}(a) = -\frac{c_{(\varepsilon\mu)1}(a)}{2\omega_1}, \qquad B_{11}(a) = -\frac{b_{(\varepsilon\mu)1}(a)}{2\omega_1 a}, \qquad (4.1.23)$$

$$A_{12}(a) = -\frac{c_{(\varepsilon\mu^2)1}(a)}{2\omega_1}, \qquad B_{12}(a) = -\frac{b_{(\varepsilon\mu^2)1}(a)}{2\omega_1 a},$$

$$A_{21}(a) = -\frac{c_{(\varepsilon^2\mu)1}(a)}{2\omega_1}, \qquad B_{21}(a) = -\frac{b_{(\varepsilon^2\mu)1}(a)}{2\omega_1 a},$$

According to (4.1.9), we get

$$\Phi(a) = \frac{da}{dt} = \varepsilon A_{10} + \varepsilon^2 A_{20} + \varepsilon^3 A_{30} + \varepsilon\mu A_{11} + \varepsilon^2 \mu A_{21} + \varepsilon\mu^2 A_{12}$$

$$+ O(\varepsilon^k \mu^l \, ; k + l = 4),$$

$$\omega(a) = \frac{d\psi}{da} = \omega_1 + \varepsilon B_{10} + \varepsilon^2 B_{20} + \varepsilon^3 B_{30} + \varepsilon\mu B_{11} + \varepsilon^2 \mu B_{21} + \varepsilon\mu^2 B_{12}$$

$$+ O(\varepsilon^k \mu^l \, ; k + l = 4),$$

at the initial conditions $a(t_0) = a_0$, $\psi(t_0) = \psi_0$.

From the first equation of (4.1.24) we obtain the dependence $a(t)$, which upon introduction into the latter equation of (4.1.24) enables us to determine the dependence $\psi\{a(t)\}$. Thanks to this, it is possible to analyse the slow transient processes leading to the steady state. The latter are analysed by assuming that $da/dt = 0$, which leads to the algebraic equation

$$G(a, \varepsilon, \mu) = A_{10} + \varepsilon A_{20} + \varepsilon^2 A_{30} + \mu A_{11} + \varepsilon \mu A_{21} + \mu^2 A_{12} = 0. \quad (4.1.24)$$

If the calculations are limited to order ε, we get from (4.1.24)

$$A_{10} = 0, \tag{4.1.25}$$

which enables us to find: (a) one isolated solution; (b) a few isolated solutions; (c) no solutions. However, sometimes the phase flow of the starting equations can be very sensitive to changes in the amplitude a and/or the parameters ε and μ. For these reasons, the full equation (4.1.24) should be taken into consideration.

Now we briefly indicate the variety of problems which can be solved using this approach, and that cannot be solved by the use of a single perturbation method [30].

A. Suppose that the parameter ε undergoes slight changes, which are impossible to avoid. We want to control such changes by treating μ as a control parameter. Inserting $a = a^0 = \text{const}$ into (4.1.24) we can find $G(\varepsilon, \mu, a^0) = G(\varepsilon, \mu) = 0$. Thus, in accordance with the changes of ε, we can find the values of μ in order to maintain a constant amplitude.

Equation (4.1.24) is transformed into the form

$$A_{30}\varepsilon^2 + 2A'_{21}\varepsilon\mu + A_{12}\mu^2 + 2A'_{20}\varepsilon + 2A'_{11}\mu + A_{10} = 0, \tag{4.1.26}$$

where

$$A'_{21} = \frac{1}{2}A_{21}, \quad A'_{20} = \frac{1}{2}A_{20}, \quad A'_{11} = \frac{1}{2}A_{11}. \tag{4.1.27}$$

Equation (4.1.26) presents implicit second-order algebraic functions if A_{30}, A'_{21} and A_{12} are not equal to zero at the same time. The form of the function is determined by the following expressions

$$W = \det \begin{vmatrix} A_{30} & A'_{21} & A'_{20} \\ A'_{21} & A_{12} & A'_{11} \\ A'_{20} & A'_{11} & A_{10} \end{vmatrix}, \quad V = \det \begin{vmatrix} A_{30} & A'_{21} \\ A'_{21} & A_{12} \end{vmatrix}, \tag{4.1.28}$$

$$S = A_{10} + A_{12}, \quad W_{22} = A_{30}A_{10} - (A'_{20})^2, \quad W_{11} = A_{12}A_{10} - (A'_{11})^2.$$

By means of shifting the origin of the coordinate system and rotating the axes, it is possible to obtain the following functional forms (the expressions W, V, S are the invariants of such shifts and rotations):

1. $V > 0$, $AW < 0$. Curve (4.1.26) is the ellipse $\varepsilon^2/A^2 + \mu^2/B^2 = 1$.
2. $V > 0$, $W = 0$. Equation (4.1.26) can be transformed to $\varepsilon^2/A^2 + \mu^2/B^2 = 0$ and the solution is the point $(0, 0)$.

3. $V > 0$, $AW > 0$. Curve (4.1.26) is an imaginary ellipse (no real curve exists).

4. $V < 0$, $W \neq 0$. Equation (4.1.26) is the equilateral hyperbola $\varepsilon^2/A^2 - \mu^2/B^2 = 1$.

5. $V < 0$, $W = 0$. The solution of (4.1.26) is the pair of intersecting lines $\varepsilon^2/A^2 - \mu^2/B^2 = 0$.

6. $V = 0$, $W \neq 0$. The curve governed by (4.1.26) is the parabola $\mu^2 = 2p\varepsilon$.

7. $V = 0$, $W = 0$, $W_{11} < 0$ or $W_{22} > 0$. Equation (4.1.26) represents a pair of parallel lines $\mu^2 - A^2 = 0$.

8. $V = 0$, $W = 0$, $W_{11} > 0$ or $W_{22} > 0$. The solutions of (4.1.26) are imaginary parallel lines $\mu^2 + A^2 = 0$ (no real curve exists).

9. $V = 0$, $W = 0$, $W_{11} = 0$ or $W_{22} = 0$. The solution of (4.1.26) is a double line $\mu^2 = 0$.

The coefficients of (4.1.26) are functions of the amplitude a and their values are determined by the functions $f_{(*)}$.

B. Suppose that we would like to have $a = a(\varepsilon)$ and the shape of $a(\varepsilon)$ should be fixed a priori. The problem is then again reduced to implicit algebraic functions of second order.

C. Different branching phenomena can be expected. We can find the hysteresis algebraic points defined by the following equations

$$G(a, \varepsilon, \mu) = 0,$$
$$G_a(a, \varepsilon, \mu) = 0, \tag{4.1.29}$$
$$G_{aa}(a, \varepsilon, \mu) = 0.$$

If it is possible to eliminate the amplitude a from one of (4.1.29), then the other two enable us to find the hysteresis points. The bifurcation and isolated variety points are defined by the following three equations

$$G(a, \varepsilon, \mu) = 0,$$
$$G_a(a, \varepsilon, \mu) = 0, \tag{4.1.30}$$
$$G_\varepsilon(a, \varepsilon, \mu) = 0.$$

As mentioned above, (4.1.30) can posses several different solutions for a. Thus, the M-multiple limit variety can be defined by the following equations

$$G(a_1, \varepsilon, \mu) = 0,$$

$$\vdots$$

$$G(a_m, \varepsilon, \mu) = 0,$$

$$\vdots$$

$$G_a(a_1, \varepsilon, \mu) = 0,$$

$$\vdots$$

$$G_a(a_m, \varepsilon, \mu) = 0.$$

Using μ as a parameter, we can control the branching phenomena mentioned above.

D. We can find the (ε, μ) set of parameters for which no real solutions of (4.1.24) exist. Thus, a domain of the assumed solution (4.1.7) can be defined in the two-parameter space.

E. Suppose that we want to change the amplitude of oscillations, but the frequency of oscillations should not undergo any changes (or it should be controlled only by a linear pair of equations). In order to fulfil these requirements we have

$$G(a, \varepsilon, \mu) = A_{10} + \varepsilon A_{20} + \varepsilon^2 A_{30} + \mu A_{11} + \varepsilon\mu A_{21} + \mu^2 A_{12} = 0,$$
$$H(a, \varepsilon, \mu) = B_{10} + \varepsilon B_{20} + \varepsilon^2 B_{30} + \mu B_{11} + \varepsilon\mu B_{21} + \mu^2 B_{12} = 0. \quad (4.1.31)$$

After eliminating a from one of (4.1.31) there remains one equation which defines an implicit algebraic function of second order in ε and μ. One can freely choose one parameter and then calculate the value of the second one. Thus, by such an appropriate choice of the parameters ε and μ, the amplitude of the one-frequency oscillations will change; however, the frequency ω_1 will always remain constant.

We consider the following example from the field of mechanics [31]. An elastic beam of constant cross-section is connected by a spring k_2 with a discrete one-degree-of-freedom system (see Fig. 4.1). We assume that the

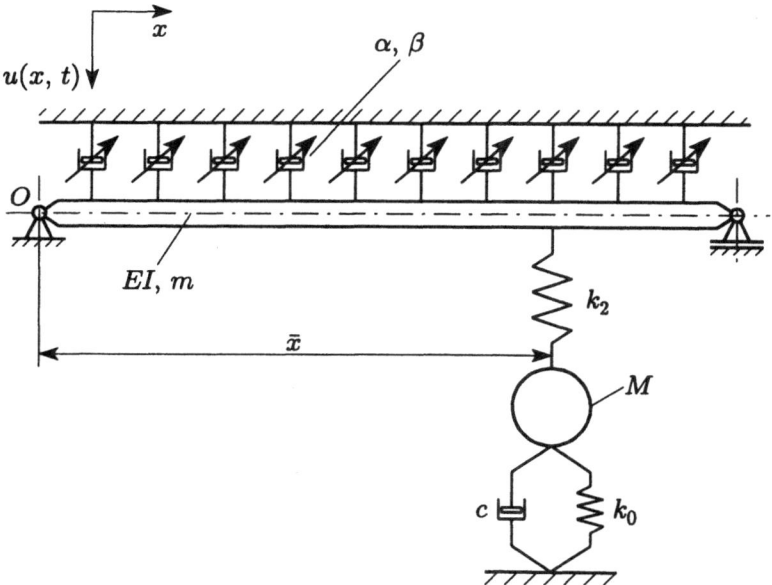

Fig. 4.1. Self-excited vibrations of a beam connected with a one degree-of-freedom system

linear coupling stiffness involves a time delay and that the nonlinearities, the time delay, and the amplitude of oscillations are small. Our system is an autonomous one, and the Van der Pol damping acting on the beam is responsible for oscillations. Within the framework of the usual assumptions of the elementary theory of bending we obtain the following set of governing equations:

$$l\,EI\frac{\partial^4 u}{\partial x^4} + ml\frac{\partial^2 u}{\partial t^2} = l(\alpha - \beta u^2)\frac{\partial u}{\partial t} - k_2\{y(t,\bar{x}) - \delta(x - \bar{x})y(t - \hat{\mu})\},$$
$$M\ddot{y} = -c\dot{y} - (k_0 + k_2) + k_2 u(t - \hat{\mu}, \bar{x}), \qquad\qquad (4.1.32)$$

where the damping coefficient α and β and the mass m are taken per unit length, and $\hat{\mu}$ is a time delay. The other standard parameters are given in Fig. 4.1. We have the following boundary conditions:

$$u(x,t)|_{x=0} = u(x,t)|_{x=l} = 0,$$
$$\left.\frac{\partial^2 u(x,t)}{\partial x^2}\right|_{x=0} = \left.\frac{\partial^2 u(x,t)}{\partial x^2}\right|_{x=l} = 0. \qquad\qquad (4.1.33)$$

In addition to the nonlinear mechanical system with a time delay, the governing delay nonlinear differential equations can be found in problems related to biology, blood circulation, and control systems. Therefore, we transform the dimensional equations (4.1.32) to nondimensional form. Thanks to this we reduce the number of valid parameters, and our further calculations are valid not only for the mechanical system shown in Fig. 4.1, but for other systems as well. The new nondimensional governing equations are:

$$\frac{\partial^2 w(\tau,\xi)}{\partial\tau^2} + \rho^4\frac{\partial^4 w(\tau,\xi)}{\partial\xi^4} = \varepsilon(1 - w^2(\tau,\xi))\frac{\partial w(\tau,\xi)}{\partial\tau} - \varepsilon A w(\tau,\bar{\xi})$$
$$+\varepsilon B\delta(\xi - \bar{\xi})\eta(\tau - \mu), \qquad\qquad (4.1.34)$$
$$\frac{d^2\eta(\tau)}{d\tau^2} = -\varepsilon C\frac{d\eta}{d\tau} + \varepsilon F w(\tau - \mu, \bar{\xi}),$$

and the new boundary conditions are

$$w(\xi,\tau)|_{\xi=0} = w(\xi,\tau)|_{\xi=l} = 0,$$
$$\left.\frac{\partial^2 w(\xi,\tau)}{\partial\tau^2}\right|_{\xi=0} = \left.\frac{\partial^2 w(\xi,\tau)}{\partial\tau^2}\right|_{\xi=l} = 0. \qquad\qquad (4.1.35)$$

The nondimensional parameters are defined as follows:

$$\tau = \Omega t, \quad w = (\beta\alpha^{-1})^{1/2}u, A = k_2\alpha^{-1}l^{-1}\Omega^{-1},$$
$$D = (k_0 + k_2)M^{-1}\Omega^{-2}, \quad \mu = \Omega\hat{\mu}, \quad \varepsilon = \alpha(m\Omega)^{-1}, \qquad\qquad (4.1.36)$$
$$B = k_2\alpha^{-1/2}\beta^{1/2}\Omega^{-1}, \quad \xi = xl^{-1}, \quad \rho^4 = EI m^{-1}\Omega^{-2}l^{-4},$$
$$C = cma^{-1}M^{-1}, \quad F = k_2 ma^{-1/2}\beta^{-1/2}\Omega^{-1}l^{-1}M^{-1}.$$

In order to avoid tedious calculations we assume that

$$\eta(\tau - \mu) = \eta(\tau) - \mu\frac{d\eta}{d\tau},$$

$$w(\tau - \mu, \bar{\xi}) = w(\tau) - \mu\frac{\partial w(\tau, \bar{\xi})}{\partial \tau}. \tag{4.1.37}$$

Taking (4.1.37) into consideration, we obtain from (4.1.33) the following set of equations:

$$\frac{\partial^2 w(\tau, \xi)}{\partial \tau^2} + \rho^4\frac{\partial^4 w(\tau, \xi)}{\partial \xi^4} = \varepsilon(1 - w^2(\tau, \xi))\frac{\partial w(\tau, \xi)}{\partial \tau} - \varepsilon A w(\tau, \bar{\xi})$$

$$+\varepsilon B\delta(\xi - \bar{\xi})\eta(\tau) - \varepsilon\mu B\delta(\xi - \bar{\xi})\frac{d\eta}{d\tau}, \tag{4.1.38}$$

$$\frac{d^2\eta(\tau)}{d\tau^2} + D\eta = -\varepsilon C\frac{d\eta}{d\tau} + \varepsilon F(\tau, \bar{\xi}) - \varepsilon\mu F\frac{\partial w(\tau, \bar{\xi})}{\partial \tau}.$$

From the first equation of (4.1.38), and for $\varepsilon = 0$, we determine the frequency $\nu = (n\pi\rho)^2$, $n \in N$. Limiting our calculations to $n = 1$, and with regard to the earlier section, the solutions of (4.1.38) are sought in the form

$$w(\xi, \tau) = a(t)\sin \pi\xi \cos \psi(\tau) + \varepsilon W_{10}(\xi, a, \psi) + \varepsilon^2 W_{20}(\xi, a, \psi)$$

$$+\varepsilon^3 W_{30}(\xi, a, \psi) + \varepsilon\mu W_{11}(\xi, a, \psi) + \varepsilon^2\mu W_{21}(\xi, a, \psi), \tag{4.1.39}$$

$$\eta(\tau) = \varepsilon\eta_{10}(a, \psi) + \varepsilon^2\eta_{20}(a, \psi) + \varepsilon^3\eta_{30}(a, \psi)$$

$$+\varepsilon\mu\eta_{11}(a, \psi) + \varepsilon^2\mu\eta_{21}(a, \psi),$$

where $W_{kl}(\xi, a, \psi)$ and $\eta_{kl}(a, \psi)$ are the limited and periodic (with regard to ψ) functions to be obtained. The unknown amplitude $a(\tau)$ and phase $\psi(\tau)$ are calculated from

$$\frac{da}{dt} = \varepsilon A_{10}(a) + \varepsilon^2 A_{20}(a) + \varepsilon^3 A_{30}(a) + \varepsilon\mu A_{11}(a) + \varepsilon^2\mu A_{21}(a), \tag{4.1.40}$$

$$\frac{d\psi}{dt} = \nu + \varepsilon B_{10}(a) + \varepsilon^2 B_{20}(a) + \varepsilon^3 B_{30}(a)$$

$$+\varepsilon\mu B_{11}(a) + \varepsilon^2\mu B_{21}(a). \tag{4.1.41}$$

Proceeding in an analogous way, we find the sequence of recurrent linear equations

$$\varepsilon : \nu^2\frac{\partial^2 W_{10}}{\partial \psi^2} + \rho^4\frac{\partial^4 W_{10}}{\partial \xi^4} - 2\nu B_{10}a \cos \psi \sin \pi\xi - 2\nu A_{10}\sin \psi \sin \pi\xi$$

$$= -a\nu \sin \pi\xi \sin \psi + \nu a^3 \sin^3 \pi\xi \sin \psi \cos^2 \psi$$

$$-aA \sin \pi\bar{\xi} \sin \psi \cos \psi, \tag{4.1.42}$$

$$\nu^2\frac{\partial^2 \eta_{10}}{\partial \psi^2} + D\eta_{10} = aF \sin \pi\bar{\xi} \cos \psi;$$

$$\varepsilon^2 \; : \; \nu^2 \frac{\partial^2 W_{20}}{\partial \psi^2} + \rho^4 \frac{\partial^4 W_{20}}{\partial \xi^4} - 2\nu B_{10} a \cos \psi \sin \pi\xi - 2\nu A_{10} \sin \psi \sin \pi\xi$$

$$= -2\nu B_{10} \frac{\partial^2 W_{10}}{\partial \psi^2} - 2\nu A_{10} \frac{\partial^2 W_{10}}{\partial a \partial \psi} - \left(A_{10} \frac{\mathrm{d}A_{10}}{\mathrm{d}a} \right.$$

$$\left. - B_{10}^2 a \right) \cos \psi \sin \pi\xi + \left(2A_{10}B_{10} + a \frac{\mathrm{d}B_{10}}{\mathrm{d}a} A_{10} \right) \sin \psi \sin \pi\xi$$

$$+ A_{10} \cos \psi \sin \pi\xi - a B_{10} \sin \psi \sin \pi\xi + \nu \frac{\partial W_{10}}{\partial \psi} \qquad (4.1.43)$$

$$- A_{10} a^2 \cos^2 \psi \sin^2 \pi\xi + 2a^2 W_{10} \nu \sin \psi \cos \psi \sin^2 \pi\xi$$

$$+ a^3 B_{10} \sin \psi \cos^2 \psi \sin^3 \pi\xi - \frac{\partial W_{10}}{\partial \psi} \nu a^2 \cos^2 \psi \sin^2 \pi\xi$$

$$- A W_{10} + B \delta(\xi - \bar\xi) \eta_{10},$$

$$\nu^2 \frac{\partial^2 \eta_{20}}{\partial \psi^2} + D\eta_{20} = -2\nu A_{10} \frac{\partial^2 \eta_{10}}{\partial a \partial \psi} - 2\nu B_{10} \frac{\partial^2 \eta_{10}}{\partial \psi^2}$$

$$- C \frac{\partial \eta_{10}}{\partial \psi} + F W_{10};$$

$$\varepsilon^3 \; : \; \nu^2 \frac{\partial^2 W_{30}}{\partial \psi^2} + \rho^4 \frac{\partial^4 W_{30}}{\partial \xi^4} - 2\nu B_{30} a \cos \psi \sin \pi\xi - 2\nu A_{30} \sin \psi \sin \pi\xi$$

$$= -\left(\frac{\mathrm{d}A_{20}}{\mathrm{d}a} A_{10} + \frac{\mathrm{d}A_{10}}{\mathrm{d}a} A_{20} - 2a B_{10} B_{20} \right) \cos \psi \sin \pi\xi$$

$$+ \left\{ 2A_{20}B_{10} + 2A_{10}B_{20} + a \left(\frac{\mathrm{d}B_{10}}{\mathrm{d}a} A_{20} \right.\right.$$

$$\left.\left. + \frac{\mathrm{d}B_{20}}{\mathrm{d}a} A_{10} \right) \right\} \sin \psi \sin \pi\xi - A_{10}^2 \frac{\partial W_{10}}{\partial a^2} - 2 \frac{\partial^2 W_{10}}{\partial a \partial \psi} (\nu A_{20}$$

$$+ A_{10} B_{10}) - A_{10} \frac{\partial W_{10}}{\partial a} \frac{\mathrm{d}A_{10}}{\mathrm{d}a} - \frac{\partial^2 W_{10}}{\partial \psi^2} (B_{10}^2 + 2\nu B_{20})$$

$$- A_{10} \frac{\partial W_{10}}{\partial \psi} \frac{\mathrm{d}B_{10}}{\mathrm{d}a} - 2\nu A_{10} \frac{\partial^2 W_{20}}{\partial a \partial \psi} - 2\nu B_{10} \frac{\partial^2 W_{20}}{\partial \psi^2}$$

$$+ A_{20} \cos \psi \sin \pi\xi - a B_{20} \sin \psi \sin \pi\xi$$

$$+ A_{10} \frac{\partial W_{10}}{\partial a} + B_{10} \frac{\partial W_{10}}{\partial \psi} + \nu \frac{\partial W_{20}}{\partial \psi}$$

$$- 2a A_{10} W_{10} \cos^2 \psi \sin^2 \pi\xi - a^2 A_{20} \cos^3 \psi \sin^3 \pi\xi \qquad (4.1.44)$$

$$+ W_{10}^2 \nu a \sin \psi \sin \pi\xi + 2\nu a^2 W_{20} \sin \psi \cos \psi \sin^2 \pi\xi$$

$$+ 2a^2 B_{10} W_{10} \sin \psi \cos \psi \sin^2 \pi\xi + a^3 B_{20} \sin \psi \cos^2 \psi \sin^3 \pi\xi$$

$$- a^2 A_{10} \frac{\partial W_{10}}{\partial a} \cos^2 \psi \sin^2 \pi\xi - 2\nu a W_{10} \frac{\partial W_{10}}{\partial \psi} \cos \psi \sin \pi\xi$$

$$- a^2 B_{10} \frac{\partial W_{10}}{\partial \psi} \cos^2 \psi \sin^2 \pi\xi - a^2 \nu \frac{\partial W_{20}}{\partial \psi} \cos^2 \psi \sin^2 \pi\xi$$

$$-AW_{20} + B\eta_{20}\delta(\xi - \bar{\xi}),$$

$$\nu^2 \frac{\partial^2 \eta_{30}}{\partial \psi^2} + D\eta_{30} = -2A_{10}^2 \frac{\partial^2 \eta_{10}}{\partial a^2} - 2\frac{\partial^2 \eta_{10}}{\partial a \partial \psi}(\nu A_{20} + A_{10}B_{10})$$

$$-A_{10}\frac{dA_{10}}{da}\frac{\partial \eta_{10}}{\partial a} - \frac{\partial^2 \eta_{10}}{\partial \psi^2}(2\nu B_{20} + B_{10}^2)$$

$$-A_{10}\frac{dB_{10}}{da}\frac{\partial \eta_{10}}{\partial \psi} - 2\nu A_{10}\frac{\partial^2 \eta_{20}}{\partial a \partial \psi} - 2\nu B_{10}\frac{\partial^2 \eta_{20}}{\partial \psi^2}$$

$$-CA_{10}\frac{\partial \eta_{10}}{\partial a} - CB_{10}\frac{\partial \eta_{10}}{\partial \psi} - C\nu\frac{\partial \eta_{20}}{\partial \psi} + FW_{20};$$

$$\varepsilon^2 \mu \; : \; \nu^2 \frac{\partial^2 W_{21}}{\partial \psi^2 W_{21}} + \rho^4 \frac{\partial^4 W_{30}}{\partial \xi^4} - 2\nu B_{21}a\cos\psi\sin\pi\xi$$

$$-2\nu A_{21}\sin\psi\sin\pi\xi = A_{11}\cos\psi\sin\pi\xi$$

$$-aB_{11}\sin\psi\sin\pi\xi + \nu\frac{\partial W_{11}}{\partial \psi} - a^2 A_{11}\cos^3\psi\sin^3\pi\xi$$

$$+2\nu a^2 W_{11}\sin\psi\cos\psi\sin^2\pi\xi + a^3 B_{11}\sin\psi\cos^2\psi\sin^3\pi\xi$$

$$-\nu a^2\frac{\partial W_{11}}{\partial \psi}\cos^2\psi\sin^2\pi\xi - AW_{11} + B\eta_{11}\delta(\xi - \bar{\xi}) \qquad (4.1.45)$$

$$-B\nu\delta(\xi - \bar{\xi})\frac{\partial \eta_{10}}{\partial \psi},$$

$$\nu^2\frac{\partial^2 \eta_{21}}{\partial \psi^2} + D\eta_{21} = -C\nu\frac{\partial \eta_{11}}{\partial \psi} + FW_{11} - FA_{10}\cos\psi\sin\pi\xi$$

$$+aFB_{10}\sin\psi\sin\pi\xi - F\nu\frac{\partial W_{10}}{\partial \psi};$$

$$\varepsilon\mu \; : \; \nu^2\frac{\partial^2 W_{11}}{\partial \psi^2} + \rho^4\frac{\partial^4 W_{11}}{\partial \xi^4} = 0, \qquad (4.1.46)$$

$$\nu^2\frac{\partial^2 \eta_{11}}{\partial \psi^2} + D\eta_{11} = a\nu F\sin\psi\sin\pi\bar{\xi}.$$

The solution of this set of equations gives:

$$W_{10} = -\frac{1}{1280\nu}a^3\sin 3\pi\xi\sin\psi - \frac{3}{128\nu}a^3\sin\pi\xi\sin 3\psi$$

$$-\frac{1}{1152\nu}a^3\sin 3\pi\xi\sin 3\psi,$$

$$\eta_{10} = \frac{Fa}{(D - \nu^2)}\sin\pi\bar{\xi}\cos\psi, \qquad \bar{\xi} \in (0,1),$$

$$W_{20} = A_1\sin 3\pi\xi\sin\psi + A_2\sin\pi\xi\sin 3\psi + A_3\sin 3\pi\xi\sin 3\psi$$

$$+A_4\sin 3\pi\xi\cos\psi + A_5\sin\pi\xi\cos 3\psi + A_6\sin 3\pi\xi\cos 3\psi,$$

$$A_1 = \frac{1}{80\nu^2}\left\{-\frac{a^3}{1280\nu}A\sin\pi\bar{\xi} - \frac{a^3}{32\nu}A\sin\pi\bar{\xi} + \frac{a^3}{1280\nu}A\right\},$$

$$A_2 = \frac{1}{8\nu^2}\left\{\frac{27a^3}{128\nu}A\sin\pi\bar{\xi} - \frac{3a^3}{32\nu}A\sin\pi\bar{\xi} - \frac{3a^3}{128\nu}A\right\},$$

$$A_3 = \frac{1}{72\nu^2}\left\{-\frac{9a^3}{1152\nu}A\sin\pi\bar{\xi} - \frac{a^3}{32\nu}A\sin\pi\bar{\xi} + \frac{a^3}{1152\nu}A\right\},$$

$$A_4 = \frac{1}{80\nu^2}\left\{\frac{a^3}{640}\right\}, \quad A_5 = \frac{1}{8\nu^2}\left\{\frac{9}{64}a^3\right\},$$

$$A_6 = \frac{1}{72\nu^2}\left\{\frac{3}{1152}a^3\right\},$$

$$\eta_{20} = \left\{2\nu A_{10}\frac{F}{(D-\nu^2)^2} + a\nu\frac{FC}{(D-\nu^2)^2}\right\}\sin\pi\bar{\xi}\sin\psi$$

$$+2\nu a B_{10}\frac{F}{(D-\nu^2)^2}\sin\pi\bar{\xi}\cos\psi - \frac{a^3 F}{1280\nu(D-\nu^2)}\sin 3\pi\bar{\xi}\sin\psi$$

$$-\frac{3a^3 F}{128\nu(D-9\nu^2)}\sin\pi\bar{\xi}\sin 3\psi - \frac{a^3 F}{1152\nu(D-9\nu^2)}\sin 3\pi\bar{\xi}\sin 3\psi,$$

$$W_{11} = 0, \quad \eta_{11} = \frac{Fa\nu}{(D-\nu^2)}\sin\pi\bar{\xi}\sin\psi, \qquad (4.1.47)$$

$$A_{10} = \frac{1}{2}a\left(1 - \frac{3}{16}a^2\right), \quad B_{10} = \frac{1}{2\nu}A\sin\pi\bar{\xi}, \quad A_{20} = 0,$$

$$B_{20} = -\frac{1}{\nu}\left\{\frac{1}{8} + \frac{3}{64}a^2 + \frac{1}{8\nu^2}A^2\sin^2\pi\bar{\xi} + \frac{BF}{2(D-\nu^2)}\sin\pi\bar{\xi}\right\},$$

$$A_{30} = -\frac{3a^3}{256\nu^4}A^2\sin^2\pi\bar{\xi} - \frac{3BFa^3}{64\nu^2(D-\nu^2)}\sin\pi\bar{\xi} + \frac{3a^3}{256\nu^2}$$

$$+\frac{BFa}{2\nu^2(D-\nu^2)}\sin\pi\bar{\xi} - \frac{BFa\left(1-\frac{3}{16}a^2\right)}{2(D-\nu^2)^2}\sin\pi\bar{\xi}$$

$$-\frac{CBFa}{2(D-\nu^2)^2}\sin\pi\bar{\xi},$$

$$B_{30} = \frac{a}{2\nu^3}\left\{\frac{1}{8} + \frac{3}{64}a^2 + \frac{1}{8\nu^2}A^2\sin^2\pi\bar{\xi} - \frac{BF}{2(D-\nu^2)}\sin\pi\bar{\xi}\right\}$$

$$-\frac{BF}{4\nu^2(D-\nu^2)^2}A^2\sin^3\pi\bar{\xi},$$

$$A_{11} = 0, \quad B_{11} = 0,$$

$$A_{21} = -\frac{BFa}{2\nu(D-\nu^2)}(\nu+1)\sin\pi\bar{\xi}, \quad B_{21} = 0.$$

In the calculations we have not taken into account harmonics of order greater than three, and we have omitted powers of the amplitudes which were greater than three. Let us consider the stationary state which leads to the following algebraic equation:

$$A_{10} + \varepsilon^2 A_{30} + \varepsilon \mu A_{21} = 0. \tag{4.1.48}$$

Because we have limited the calculations to the first power of μ in our example, we can use the general discussion given earlier by substituting μ for a and considering further the implicit function (4.1.48) with regard to a and ε.

From (4.1.48) one obtains

$$\varepsilon^2 \frac{BF}{\nu^2(D - \nu^2)} \{D - \nu^2(C + 2)\} \sin \pi \bar{\xi}$$

$$-\varepsilon \frac{BF(\nu + 1)\mu}{\nu(D - \nu^2)} \sin \pi \bar{\xi} - \frac{1}{16} a^2 + 1 = 0. \tag{4.1.49}$$

We have also determined

$$W = \frac{3BF}{16\nu^2(D - \nu^2)^2} \sin \pi \bar{\xi} \left\{ (\nu^2(C + 2) - D) + \mu^2 \frac{BF(\nu + 1)^2}{4} \sin \pi \bar{\xi} \right\},$$

$$V = \frac{3BF}{16\nu^2(D - \nu^2)^2} \{\nu^2(C + 2) - D\} \sin \pi \bar{\xi}. \tag{4.1.50}$$

These results allow us to come to some important conclusions from the point of view of possible applications.

If $\nu^2(C + 2) > D$, then in the considered system, a one-frequency periodic solution does not exist, because $W > 0$.

If $\nu^2(C + 2) = D$, then $W > 0$, and the curve $\varepsilon(a)$ is a parabola.

If $\nu^2(C + 2) < D$ and $W \neq 0$, then the curve $\varepsilon(a)$ is an equilateral hyperbola.

If $\nu^2(C + 2) < D$ and $W = 0$, we have two intersecting lines.

4.2 Simple Perturbation Technique

To introduce the reader to a simple perturbation technique applied to discrete–continuous systems we consider equations of the form [29]

$$\frac{\partial^2 u(t_1, x)}{\partial t_1^2} = c^2 \frac{\partial^2 u(t_1, x)}{\partial x^2}$$

$$+ f\left(\varepsilon, x, u(t_1, x), \frac{\partial u(t_1, x)}{\partial t_1}, \frac{\partial u(t_1, x)}{\partial x}, y(t_1 - \tau)\right), \tag{4.2.1}$$

$$L_1[y(t), \tau_r] = \varphi[\varepsilon, y(t_1), u(t_1 - \tau, \xi)]$$

with the homogeneous boundary conditions

$$u(t_1, 0) = u(t_1, l) = 0, \tag{4.2.2}$$

where:
f is a certain nonlinear function assuming to be zero for $x = 0$ and $x = l$;

$L_1(y, \tau_r)$ is a linear differential operator with delay of the form

$$L_1[y, \tau_r] = \sum_{p=0}^{P} \sum_{r=0}^{R} a_{pr} y^{(p)}(t_1 - \tau_r), \quad \tau_0 = 0, \quad \tau_r > 0;$$

φ is a nonlinear differential operator with a time delay.

Next, we assume that the nonlinear operators f and φ have continuous first derivatives, considering the other arguments in a certain sufficiently large range of their variations. Moreover, we assume that the delays occurring in the system are small. Thus, we have

$$y(t_1 - \tau) = y(t_1) - \tau \frac{dy(t_1)}{dt_1} + \frac{1}{2}\tau^2 \frac{d^2 y(t_1)}{dt_1^2} \cdots,$$

$$u(t_1 - \tau, \xi) = u(t_1, \xi) - \tau \frac{\partial u(t_1, \xi)}{\partial t_1} + \frac{1}{2}\tau^2 \frac{\partial^2 y(t_1, \xi)}{\partial t_1^2} \cdots. \tag{4.2.3}$$

Further calculations will be limited only to the first three terms of series (4.2.3) in (4.2.1) and we obtain

$$\frac{\partial^2 u(t_1, x)}{\partial t_1^2} = c^2 \frac{\partial^2 u(t_1, x)}{\partial x^2}$$

$$+ f_1 \left(\varepsilon, x, u(t_1, x), \frac{\partial u(t_1, x)}{\partial t_1}, \frac{\partial u(t_1, x)}{\partial x}, \tau, y, \frac{dy}{dt_1}, \frac{d^2 y}{dt_1^2} \right),$$

$$L_1[y(t_1), \tau_r] = \varphi_1 \left(\varepsilon, y(t_1), \tau, u(t_1, \xi), \frac{\partial u(t_1, \xi)}{\partial t_1}, \frac{\partial^2 u(t_1, \xi)}{\partial t_1^2} \right), \tag{4.2.4}$$

where the functions f_1 and φ_1 are obtained respectively from f and φ, considering (4.2.3).

Let the nonlinear functions f_1 and φ_1 vanish when $\varepsilon = \tau = 0$, which means that the nonlinear system of differential equations (4.2.4) is then reduced to a linear system.

Then, the problem lies in the analysis of the system of equations (4.2.4) with two independent small parameters τ and ε.

Let us further assume that the characteristic equation adequate for the linear part of the second equation of the system is of the form

$$\Theta(\rho) = \sum_{p=0}^{P} \sum_{r=0}^{R} a_{pr} \rho^p e^{-\tau_r \rho}, \tag{4.2.5}$$

and that its eigenvalues are different from zero and have purely imaginary values. This means that oscillations are not generated by the discrete system. The starting solution for the analytical approximate method, with $\varepsilon = 0$, $\tau = 0$, is of the form

$$U_{(*)}^0(t_1, x) = \sum_{n=1}^{\infty} \sin \frac{\pi x}{l} \left[a_{(*)n}^0 \cos(n\alpha_0 t_1) + b_{(*)n}^0 \sin(n\alpha_0 t_1) \right],$$

$$y_{(*)}^0(t_1) = 0, \tag{4.2.6}$$

where the operator $(*)$ denotes τ or ε, $a^0_{(*)n}$, and $b^0_{(*)n}$ are the amplitudes, and $T_0 = 2\pi/\alpha_0 = 2l/c$ is the period of oscillations of the linear part of the system described by the first equation in (4.2.4). For $\varepsilon \neq 0$ and $\tau \neq 0$ in a satisfactorily close neighbourhood of zero, we seek the periodic solution of system (4.2.4) a little different from (4.2.6). Generally, the contribution of higher harmonics to the solution quickly decreases, and it is sufficient to consider only a few of the first harmonics in the calculations. The period sought is equal to

$$T = T_0[1 + \eta(\varepsilon, \tau)] \tag{4.2.7}$$

and evidently depends on both of the perturbation parameters.

Let us introduce a new dimensionless time t according to the equation

$$t_1 = \frac{1 + \eta(\varepsilon, \tau)}{\alpha_0} t, \tag{4.2.8}$$

which allows us to seek a periodic solutions with period 2π.

Substituting (4.2.8) in (4.2.4), we obtain the equation

$$\frac{\partial^2 u(t, x)}{\partial t^2} = c^2 \frac{\partial^2 u(t, x)}{\partial x^2}$$
$$+ F\left(\varepsilon, x, u(t, x), \frac{\partial u(t, x)}{\partial t}, \frac{\partial u(t, x)}{\partial x}, \tau, y, \frac{dy}{dt}, \frac{d^2 y}{dt^2}\right),$$
$$L_1[y(t), \tau_r] = \phi_1\left(\varepsilon, y(t), t, u(t, \xi), \frac{\partial u(t, \xi)}{\partial t}, \frac{\partial^2 u(t, \xi)}{\partial t^2}\right), \tag{4.2.9}$$

where:

$$F = \frac{(1 + \eta^2) l^2}{\pi^2 c^2} f_1, \tag{4.2.10}$$

$$L = \sum_{p=0}^{P} \sum_{r=0}^{R} a_{pr} \left\{ \left[\frac{l(1+\eta)}{\pi c}\right]^2 y^{(p)}(t) - \tau_r \frac{l(1+\eta)}{\pi c} y^{(p+1)}(t) \right.$$
$$\left. + \frac{1}{2} \tau_r^2 y^{(p+2)}(t) \right\},$$

$$\phi = \left[\frac{l(1+\eta)}{\pi c}\right]^2 \varphi.$$

The nonlinear functions ϕ and F as well as the solution sought, y, u and η, are presented in the form of power series:

$$\phi = \phi_0 + \varepsilon \phi_\varepsilon + \varepsilon^2 \phi_{\varepsilon\varepsilon} + \ldots + \tau \phi_\tau + \tau^2 \phi_{\tau\tau} + \ldots + \tau\varepsilon \phi_{\tau\varepsilon} + \ldots,$$
$$F = F_0 + \varepsilon F_\varepsilon + \varepsilon^2 F_{\varepsilon\varepsilon} + \ldots + \tau F_\tau + \tau^2 F_{\tau\tau} + \ldots + \tau\varepsilon F_{\tau\varepsilon} + \ldots,$$
$$y = y_0 + \varepsilon y_\varepsilon + \varepsilon^2 y_{\varepsilon\varepsilon} + \ldots + \tau y_\tau + \tau^2 y_{\tau\tau} + \ldots + \tau\varepsilon y_{\tau\varepsilon} + \ldots, \tag{4.2.11}$$
$$u = u_0 + \varepsilon u_\varepsilon + \varepsilon^2 u_{\varepsilon\varepsilon} + \ldots + \tau u_\tau + \tau^2 u_{\tau\tau} + \ldots + \tau\varepsilon u_{\tau\varepsilon} + \ldots,$$
$$\eta = \eta_0 + \varepsilon \eta_\varepsilon + \varepsilon^2 \eta_{\varepsilon\varepsilon} + \ldots + \tau \eta_\tau + \tau^2 \eta_{\tau\tau} + \ldots + \tau\varepsilon \eta_{\tau\varepsilon} + \ldots.$$

Having substituted (4.2.11) into (4.2.9), and having equated the expression representing the same powers of the small parameters τ and ε as well as the same powers of their products $\tau^m \varepsilon^l$ $(m, l = 1, 2 \ldots)$, the recurrent systems of linear equations are obtained. While solving the subsequent equations of the system, we use the harmonics balance method. Let us assume that we have determined the first system of recurrent equations, where the operator $(*)$ means τ or ε. Having substituted the solutions (4.2.6) for the nonlinear functions $F_{(*)}$ and $\phi_{(*)}$ (this time for the equation we assume $t_1 = t$ and $\alpha_0 = 1$) and having developed these functions into a Fourier series, we obtain

$$F_{(*)}(t, x) = \sum_{p=0}^{\infty} \sum_{k=0}^{\infty} \sin \frac{n\pi x}{l} [A_{nk}^{(*)} \cos(kt) + B_{nk}^{(*)} \sin(kt)],$$

$$\phi_{(*)}(t) = \sum_{k=0}^{\infty} [C_k^{(*)} \cos(kt) + D_k^{(*)} \sin(kt)], \tag{4.2.12}$$

where:

$$A_{nk}^{(*)} = \frac{2}{\pi l} \int_0^l \int_0^{2\pi} F_{(*)}(t, x) \sin \frac{n\pi x}{l} \cos(kt) \, dt \, dx,$$

$$B_{nk}^{(*)} = \frac{2}{\pi l} \int_0^l \int_0^{2\pi} F_{(*)}(t, x) \sin \frac{n\pi x}{l} \sin(kt) \, dt \, dx,$$

$$C_k^{(*)} = \frac{1}{\pi} \int_0^{2\pi} \phi_{(*)}(t) \cos(kt) \, dt, \tag{4.2.13}$$

$$D_k^{(*)} = \frac{1}{\pi} \int_0^{2\pi} \phi_{(*)}(t) \sin(kt) \, dt.$$

We seek the solutions of the system of equations formed by comparison of the expression next to $(*)$ in the form of

$$U_{(*)}(t, x) = \sum_{n=1}^{N} \sum_{k=0}^{K} \sin \frac{n\pi x}{l} [a_{(*)nk} \cos(kt) + b_{(*)nk} \sin(kt)],$$

$$y_{(*)}(t) = \sum_{k=0}^{K} [c_{(*)k} \cos(kt) + d_{(*)k} \sin(kt)]. \tag{4.2.14}$$

The solution of the first equation of system (4.2.9) is explicitly determined only when

$$P_{(*)n}(a_{(*)s}^0, b_{(*)s}^0, \eta_\varepsilon) = \int_0^l \int_0^{2\pi} F_{(*)}(t, x) \sin \frac{n\pi x}{l} \cos(nt) \, dt \, dx = 0,$$

$$Q_{(*)n}(a^0_{(*)s}, b^0_{(*)s}, \eta_\varepsilon) = \int\limits_0^l \int\limits_0^{2\pi} F_{(*)}(t, x) \sin \frac{n\pi x}{l} \sin(nt) \, dt \, dx = 0. \quad (4.2.15)$$

Conditions (4.2.15) allow us to neglect the resonance terms which exponentially grow with time. Thus we obtain $2N$ of the equations, whereas the unknowns $a^0_{(*)s}$, $b^0_{(*)s}$ and η_ε are $2N + 1$. In this case, however, dealing with an autonomous system, we may assume that $b_{(*)N} = 0$.

As an example let us consider the discrete-continuous system described by the equations

$$\frac{\partial^2 u}{\partial t_1^2} = \left(\frac{30}{\pi}\right)^2 \frac{\partial^2 u(t_1, x)}{\partial x^2} + \varepsilon[0.003 - u^2(t_1, x)]\frac{\partial u(t_1, x)}{\partial t_1}$$

$$+ \varepsilon\delta(x - \bar{x})y(t_1) - \varepsilon\tau\delta(x - \bar{x})\frac{dy}{dt_1}$$

$$+ \tau\left(\frac{\partial u(t_1, x)}{\partial t_1} - \frac{\partial^2 u(t_1, x)}{\partial x^2}\right) - \tau\left(\frac{\partial u(t_1, x)}{\partial t_1}\right)^3, \quad (4.2.16)$$

$$\frac{d^2 y}{dt_1^2} + 10\varepsilon\frac{dy}{dt_1} + 400y(t_1) = 10\varepsilon u(t_1, x), \quad u(t_1, 0) = u(t_1, 1) = 0,$$

where for the sake of simplification of the calculations, the delay τ and the small parameter ε are in the evident form (4.2.1) and $\bar{x} \in [0, 1]$ is the association point of the discrete system with the continuous one. In the discrete system described by the second equation of system (4.2.16) accompanied by the lack of interaction on the side of the continuous system and as a result of damping in the system, oscillations cannot occur. The oscillations are excited in the continuous system because of damping of the Van der Pol type described by the second term on the right-hand side of the equality sign. For $\tau = \varepsilon = 0$ the period of this solution is equal to $T_0 = \pi/15$. We seek the periodic solution of system (4.2.16) with period T, insignificantly different from the period T_0. Accordingly, let us first make use of the independent variable

$$t_1 = \frac{1 + \eta(\varepsilon, \tau)}{30}t. \quad (4.2.17)$$

Having substituted (4.2.17) into (4.2.16), we obtain

$$\frac{\partial^2 u(t, x)}{\partial t^2} = \frac{1}{\pi^2}(1 + \eta)^2\frac{\partial^2 u(t, x)}{\partial x^2} + \varepsilon\frac{(1 + \eta)}{30}[0.003 - u^2(t, x)]\frac{\partial u(t, x)}{\partial t}$$

$$+ \varepsilon\frac{(1 + \eta)^2}{900}\delta(x - \bar{x})y(t) - \varepsilon\tau\frac{1 + \eta}{30}\delta(x - \bar{x})\frac{dy}{dt}$$

$$+ \tau\frac{1 + \eta}{30}\frac{\partial u(t, x)}{\partial t} - \tau\frac{(1 + \eta)^2}{900}\frac{\partial^2 u(t, x)}{\partial x^2}$$

$$- \tau\frac{30}{1 + \eta}\left(\frac{\partial u(t, x)}{\partial t}\right)^3; \quad (4.2.18)$$

$$\frac{d^2 y}{dt^2} + \varepsilon\frac{1}{3}(1 + \eta)\frac{dy}{dt} + \frac{4}{9}(1 + \eta)^2 y(t) = \frac{\varepsilon}{90}(1 + \eta)^2 u(t, \bar{x}).$$

The parameters τ and ε are treated as independent. Assuming one of them to be equal to zero, the problem is reduced to the classical perturbation method.

We assume the starting solution in the form of

$$u^{(0)} = u_\tau^{(0)} + u_\varepsilon^{(0)} = a_\varepsilon^{(0)} \sin \pi x \cos t + a_\tau^{(0)} \sin \pi x \cos t,$$
$$y^{(0)} = 0. \tag{4.2.19}$$

The amplitude sought, $a_\varepsilon^{(0)}$, will be determined from the first recurrent equation formed by the comparsion of expressions that are coefficients of the parameter ε, whereas the amplitude $a_\tau^{(0)}$ will be determined from the first recurrent equation formed by the comparsion of expressions that are coefficients of the parameter τ.

From the first equation of system (4.2.18), having equated the expressions that are coefficients of the parameters τ, we obtain

$$\frac{\partial^2 u}{\partial t^2} = \frac{1}{\pi^2} \frac{\partial^2 u}{\partial x^2} + \frac{2\eta\tau}{\pi^2} \frac{\partial^2 u_\tau^{(0)}}{\partial x^2} + \frac{1}{30} \frac{\partial u_\tau^{(0)}}{\partial t}$$
$$- \frac{1}{900} \frac{\partial^2 u_\tau^{(0)}}{\partial x^2} - 30 \left(\frac{\partial u_\tau^{(0)}}{\partial t} \right)^3. \tag{4.2.20}$$

Having equated the resonance terms to zero, we obtain

$$\eta_\tau = 0.0055,$$
$$a_\tau^{(0)} = 0.044. \tag{4.2.21}$$

The solution of (4.2.20) is

$$u_\tau = \frac{3}{128} \left(a_\tau^{(0)} \right)^3 \sin 3\pi x \sin t - \frac{3}{128} \left(a_\tau^{(0)} \right)^3 \sin \pi x \sin 3t$$
$$+ a_\tau \sin \pi x \cos t, \tag{4.2.22}$$

where the amplitude a_τ will be determined from the subsequent reccurent equation. This equation is of the form

$$\frac{\partial^2 u_{\tau\tau}}{\partial t^2} = \frac{1}{\pi^2} \frac{\partial^2 u_{\tau\tau}}{\partial x^2} + \frac{2}{\pi^2} \eta_\tau \frac{\partial^2 u_\tau}{\partial x^2} + \frac{2}{\pi^2} \eta_{\tau\tau} \frac{\partial^2 u_\tau^{(0)}}{\partial x^2}$$
$$+ \frac{1}{30} \eta_\tau \frac{\partial u_\tau^{(0)}}{\partial t} + \frac{1}{30} \frac{\partial u_\tau}{\partial t} - \frac{1}{900} \frac{\partial^2 u_\tau}{\partial x^2} - \frac{2}{900} \eta_\tau \frac{\partial^2 u_\tau^{(0)}}{\partial x^2}$$
$$- 90 \frac{\partial (u^{(0)})^2 u_\tau}{\partial t} + 30 \eta_\tau \frac{\partial (u_\tau^{(0)})^3}{\partial t}. \tag{4.2.23}$$

From (4.2.23), having equated the resonance terms to zero, we obtain

$$-\frac{1}{30} \eta_\tau a_\tau^{(0)} - \frac{1}{30} a_\tau - \frac{9}{16} (a_\tau^{(0)})^2 a_\tau = 0,$$
$$-2\eta_\tau a_\tau - a_\tau^{(0)} \eta_{\tau\tau} + \frac{1}{900} a_\tau \pi^2 + \frac{2}{900} \eta_\tau a_\tau^{(0)} \pi^2 \tag{4.2.24}$$
$$+ \frac{135}{206} (a_\tau^{(0)})^5 - \frac{135}{8} \eta_\tau (a_\tau^{(0)})^3 = 0.$$

Solving the system of equations (4.2.24) we get:

$$a_\tau = 0.00002, \qquad \eta_{\tau\tau} = -0.00006. \tag{4.2.25}$$

From the second equation of system (4.2.18), we obtain

$$y_\tau = 0. \tag{4.2.26}$$

Let us now determine the perturbation equations formed owing to the comparison of the expressions that are coefficients of the parameter ε.

From the first equation of system (4.2.18), we obtain

$$\frac{\partial^2 u_\varepsilon}{\partial t^2} = \frac{1}{\pi^2}\frac{\partial^2 u_\varepsilon}{\partial x^2} + \frac{1}{\pi^2}2\eta_\varepsilon\frac{\partial^2 u_\varepsilon^{(0)}}{\partial x^2} + \frac{1}{30}\big[0.003 - (u_\varepsilon^{(0)})^2\big]\frac{\partial u_\varepsilon^{(0)}}{\partial t}, \tag{4.2.27}$$

and having equated the resonance terms to zero, we obtain a system of algebraic equations. Solving it, we have

$$a_\varepsilon^{(0)} = 0.12649, \qquad \eta_\varepsilon = 0. \tag{4.2.28}$$

The solution of (4.2.27) is

$$u_\varepsilon = a_\varepsilon \sin(\pi x)\cos t - 5\cdot 10^{-7}\sin 3\pi x \cos t. \tag{4.2.29}$$

From the second equation of system (4.2.18), we obtain

$$\frac{d^2 y}{dt^2} + \frac{4}{9}y_\varepsilon = \frac{1}{90}u_\varepsilon^{(0)}(t,x). \tag{4.2.30}$$

We seek the solution of (4.2.30) in the form

$$y_\varepsilon = b_\varepsilon \cos t + c_\varepsilon \sin t. \tag{4.2.31}$$

Having substituted (4.2.31) into (4.2.30), we find

$$b_\varepsilon = -0.088\sin(\pi\bar{x}), \qquad c_\varepsilon = 0. \tag{4.2.32}$$

From the second equation of system (4.2.18), having equated the coefficients of ε^2, we obtain

$$\frac{d^2 y_{\varepsilon\varepsilon}}{dt^2} + \frac{4}{9}y_{\varepsilon\varepsilon} = \frac{1}{3}b_\varepsilon \sin t + \frac{1}{90}a_\varepsilon \sin(\pi\bar{x})\cos t. \tag{4.2.33}$$

We seek the solution of (4.2.33) in the form

$$y_{\varepsilon\varepsilon} = b_{\varepsilon\varepsilon}\cos t + c_{\varepsilon\varepsilon}\sin t. \tag{4.2.34}$$

Having substituted (4.2.34) into (4.2.33), we calculate:

$$b_{\varepsilon\varepsilon} = -\frac{\partial_\varepsilon}{50}\sin(\pi\bar{x}), \qquad c_{\varepsilon\varepsilon} = -0.0048\sin\pi x. \tag{4.2.35}$$

From the first equation of system (4.2.18), having equated the coefficients of ε^2, we obtain

$$\frac{\partial^2 u_{\varepsilon\varepsilon}}{\partial t^2} = \frac{1}{\pi^2}\frac{\partial^2 u_{\varepsilon\varepsilon}}{\partial x^2} + \frac{2}{\pi^2}\eta_{\varepsilon\varepsilon}\frac{\partial^2 u_\varepsilon^{(0)}}{\partial x^2} + 0.0001\frac{\partial u_\varepsilon}{\partial t} - \frac{1}{30}(u_\varepsilon^{(0)})^2\frac{\partial u_\varepsilon}{\partial t}$$
$$- \frac{1}{15}u_\varepsilon^{(0)}u_\varepsilon\frac{\partial u_\varepsilon^{(0)}}{\partial t} + y_\varepsilon\delta(x - \bar{x}). \tag{4.2.36}$$

From (4.2.36) we finally calculate:

$$a_\varepsilon = 0, \qquad \eta_{\varepsilon\varepsilon} = -0.01 \sin^2 \pi\bar{x}. \tag{4.2.37}$$

By means of analogous calculations it is possible to determine the recurrent equations occurring with the combinations $\varepsilon^k \tau^l$, where $k \geq 1$ and $l \geq 1$.

In the case when the characteristic equation (4.1.1) does not have imaginary eigenvalues, the periodic solution is sought in the form [37]

$$\nu(t, x) = \sum_{k=1}^{K} \sum_{l=0}^{L} \varepsilon^k \mu^l V_{kl}\{x, a(t), \psi(t)\},$$

$$y(t) = a(t)\{\alpha e^{i\psi(t)} + \bar{\alpha} e^{-i\psi(t)}\} + \sum_{k=1}^{K} \sum_{l=0}^{L} \varepsilon^k \mu^l y_{kl}\{a(t), \psi(t)\}. \tag{4.2.38}$$

where

$$\frac{da}{dt} = \sum_{k=1}^{K} \sum_{l=0}^{L} \varepsilon^k \mu^l A_{kl}\{a(t)\},$$

$$\frac{d\psi}{dt} = \omega + \sum_{k=1}^{K} \sum_{l=0}^{L} \varepsilon^k \mu^l B_{kl}\{a(t)\}. \tag{4.2.39}$$

α and $\bar{\alpha}$ are determined from the equations

$$\sum_{p=0}^{P} \left(A_p e^{-\tau_p \omega i} - E\omega i \right) \alpha = 0, \quad \sum_{p=0}^{P} \left(A_p e^{\tau_p \omega i} + E\omega i \right) \bar{\alpha} = 0. \tag{4.2.40}$$

The eigenvectors β and $\bar{\beta}$ of the adjoint set of equations are obtained from

$$\sum_{p=0}^{P} \left(A_p^* e^{\tau_p \omega i} + E\omega i \right) \beta = 0, \quad \sum_{p=0}^{P} \left(A_p^* e^{-\tau_p \omega i} - E\omega i \right) \bar{\beta} = 0, \tag{4.2.41}$$

where A_p^* are the matrices conjugate to the A_p matrices. From the first of equations (4.2.38) we have

$$\frac{\partial \nu}{\partial t} = \sum_{k=1}^{K} \sum_{l=0}^{L} \varepsilon^k \mu^l \left\{ \frac{\partial V_{kl}}{\partial a} \left(\frac{da}{dt} \right) + \frac{\partial V_{kl}}{\partial \psi} \left(\frac{d\psi}{dt} \right) \right\},$$

$$\frac{\partial^2 \nu}{\partial t^2} = \sum_{k=1}^{K} \sum_{l=0}^{L} \varepsilon^k \mu^l \left\{ \frac{\partial^2 V_{kl}}{\partial a^2} \left(\frac{da}{dt} \right)^2 + 2 \frac{\partial^2 V_{kl}}{\partial a \partial \psi} \frac{d\psi}{dt} \frac{da}{dt} \right. \tag{4.2.42}$$

$$\left. + \frac{\partial V_{kl}}{\partial a} \left(\frac{d^2 a}{dt^2} \right) + \frac{\partial^2 V_{kl}}{\partial a^2} \left(\frac{d\psi}{dt} \right)^2 + \frac{\partial V_{kl}}{\partial \psi} \left(\frac{d^2 \psi}{dt^2} \right) \right\}.$$

From the first of equations (4.1.1) we obtain

$$\frac{\partial^2 \nu}{\partial t^2} - L_x^{(2m)} \{\nu(t, x)\} = \sum_{k=1}^{K} \sum_{l=0}^{L} \varepsilon^k \mu^l \left\{ \frac{\partial^2 V_{kl}}{\partial a^2} \left(\frac{da}{dt} \right)^2 \right.$$

$$+2\frac{\partial^2 V_{kl}}{\partial a \partial \psi}\frac{\mathrm{d}\psi}{\mathrm{d}t}\frac{\mathrm{d}a}{\mathrm{d}t} + \frac{\partial V_{kl}}{\partial a}\left(\frac{\mathrm{d}^2 a}{\mathrm{d}t^2}\right) + \frac{\partial^2 V_{kl}}{\partial a^2}\left(\frac{\mathrm{d}\psi}{\mathrm{d}t}\right)^2$$

$$+\frac{\partial V_{kl}}{\partial \psi}\left(\frac{\mathrm{d}^2\psi}{\mathrm{d}t^2}\right) - L_x^{(2m)}\{V_{kl}\}\Big\}, \tag{4.2.43}$$

while from the second one we obtain

$$\frac{\mathrm{d}y}{\mathrm{d}t} = \frac{\mathrm{d}a}{\mathrm{d}t}\left(\alpha e^{i\psi(t)} + \bar{\alpha}e^{-i\psi(t)}\right) + ia(t)\frac{\mathrm{d}\psi}{\mathrm{d}t}\left(\alpha e^{i\psi(t)} - \bar{\alpha}e^{-i\psi(t)}\right)$$

$$+\sum_{k=1}^{K}\sum_{l=0}^{L}\varepsilon^k \mu^l\left\{\frac{\partial y_{kl}}{\partial a}\frac{\mathrm{d}a}{\mathrm{d}t} + \frac{\partial y_{kl}}{\partial \psi}\frac{\mathrm{d}\psi}{\mathrm{d}t}\right\}. \tag{4.2.44}$$

From (4.2.39) it follows that

$$\frac{\mathrm{d}^2 a}{\mathrm{d}t^2} = \varepsilon^2 A_{10}\frac{\mathrm{d}A_{10}}{\mathrm{d}a} + \varepsilon^2\mu\left(\frac{\mathrm{d}A_{10}}{\mathrm{d}a}A_{11} + \frac{\mathrm{d}A_{11}}{\mathrm{d}a}A_{10}\right)$$

$$+\varepsilon^3\left(\frac{\mathrm{d}A_{20}}{\mathrm{d}a}A_{10} + \frac{\mathrm{d}A_{10}}{\mathrm{d}a}A_{20}\right) + O(\varepsilon^k\mu^l\,;k+l=4), \tag{4.2.45}$$

$$\frac{\mathrm{d}a}{\mathrm{d}t}\frac{\mathrm{d}\psi}{\mathrm{d}t} = \varepsilon\omega A_{10} + \varepsilon^2(\omega A_{20} + A_{10}B_{10}) + \varepsilon\mu\omega A_{11} + \varepsilon^2\mu(\omega A_{21}$$

$$+A_{11}B_{10} + A_{10}B_{11}) + \varepsilon\mu^2 A_{12}\omega + \varepsilon^3(A_{30}\omega \tag{4.2.46}$$

$$+A_{20}B_{10} + A_{10}B_{20}) + O(\varepsilon^k\mu^l\,;k+l=4),$$

$$\frac{\mathrm{d}^2\psi}{\mathrm{d}t^2} = \varepsilon^2\frac{\mathrm{d}B_{10}}{\mathrm{d}a}A_{10} + \varepsilon^2\mu\left\{\frac{\mathrm{d}B_{10}}{\mathrm{d}a}A_{11} + \frac{\mathrm{d}B_{11}}{\mathrm{d}a}A_{10}\right\}$$

$$+\varepsilon^3\left\{\frac{\mathrm{d}B_{10}}{\mathrm{d}a}A_{20} + \frac{\mathrm{d}B_{20}}{\mathrm{d}a}A_{10}\right\} + O(\varepsilon^k\mu^l\,;k+l=4), \tag{4.2.47}$$

$$\left(\frac{\mathrm{d}a}{\mathrm{d}t}\right)^2 = \varepsilon^2 A_{10}^2 + 2\varepsilon^2\mu A_{10}A_{11} + 2\varepsilon^3 A_{20}A_{10}$$

$$+O(\varepsilon^k\mu^l\,;k+l=4), \tag{4.2.48}$$

$$\left(\frac{\mathrm{d}\psi}{\mathrm{d}t}\right)^2 = \omega^2 + 2\varepsilon\omega B_{10} + 2\varepsilon\mu B_{10}\omega + \varepsilon^2(2\omega B_{20} + B_{10}^2)$$

$$+2\varepsilon\mu^2 B_{12}\omega + 2\varepsilon^2\mu(\omega B_{21} + B_{11}B_{10}) \tag{4.2.49}$$

$$+2\varepsilon^3(\omega B_{30} + B_{20}B_{10}) + O(\varepsilon^k\mu^l\,;k+l=4).$$

Since y and ν can be expressed as power series

$$y(t-\mu) = \sum_{n=0}^{N}\frac{1}{n!}\frac{\mathrm{d}^n y(t)}{\mathrm{d}t^n}(-\mu)^n,$$

$$\nu(t-\mu,\xi) = \sum_{n=0}^{N}\frac{1}{n!}\frac{\mathrm{d}^n\nu(t,\xi)}{\mathrm{d}t^n}(-\mu)^n, \tag{4.2.50}$$

then the function εf and εF can be expanded in a power series of the small parameters μ and ε. Further calculations were carried out for $n = 1$ ($\dot{y}_1 = -\mu(dy/dt)$ and $\dot{\nu}_1 = -\mu(d\nu/dt)$) and under the assumption that

$$\dot{\nu}_1 = -\mu \sum_{k=1}^{N} \sum_{l=0}^{L} \varepsilon^k \mu^l \left\{ \frac{\partial V_{kl}}{\partial a} \frac{da}{dt} + \frac{\partial V_{kl}}{\partial \psi} \frac{d\psi}{dt} \right\}, \tag{4.2.51}$$

$$\dot{y}_1 = -\mu \left\{ \frac{da}{dt} \left(\alpha e^{i\psi(t)} + \bar{\alpha} e^{-i\psi(t)} \right) - \mu a(t) \left\{ \frac{d\psi}{dt} i \left(\alpha e^{i\psi(t)} - \bar{\alpha} e^{-i\psi(t)} \right) \right. \right.$$
$$\left. \left. - \mu \sum_{k=1}^{K} \sum_{l=0}^{L} \varepsilon^k \mu^l \left(\frac{\partial y_{kl}}{\partial a} \frac{da}{dt} + \frac{\partial y_{kl}}{\partial \psi} \frac{d\psi}{dt} \right) \right\} \right\}.$$

The necessary derivatives of the functions f and F were calculated at the point $\mu = \varepsilon = 0$ and $\nu_0 = 0$, $y_0(t) = a(t)\{\alpha e^{i\psi(t)} + \bar{\alpha} e^{-i\psi(t)}\}$. The sequence of recurrent linear differential equations obtained are of the form

$$\varepsilon \; : \; \omega^2 \frac{\partial^2 V_{10}(x, a, \psi)}{\partial \psi^2} = L_x^{(2m)}\{V_{10}\} + f_\varepsilon,$$

$$\omega \frac{\partial y_{10}(a, \psi)}{\partial \psi} = \sum_{p=0}^{P} A_p y_{10}(a, \psi - \tau_p \omega) - A_{10}(\alpha e^{i\psi} + \bar{\alpha} e^{-i\psi})$$

$$-ia B_{10}(\alpha e^{i\psi} - \bar{\alpha} e^{-i\psi}) - \sum_{p=0}^{P} \tau_p A_p \left\{ A_{10}(\alpha e^{i(\psi - \tau_p \omega)} \right. \tag{4.2.52}$$

$$\left. + \bar{\alpha} e^{-i(\psi - \tau_p \omega)}) - ia B_{10}(\alpha e^{i(\psi - \tau_p \omega)} - \bar{\alpha} e^{-i(\psi - \tau_p \omega)}) \right\} + F_\varepsilon;$$

$$\varepsilon^2 \; : \; \omega^2 \frac{\partial^2 V_{20}(x, a, \psi)}{\partial \psi^2} = L_x^{(2m)}\{V_{20}\} + f_{\varepsilon\varepsilon},$$

$$\omega \frac{\partial y_{20}(a, \psi)}{\partial \psi} = \sum_{p=0}^{P} A_p y_{20}(a, \psi - \tau_p \omega) - A_{20}(\alpha e^{i\psi} + \bar{\alpha} e^{-i\psi})$$

$$-ia B_{20}(\alpha e^{i\psi} - \bar{\alpha} e^{-i\psi}) - \sum_{p=0}^{P} A_p \left\{ \tau_p A_{20}(\alpha e^{i(\psi - \tau_p \omega)} \right. \tag{4.2.53}$$

$$\left. + \bar{\alpha} e^{-i(\psi - \tau_p \omega)}) + ia B_{20}(\alpha e^{i(\psi - \tau_p \omega)} - \bar{\alpha} e^{-i(\psi - \tau_p \omega)}) \right\} + F_{\varepsilon\varepsilon};$$

$$\varepsilon\mu \; : \; \omega^2 \frac{\partial^2 V_{11}(x, a, \psi)}{\partial \psi^2} = L_x^{(2m)}\{V_{11}\} + f_{\varepsilon\mu},$$

$$\omega \frac{\partial y_{11}(a, \psi)}{\partial \psi} = \sum_{p=0}^{P} A_p y_{11}(a, \psi - \tau_p \omega) - A_{11}(\alpha e^{i\psi} + \bar{\alpha} e^{-i\psi})$$

$$-ia B_{11}(\alpha e^{i\psi} - \bar{\alpha} e^{-i\psi}) - \sum_{p=0}^{P} \tau_p A_p \left\{ A_{11}(\alpha e^{i(\psi - \tau_p \omega)} \right. \tag{4.2.54}$$

$$\left. + \bar{\alpha} e^{-i(\psi - \tau_p \omega)}) - ia B_{11}(\alpha e^{i(\psi - \tau_p \omega)} - \bar{\alpha} e^{-i(\psi - \tau_p \omega)}) \right\} + F_{\varepsilon\mu};$$

$$\varepsilon^3 \; : \; \omega^2 \frac{\partial^2 V_{30}(x, a, \psi)}{\partial \psi^2} = L_x^{(2m)}\{V_{30}\} + f_{\varepsilon^3},$$

$$\omega \frac{\partial y_{30}(a, \psi)}{\partial \psi} = \sum_{p=0}^{P} A_p y_{30}(a, \psi - \tau_p \omega) - A_{30}(\alpha e^{i\psi} + \bar{\alpha} e^{-i\psi})$$

$$-iaB_{30}(\alpha e^{i\psi} - \bar{\alpha} e^{-i\psi}) - \sum_{p=0}^{P} \tau_p A_p \Big\{ A_{30}(\alpha e^{i(\psi - \tau_p \omega)} \qquad (4.2.55)$$

$$+\bar{\alpha} e^{-i(\psi - \tau_p \omega)}) + iaB_{12}(\alpha e^{i(\psi - \tau_p \omega)} - \bar{\alpha} e^{-i(\psi - \tau_p \omega)}) \Big\} + F_{\varepsilon^3};$$

$$\varepsilon \mu^2 \; : \; \omega^2 \frac{\partial^2 V_{12}(x, a, \psi)}{\partial \psi^2} = L_x^{(2m)}\{V_{12}\} + f_{\varepsilon \mu^2},$$

$$\omega \frac{\partial y_{12}(a, \psi)}{\partial \psi} = \sum_{p=0}^{P} A_p y_{12}(a, \psi - \tau_p \omega) - A_{12}(\alpha e^{i\psi} + \bar{\alpha} e^{-i\psi})$$

$$-iaB_{12}(\alpha e^{i\psi} - \bar{\alpha} e^{-i\psi}) - \sum_{p=0}^{P} A_p \Big\{ \tau_p A_{12}(\alpha e^{i(\psi - \tau_p \omega)} \qquad (4.2.56)$$

$$+\bar{\alpha} e^{-i(\psi - \tau_p \omega)}) + iaB_{12}(\alpha e^{i(\psi - \tau_p \omega)} - \bar{\alpha} e^{-i(\psi - \tau_p \omega)}) \Big\} + F_{\varepsilon \mu^2};$$

$$\varepsilon^2 \mu \; : \; \omega^2 \frac{\partial^2 V_{21}(x, a, \psi)}{\partial \psi^2} = L_x^{(2m)}\{V_{21}\} + f_{\varepsilon^2 \mu},$$

$$\omega \frac{\partial y_{21}(a, \psi)}{\partial \psi} = \sum_{p=0}^{P} A_p y_{21}(a, \psi - \tau_p \omega) - A_{21}(\alpha e^{i\psi} + \bar{\alpha} e^{-i\psi})$$

$$-iaB_{21}(\alpha e^{i\psi} - \bar{\alpha} e^{-i\psi}) - \sum_{p=0}^{P} A_p \Big\{ \tau_p A_{21}(\alpha e^{i(\psi - \tau_p \omega)} \qquad (4.2.57)$$

$$+\bar{\alpha} e^{-i(\psi - \tau_p \omega)}) + iaB_{21}(\alpha e^{i(\psi - \tau_p \omega)} - \bar{\alpha} e^{-I(\psi - \tau_p \omega)}) \Big\} + F_{\varepsilon^2 \mu}.$$

Here

$$f_\varepsilon = f(x, y_0), \qquad F_\varepsilon = F(x, \dot{\nu}_1, y, \dot{y}_1),$$

$$f_{\varepsilon^2} = -2\omega B_{10} \frac{\partial^2 V_{10}}{\partial \psi^2} - 2\omega A_{10} \frac{\partial^2 V_{10}}{\partial a \partial \psi} + \frac{\partial f}{\partial \nu} V_{10} + \sum_{l=1}^{m} \frac{\partial f}{\partial y_l} y_{(10)l},$$

$$F_{\varepsilon^2} = -\frac{\partial y_{10}}{\partial a} A_{10} - \frac{\partial y_{10}}{\partial \psi} B_{10} - \frac{\partial F}{\partial \nu} V_{10} + \sum_{l=1}^{m} \frac{\partial F}{\partial y_l} y_{(10)l},$$

$$f_{\varepsilon \mu} = 0, \qquad F_{\varepsilon \mu} = -\sum_{n=1}^{m} \frac{\partial F}{\partial \dot{y}_{1n}} a\omega i(\alpha e^{i\psi} - \bar{\alpha} e^{-i\psi}),$$

$$f_{\varepsilon^3} = \frac{\partial^2 f}{\partial \nu^2} V_{10}^2 + \frac{\partial f}{\partial \nu} V_{20} + \sum_{l=1}^{m} \frac{\partial^2 f}{\partial y_l^2} y_{(10)l}^2 + \sum_{l=1}^{m} \frac{\partial f}{\partial y_l} y_{(20)l}$$

$$+2\sum_{l=1}^{m}\frac{\partial^2 f}{\partial\nu\partial y_l}V_{10}y_{(10)l} - 2\omega B_{10}\frac{\partial^2 V_{20}}{\partial\psi^2} - (2\omega B_{20}+B_{10}^2)\frac{\partial^2 V_{10}}{\partial\psi^2}$$

$$-A_{10}\frac{\mathrm{d}B_{10}}{\mathrm{d}a}\frac{\partial V_{10}}{\partial\psi} - A_{10}\frac{\mathrm{d}A_{10}}{\mathrm{d}a}\frac{\partial V_{10}}{\partial a} - 2\omega A_{10}\frac{\partial^2 V_{20}}{\partial a\partial\psi}$$

$$-2(\omega A_{20}+A_{10}B_{10})\frac{\partial^2 V_{10}}{\partial\psi\partial a} - A_{10}^2\frac{\partial^2 V_{10}}{\partial a^2},$$

$$F_{\varepsilon^3} = -\frac{\partial y_{10}}{\partial a}A_{20} - \frac{\partial y_{20}}{\partial a}A_{10} - \frac{\partial y_{20}}{\partial\psi}B_{20} + \frac{\partial^2 F}{\partial\nu^2}V_{10}^2 + \frac{\partial f}{\partial\nu}V_{20} \qquad (4.2.58)$$

$$+\sum_{l=1}^{m}\frac{\partial^2 F}{\partial y_l^2}y_{(10)l} + \sum_{l=1}^{m}\frac{\partial F}{\partial y_l}y_{(10)l}^2 + \sum_{l=1}^{m}\frac{\partial F}{\partial y_l}y_{(20)l}$$

$$+2\sum_{l=1}^{m}\frac{\partial^2 F}{\partial\nu\partial y_l}V_{10}y_{(10)l},$$

$$f_{\varepsilon\mu^2} = \sum_{n=1}^{m}\frac{\partial^2 f}{\partial y_{1n}^2}a^2\omega^2(\alpha e^{\mathrm{i}\psi} - \bar\alpha e^{-\mathrm{i}\psi})^2,$$

$$F_{\varepsilon\mu^2} = \sum_{n=1}^{M}\frac{\partial^2 F}{\partial y_{1n}^2}a^2\omega^2(\alpha e^{\mathrm{i}\psi} - \bar\alpha e^{-\mathrm{i}\psi})^2,$$

$$f_{\varepsilon^2\mu} = 2\frac{\partial f}{\partial\nu}V_{11} + 2\sum_{l=1}^{m}\frac{\partial f}{\partial y_{(11)l}} + 2\sum_{n=1}^{m}\frac{\partial f}{\partial y_{1n}}\{-A_{10}(\alpha e^{\mathrm{i}\psi}+\bar\alpha e^{-\mathrm{i}\psi})$$

$$-aB_{10}\mathrm{i}(\alpha e^{\mathrm{i}\psi}-\bar\alpha e^{-\mathrm{i}\psi})\} - \frac{\partial y_{11}}{\partial\psi}\omega$$

$$-V_{10}a\omega\mathrm{i}\sum_{n=1}^{m}\frac{\partial^2 f}{\partial\nu\partial\dot y_{1n}}(\alpha e^{\mathrm{i}\psi}-\bar\alpha e^{-\mathrm{i}\psi})$$

$$-y_{10}a\omega\mathrm{i}(\alpha e^{\mathrm{i}\psi}-\bar\alpha e^{-\mathrm{i}\psi})\sum_{n=1}^{m}\sum_{l=1}^{m}\frac{\partial^2 f}{\partial y_l\partial y_{1n}} - 2\omega B_{10}\frac{\partial^2 V_{11}}{\partial\psi^2}$$

$$-2\omega B_{11}\frac{\partial^2 V_{10}}{\partial\psi^2} - 2\omega A_{10}\frac{\partial^2 V_{11}}{\partial a\partial\psi} - 2\omega A_{11}\frac{\partial^2 V_{10}}{\partial a\partial\psi},$$

$$F_{\varepsilon^2\mu} = -A_{10}\frac{\partial y_{11}}{\partial a} - A_{11}\frac{\partial y_{10}}{\partial a} - B_{10}\frac{\partial y_{11}}{\partial a} - B_{11}\frac{\partial y_{10}}{\partial\psi}$$

$$-\sum_{n=1}^{m}\frac{\partial^2 F}{\partial\nu\partial\dot y_{1n}}V_{10}a\omega\mathrm{i}(\alpha e^{\mathrm{i}\psi}+\bar\alpha e^{-\mathrm{i}\psi})$$

$$-\sum_{l=1}^{m}\sum_{n=1}^{m}\frac{\partial^2 F}{\partial y_l\partial\dot y_{1n}}y_{10}a\omega\mathrm{i}(\alpha e^{\mathrm{i}\psi}+\bar\alpha e^{-\mathrm{i}\psi})$$

$$+2\sum_{l=1}^{M}\frac{\partial F}{\partial y_l}y_{(11)l} + 2\frac{\partial F}{\partial\nu}V_{11} - \omega\frac{\partial F}{\partial\dot\nu_1}\frac{\partial V_{11}}{\partial\psi}$$

$$+2\sum_{n=1}^{m}\frac{\partial F}{\partial \ddot{y}_{1n}}\left\{-A_{10}(\alpha e^{i\psi}+\bar{\alpha}e^{-i\psi})\right.$$

$$\left.-aB_{10}(\alpha e^{i\psi}+\bar{\alpha}e^{-i\psi})-\frac{\partial y_{11}}{\partial \psi}\omega\right\}.$$

To achieve a complete ordering of all the recurrent equations we take the additional condition that $\varepsilon^{i-1}<\mu^{i}$, where i is a positive integer. After expanding the function $f_{(*)}$ into a Fourier series, one obtains

$$f_{(*)}=\sum_{s=1}^{\infty}\sum_{n=1}^{\infty}\left\{b_{(*)sn}(a)\cos n\psi+c_{(*)sn}(a)\sin n\psi\right\}X_{s}(x), \tag{4.2.59}$$

where

$$b_{(*)sn}(a)=\frac{1}{2\pi l}\int_{0}^{l}\mathrm{d}x\int_{0}^{2\pi}f_{(*)}(x,a,\psi)X_{s}(x)\cos n\psi\,\mathrm{d}\psi,$$

$$c_{(*)sn}(a)=\frac{1}{2\pi l}\int_{0}^{l}\mathrm{d}x\int_{0}^{2\pi}f_{(*)}(x,a,\psi)X_{s}(x)\sin n\psi\,\mathrm{d}\psi. \tag{4.2.60}$$

The functions $F_{(*)}$ are expanded into a complex Fourier series

$$F_{(*)}=\sum_{n=-\infty}^{\infty}C_{(*)n}(a)e^{in\psi}, \tag{4.2.61}$$

where

$$C_{(*)n}(a)=\frac{1}{2\pi}\int_{0}^{2\pi}F_{(*)}e^{in\psi}\,\mathrm{d}\psi,\quad n=\pm1,\pm2,\ldots \tag{4.2.62}$$

We describe a procedure of solving the recurrent set of ODE's based on (4.2.52). In order to avoid terms ascending unrestrictedly in time in these equations, the following conditions must be satisfied:

$$-\{A_{10}(a)+iaB_{10}(a)\}\left\{(\alpha,\beta)+\sum_{p=1}^{P}\tau_{p}(A_{p}\alpha,\beta)e^{-\tau_{p}\omega i}\right\}$$

$$+\left(C_{(10)l}(a),\beta\right)=0. \tag{4.2.63}$$

Here (a,b) denotes the scalar product. By equating to zero the real and imaginary parts of (4.2.61), we obtain two equations to determine the quantities $A_{10}(a)$ and $B_{10}(a)$. Then we can find V_{10} and y_{10}, and further successfully solve the recurrent set of equations.

Analytical conditions for the existence of a two-parameter family of periodic and quasiperiodic orbits in autonomous and non-autonomous systems can be found in [32, 35, 36, 38].

4.3 Nonlinear Behaviour
of Electromechanical Systems

4.3.1 Introduction

Usually the dynamics of nonlinear discrete–continuous systems governed by ordinary and partial differential equations (the case considered here) causes some difficulties in nonlinear analysis. It is often brought about by the time-consuming numerical techniques used to find the solution of the partial differential equations. Additionally, real physical systems possess many parameters which can be changed over wide ranges and in practice direct simulation of the governing equations is costly and tedious.

For simple dynamical systems, averaging formulas can be derived without computers. However, in the case of complex systems (such as the example considered in this book) this classical approach leads to serious difficulties. Therefore, the idea of applying the averaging technique has been supported by symbolic computation with the use of the *Mathematica* package. A program has been written in the it Mathematica language which has yielded the averaged equations. This set of equations has been transformed (using one of the *Mathematica* options) to Fortran expressions, and further, a numerical analysis has been carried out.

Admittedly, in this section, on the one hand, the electromechanical system serves as an example for a systematic strategy of solving many other relative problems, which can be found in nonlinear mechanical, biological or chemical dynamical systems. It consists of a few steps: 1. Dynamical equations are derived. 2. The averaging method is proposed and the program for symbolic computation yields the averaged equations (AVE). 3. Further systematic study of the obtained ODE's is developed.

The model was first discussed by Rubanik [137], where attention was focused on the averaging procedure only starting with the governing equations. Here, the system under consideration is discussed in some detail, and the symbolic computation is used to obtain the differential amplitude equations based on the application of the *Mathematica* package. Contrary to the approach in [137], also the numerical analysis of the averaged differential equations found is carried out to show interesting nonlinear phenomena.

The averaging method, also based on the symbolic computation, has been proposed by one of the authors earlier [21d, 22d]. Here two parts of the project are clarified. First, more attention is paid to the discussion of the electromechanical nonlinear system including its electrical model. Second, instead of an approximation of the time delay function by a Taylor series, a Galerkin method is applied. As has been pointed out by Elsgolc [33d], the truncation of the Taylor series to one term, as well as the use of more Taylor terms in the function with the time delay approximation, does not lead to an improvement in accuracy of the numerical calculations. For this reason a new approach, based on a Galerkin approximation, which does not possess any limitations because of the delay magnitude, is used. Third, the qualitatively new numerical results are discussed and illustrated now in comparison to the author's previous works [34, 24d, 25d].

4.3.2 Dynamics Equations

A string (a continuous system) is embedded in a magnetic field. For a certain set of parameters the stationary position of the string becomes unstable and the string starts to vibrate. Because the string possesses an inductance L, a resistance R and a capacitance C the movement in the magnetic field causes the occurrence of voltage and current. The amplifier controls the change of the current amplitudes with the time delay as the experiment shows. Figure 4.2 presents the electromechanical model and its electric scheme.

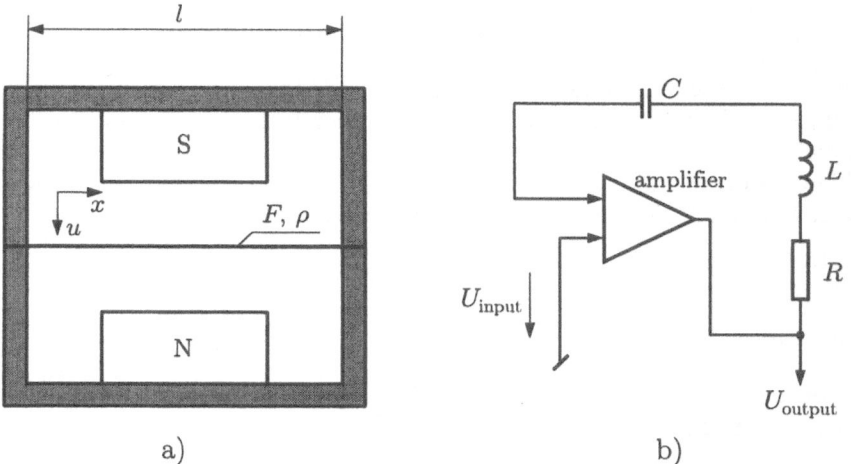

a)   b)

Fig. 4.2. Scheme of a string embedded in a magnetic field (**a**) and its electrical model including an amplifier (**b**)

The magnetic induction $B(x)$ acting along the string generates voltage at the ends of the string according to the following equation

$$U_{\text{input}}(t) = \int\limits_0^l B(x)\frac{\partial u(t,x)}{\partial t}\,dx, \tag{4.3.1}$$

where x is the spatial coordinate, t denotes time, $u(t,x)$ is the amplitude of oscillations of the string at the point (t,x) and l is the length of the string. The amplifier gives the output voltage

$$\tilde{U}_{\text{output}}(t) = \tilde{h}_1 U_{\text{input}}(t) - \tilde{h}_2 U_{\text{input}}(t), \tag{4.3.2}$$

where $\tilde{h}_i$ $(i = 1, 2)$ are constant coefficients.

The current oscillations including the time delay in the amplifier are governed by the equation

$$\ddot{I}(t) + 2\lambda\dot{I}(t) + kI(t) = \dot{U}_{\text{output}}(t - \nu) \tag{4.3.3}$$

where:

$$2\lambda = RL^{-1}, \quad k = (LC)^{-1},$$
$$U_{\text{output}}(t) = h_1 U_{\text{input}}(t) - h_2 U_{\text{input}}^3(t) \tag{4.3.4}$$
$$h_1 = L^{-1}\tilde{h}_1, \quad h_2 = L^{-1}\tilde{h}_2.$$

In the above expresion, the "dot" denotes differentiation with respect to t, $I(t)$ denotes the changes of the current and ν is the time delay. The changes in time of $I(t)$ and the changes in x of $B(x)$ play the role of the force acting on the string, whose oscillations are governed by the equation

$$\frac{\partial^2 u(t,x)}{\partial t^2} - c^2 \frac{\partial^2 u(t,x)}{\partial x^2} = \frac{\varepsilon}{\rho}\left(2h_0 \frac{\partial u(t,x)}{\partial t} - B(x)I(x)\right), \tag{4.3.5}$$

where h_0 is the external damping coefficient, ρ is the mass density per unit length, $c^2 = F/\rho$, F denotes the string's tension and ε is a small positive parameter.

The frequencies of free oscillations of the string are given by $\omega_s = \pi cs/l$ ($s = 1, 2, \ldots$), and the homogeneous boundary conditions are as follows:

$$u(t,0) = u(t,l) = 0. \tag{4.3.6}$$

4.3.3 Averaging

Our considerations are limited to first-order averaging. For $\varepsilon = 0$ the solution to (4.3.5) is given by

$$u_0 = a_1 \cos(\omega_1 t + \theta_1) \sin\left(\frac{\pi x}{l}\right) + a_3 \cos(3\omega_1 t + \theta_3) \sin\left(\frac{3\pi x}{l}\right) \tag{4.3.7}$$

where a_1, a_3 are the amplitudes and θ_1, θ_3 are the phases.

For $\varepsilon \neq 0$, yet small enough, the solution to (4.3.5) is expected to be of the form

$$u = u_0 + \varepsilon u_1(x, a_1, a_3, \theta_1, \theta_3) + \text{h.o.t.} \tag{4.3.8}$$

where "h.o.t." denotes higher-order terms. Supposing that $B(x)$ is symmetric with respect to the ends of the string, i.e., $B(x) = B(l - x)$, we take

$$B = B_1 \sin\left(\frac{\pi x}{l}\right) + B_3 \sin\left(\frac{3\pi x}{l}\right). \tag{4.3.9}$$

From (1) we obtain

$$U_{\text{input}} = -\frac{1}{2}B_1 a_1 \omega_1 l \sin(\omega_1 t + \theta_1) - \frac{3}{2}B_3 a_3 \omega_1 l \sin(3\omega_1 t + \theta_3) \tag{4.3.10}$$

and the right-hand side of (4.3.3) is calculated using the symbolic calculation:

$$U_{\text{output}}(t - \nu) = -\frac{1}{2}B_1 a_1 h_1 l \omega_1^2 \cos(\Psi_{10}) - \frac{9}{2}B_3 a_3 h_1 l \omega_1^2 \cos(\Psi_{30})$$

$$+\frac{3}{8}B_1^3 a_1^3 h_2 l^3 \omega_1^4 \cos(\Psi_{10}) \sin^2(\Psi_{30}) \tag{4.3.11}$$

$$+\frac{27}{8}B_1^2 B_3 a_1^2 a_3 h_2 l^3 \omega_1^4 \cos(\Psi_{30}) \sin^2(\Psi_{10})$$

$$+\frac{9}{8}B_1^2 B_3 a_1^2 a_3 h_2 l^3 \omega_1^4 \sin(2\Psi_{10}) \sin(\Psi_{30})$$

$$+\frac{81}{8}B_1 B_3^2 a_1 a_3^2 h_2 l^3 \omega_1^4 \sin(2\Psi_{30}) \sin(\Psi_{10})$$

$$+\frac{27}{8}B_1 B_3^2 a_1 a_3^2 h_2 l^3 \omega_1^4 \cos(\Psi_{10}) \sin^2(\Psi_{30})$$

$$+\frac{243}{8}B_3^3 a_3^2 h_2 l^3 \omega_1^4 \cos(\Psi_{30}) \sin^2(\Psi_{30}),$$

where:

$$\Psi_{10} = \omega t + \theta_1 - \nu t,$$
$$\Psi_{30} = 3\omega t + \theta_3 - 3\nu t. \tag{4.3.12}$$

From (4.3.11) we take only the harmonics $\sin(i\omega t)$, $\cos(i\omega t)$ $(i = 1, 3)$, and therefore

$$\dot{U}_{\text{output}}(t - \nu) = b_{1c} \cos \omega t + b_{1s} \sin \omega t + b_{3c} \cos 3\omega t + b_{3s} \sin 3\omega t \tag{4.3.13}$$

where:

$$b_{1c} = A \cos \theta_1^* - \frac{9}{32}B_1^2 B_3 a_1^2 a_3 h_2 l^3 \omega_1^4 \cos(2\theta_1^* - \theta_3^*),$$

$$b_{1s} = -A \sin \theta_1^* - \frac{45}{64}B_1^2 B_3 a_1^2 a_3 h_2 l^3 \omega_1^4 \cos(2\theta_1^* - \theta_3^*),$$

$$b_{3c} = A \cos \theta_3^* - \frac{3}{32}B_1^2 B_3 a_1^3 h_2 l^3 \omega_1^4 \cos(3\theta_1^*),$$

$$b_{3s} = -A \sin \theta_3^* - \frac{3}{32}B_1^3 a_1^3 h_2 l^3 \omega_1^4 \sin(3\theta_1^*),$$

$$A = -\frac{1}{2}B_1 a_1 h_1 l \omega_1^2 + \frac{3}{12}B_1^3 a_1^3 h_2 l^3 \omega_1^4 + \frac{27}{16}B_1 B_3^2 a_1 a_3^2 h_2 l^3 \omega_1^4,$$

$$C = -\frac{9}{2}B_3 a_3 h_1 l \omega_1^2 + \frac{27}{16}B_1^2 a_1^2 a_3 h_2 l^3 \omega_1^4 + \frac{243}{32}B_3^3 a_3^3 h_2 l^3 \omega_1^4, \tag{4.3.14}$$

$$\theta_1^* = \theta_1 - \nu t, \qquad \theta_3^* = \theta_3 - 3\nu t.$$

The solution to the linear equation (4.3.3) has the form

$$I_0(t) = \sum_{i=1,3} \{(b_{ic}M_i - b_{is}N_i) \cos i\omega t$$

$$+ (b_{ic}N_i - b_{is}M_i) \sin i\omega t\} + \text{h.h.} \tag{4.3.15}$$

where the abbreviation "h.h." denotes higher harmonics which are not taken into account, and M_i, N_i are given below:

$$M_i = \frac{k - i^2\omega_1^2}{(k - i^2\omega_1^2)^2 + 4i^2\lambda^2\omega_1^2},$$

$$N_i = \frac{2\lambda i\omega_1}{(k - i^2\omega_1^2)^2 + 4i^2\lambda^2\omega_1^2}. \tag{4.3.16}$$

The further analysis is straightforward for the perturbation technique. Because $B(x)$ and $I(t)$ are defined, therefore (4.3.5) can be solved using a modified classical perturbation approach (it is assumed that $u_1(x, a_1, a_3, \theta_1, \theta_3)$ is a limited and periodic function).

Substituting (4.3.8) for (4.3.4) and taking into account that $a_i = a_i(t)$ and $\theta_i = \theta_i(t)$ $(i = 1, 3)$ slowly change in time, the following resonance terms are calculated from the right-hand side of (4.3.5) (further referred to as R_i)

$$R_{ic} = \frac{2}{\pi l} \int_0^l \int_0^{2\pi} R \sin\frac{\pi i x}{l} \cos \Psi_{i0}\, d\Psi_{i0}$$

$$R_{is} = \frac{2}{\pi l} \int_0^l \int_0^{2\pi} R \sin\frac{\pi i x}{l} \sin \Psi_{i0}\, d\Psi_{i0} \tag{4.3.17}$$

$$\Psi_{i0} = i\omega + \theta_i, \quad i = 1, 3 \tag{4.3.18}$$

where R_{ic}, R_{is} correspond to the coefficients of $\cos i\Psi_{i0}$ and $\sin i\Psi_{i0}$, respectively. The comparison of the coefficients of $\cos i\Psi_{i0}$ and $\sin i\Psi_{i0}$ and generated by the left-hand side of (4.3.5) to those defined by (4.3.17) leads to the following averaged amplitude equations

$$\dot{a}_1 = -\frac{\varepsilon h_0 a_1}{\rho} - \frac{\varepsilon B_1}{2\rho\omega}\left\{(b_{1c}M_1 - b_{1s}N_1)\sin\theta_1 + (b_{1c}N_1 - b_{1s}M_1)\cos\theta_1\right\},$$

$$\dot{a}_3 = -\frac{\varepsilon h_0 a_3}{\rho} - \frac{\varepsilon B_3}{2\rho\omega}\left\{(b_{3c}M_3 - b_{3s}N_3)\sin\theta_3 + (b_{3c}N_3 - b_{3s}M_3)\cos\theta_3\right\},$$

$$\dot{\theta}_1 = -\frac{\varepsilon B_1}{2a_1\rho\omega}\left\{(b_{1c}M_1 - b_{1s}N_1)\cos\theta_1 + (b_{1c}N_1 - b_{1s}M_1)\sin\theta_1\right\}, \tag{4.3.19}$$

$$\dot{\theta}_3 = -\frac{\varepsilon B_3}{2a_3\rho\omega}\left\{(b_{3c}M_3 - b_{3s}N_1)\cos\theta_3 + (b_{3c}N_3 - b_{3s}M_3)\sin\theta_3\right\}.$$

The analysed set of equations has some properties which can cause difficulties during numerical analysis. First of all, this is a stiff set of equations (note the occurrence of a_3 in the denominator of the last equation of (4.3.19)). As is assumed by the averaging procedure, the amplitudes a_i and the phases θ_i change with time very slowly, and a long integration to trace the behaviour of the system is required.

For the further analysis of the time dependent solutions we transform (4.3.19) into the amplitude equations. For this aim we assume

$$u_0 = (Y_1 \cos \omega_1 t + Y_2 \sin \omega_1 t) \sin \left(\frac{\pi x}{l}\right)$$

$$+ (Y_3 \cos \omega_3 t + Y_4 \sin \omega_3 t) \sin \left(\frac{3\pi x}{l}\right). \tag{4.3.20}$$

Comparison with (4.3.7) yields the following relations:

$$\begin{aligned}
Y_1(t) &= a_1(t) \cos \theta_1(t) \\
Y_2(t) &= -a_1(t) \sin \theta_1(t) \\
Y_3(t) &= a_3(t) \cos \theta_3(t) \\
Y_4(t) &= -a_3(t) \sin \theta_3(t).
\end{aligned} \tag{4.3.21}$$

In what follows, the set of the amplitude differential equations has the form

$$\begin{aligned}
\dot{Y}_1(t) &= \dot{a}_1(t) \cos \theta_1(t) - a_1(t)\dot{\theta}_1(t) \sin \theta_1(t), \\
\dot{Y}_2(t) &= -\dot{a}_1(t) \sin \theta_1(t) - a_1(t)\dot{\theta}_1(t) \cos \theta_1(t), \\
\dot{Y}_3(t) &= \dot{a}_3(t) \cos \theta_3(t) - a_3(t)\dot{\theta}_3(t) \sin \theta_3(t), \\
\dot{Y}_4(t) &= -\dot{a}_3(t) \sin \theta_3(t) - a_3(t)\dot{\theta}_3(t) \cos \theta_3(t)
\end{aligned} \tag{4.3.22}$$

where a_i and θ_1 are given by (4.3.19) and

$$\theta_1 = \arctan\left(-\frac{Y_2}{Y_1}\right), \quad \theta_3 = \arctan\left(-\frac{Y_4}{Y_3}\right), \tag{4.3.23}$$

$$a_1 = (Y_1^2 + Y_2^2)^{1/2}, \quad a_3 = (Y_3^2 + Y_4^2)^{1/2}.$$

4.3.4 Numerical Results

We consider both time-dependent and time-independent solutions. In order to get the stationary solutions we solve the nonlinear algebraic equations obtained from (4.3.19). For this a Powell hybrid method and variation of Newton's method have been used. It takes a finite-difference approximation to the Jacobian with high precision.

Figure 4.3 presents an example of numerical calculations for the following fixed parameters: $l = 0.1$, $\omega = 30.0$, $\lambda = 0.01$, $k = 35.0$, $h_2 = 5.0$, $\rho = 1.0$, $h_0 = 0.005$, $\nu = 5.0$, $B_1 = 5.0$, $B_3 = 1.0$.

h_1 has been taken as the control parameter. The increase of h_1 damps the value of the first harmonic oscillations. However, during the change of h_1 the amplitude a_3 as well as the phases θ_1 and θ_3 remain constant.

Now (4.3.22) and their time-dependent solutions will be analysed. The system of equations is stiff and the Gear method routine from the IMSL Library is used to solve the problem. Let us consider the following set of parameters: $l = 0.1$, $\omega = 190.0$, $\lambda = 0.1$, $h_1 = 0.58$, $h_2 = 0.02$, $\nu = 0.1$, $\rho = 1.0$, $B_1 = 6.3$, $B_3 = 0.08$, $\varepsilon = 0.05$. Calculations have been performed with TOL $= 10^{-8}$, which is proportional to the calculation step. As can be seen from Figure 4.3a, the variables $Y_{1,2}(t)$ change in an oscillatory manner, whereas $Y_{3,4}(t)$ decay exponentially. This allows the interpretation that the ampli-

a)

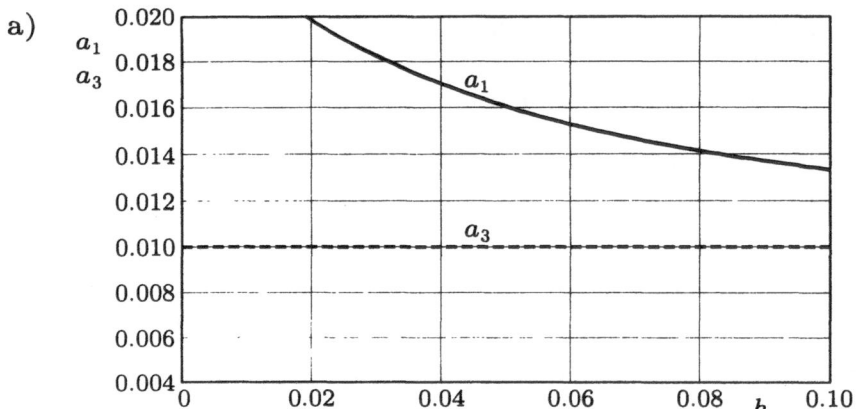

b)

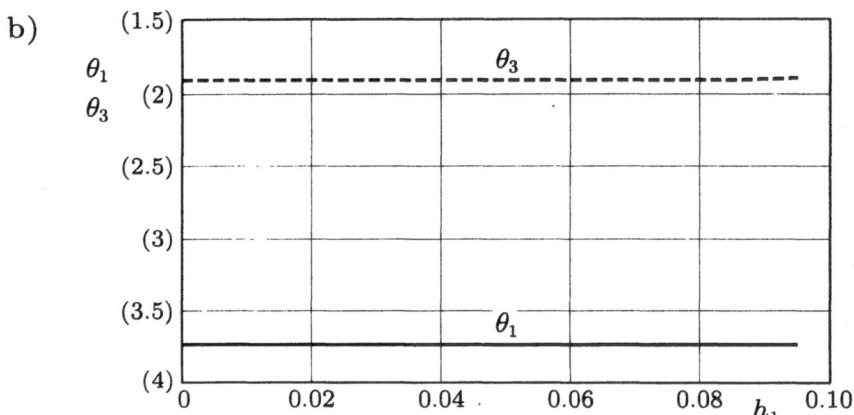

Fig. 4.3. Amplitudes (a) and phases (b) versus parameter h_1

tudes of the first and the third modes of the string behave quite differently. Furthermore, it implies that quite different analytical types of solutions can be assumed from the mathematical point of view.

The other figures illustrate the change of the mode amplitudes with the increase of the coefficient h_2. As can be seen from these figures, the increase in the nonlinear term h_2 leads to the extension of transient periodic oscillations.

Finally, let us consider the following fixed parameters: $l = 0.1$, $\omega = 900.0$, $k = 7250$, $\lambda = 0.1515$, $h_1 = 5.48$, $h_2 = 65.0$, $\rho = 1.0$, $B_1 = 0.65$, $h_0 = 0.00001$, $B_3 = 0.089$, $\varepsilon = 0.009$, $\nu = 0.00001$. Strong nonlinear behaviour is observed. In the beginning all variables do not exhibit oscillatory behaviour. After the time of about 20000 units a sudden occurrence of strong nonlinear oscillations of $Y_{1,2}(t)$ can be seen, whereas the variables $Y_{3,4}(t)$ do not change in an oscillatory manner again (Figure 4.4a). Increasing the time delay a strange

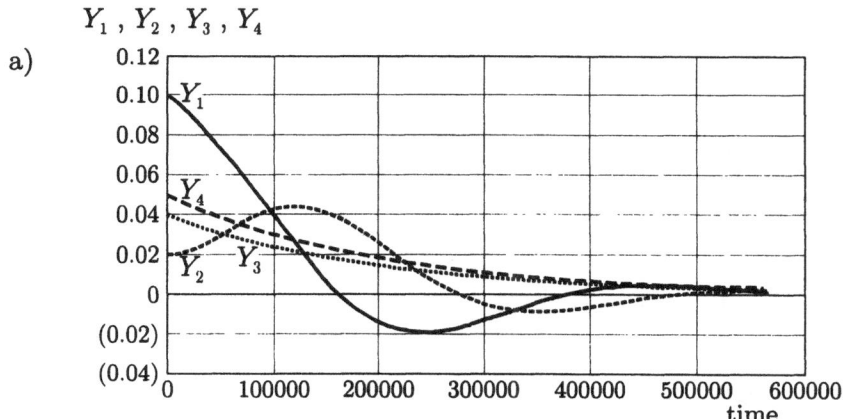

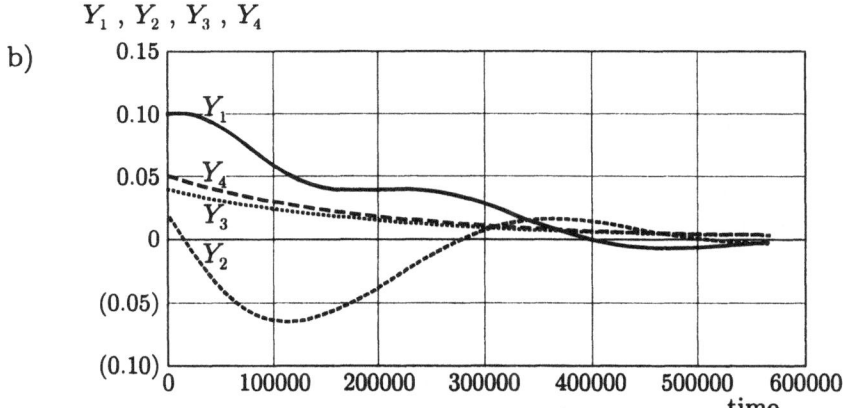

Fig. 4.4. Time evolution of the amplitudes with an increase of coefficient h_2: (a) $h_2 = 0.02$; (b) $h_2 = 0.04$

transitional state is observed: strong nonlinear oscillations $Y_{1,2}(t)$ vanishing in time are shown. The amplitude Y_3 decreases linearly, and its derivative remains constant. This means that the first and the third mode amplitudes behave qualitatively differently.

To summarize, the analysis is focused on the numerical observation of the averaged differential equations derived from the dynamical examination of the string-type electromechanical generator. Considerable attention is paid to the derivation of the averaged equations through the application of a modified perturbation technique supported by symbolic computations. The obtained set of equations is nonlinear and stiff. The numerical calculations based on solving the initial value problem have been performed to reveal some interesting results, which are here briefly summarized.

Y_1, Y_2, Y_3, Y_4

c)

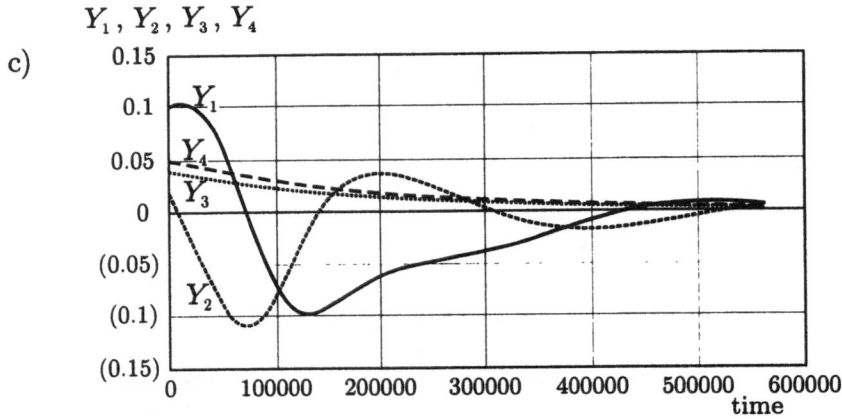

Y_1, Y_2, Y_3, Y_4

d)

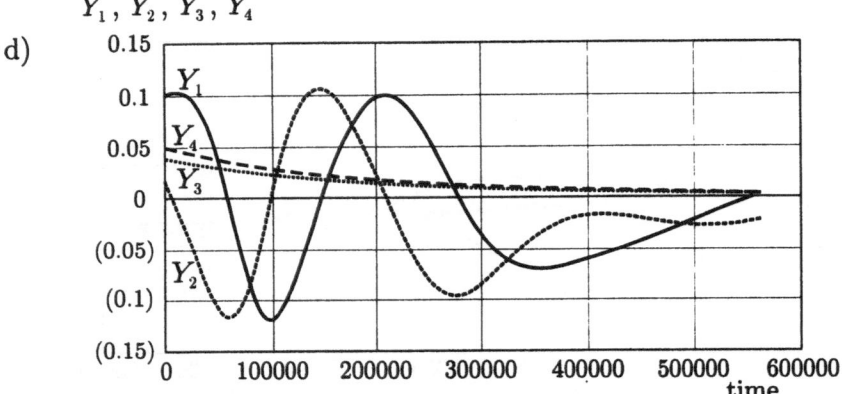

Y_1, Y_2, Y_3, Y_4

e)

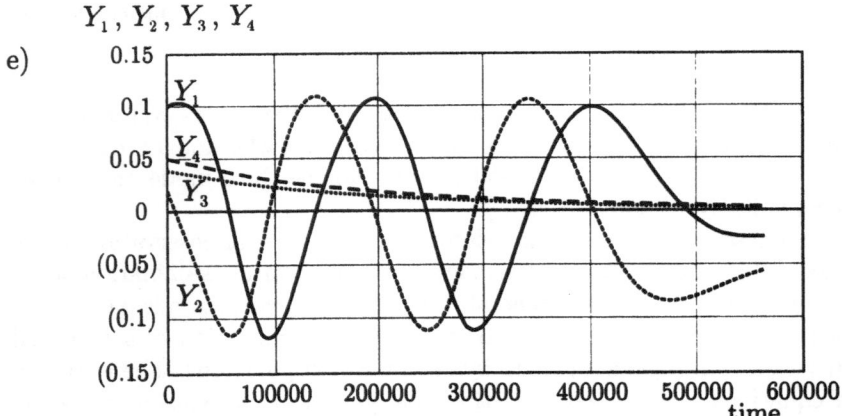

Fig. 4.4(continued). Time evolution of the amplitudes with an increase of the coefficient h_2: (c) $h_2 = 0.05$; (d) $h_2 = 0.055$; (e) $h_2 = 0.0554$

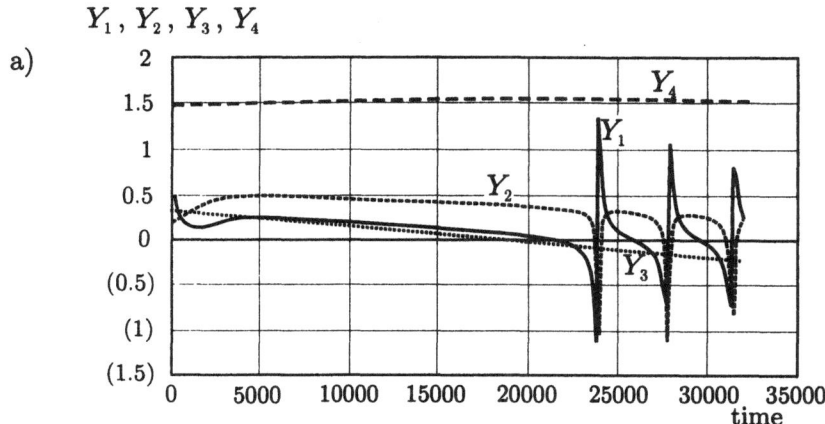

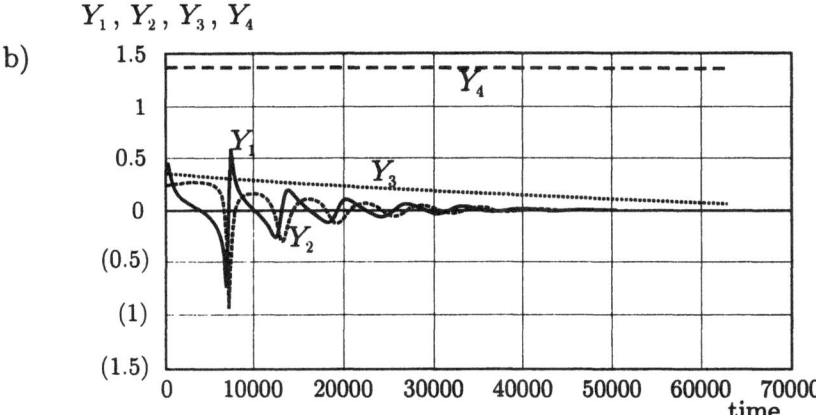

Fig. 4.5. Strong nonlinear amplitudes oscillations: (a) $\nu = 0.00001$; (b) $\nu = 0.0001$

It has been shown that in the case of stationary solutions the increase in the control parameter h_1 damps the first harmonic oscillations. The amplitude a_3 and the phases remain constant during the change of the control parameter.

In the case of time-dependent solutions interesting nonlinear behaviour has been reported in general. Amplitudes of the first two modes behave in a qualitatively different manner. $Y_{1,2}(t)$ change in time with oscillations, whereas $Y_{3,4}(t)$ decay exponentially.

Furthermore, strange nonlinear phenomenon has been exhibited. After a long nonoscillatory transitional state strong nonlinear oscillations suddenly occur.

General References

[1] Abdel-Ghaffar A.M., Rubin L.I., Nonlinear free vibrations of suspension bridges: theory and applications. *J. Engn. Mech. Amer. Sol. Civil Engn*, vol. 109, No 1, 1983, 313-345.

[2] Ambartzumyan S.A., *General theory of anysotropic shells*. Nauka, Moscow, 1974 (in Russian).

[3] *An album of fluid motion*. Van Dyke M. (ed.). Parabolic Press, Stanford, California, 1982.

[4] Andersen C.M., Geer J.F., Power series expansion for frequency and period of the limit cycle of the Van der Pol equation. *SIAM J. Appl. Math*, vol. 42, No 3, 1982, 678-693.

[5] Andersen C.M., Geer J.F., Dadfar M.B., Perturbation analysis of the limit cycle of the free Van der Pol equation. *SIAM J. Appl. Math*, vol. 44, No 5, 1984, 881-895.

[6] Andrianov I.V., On the theory of Berger plates. *PMM, J. Appl. Maths Mech*, vol. 47, No 1, 1983, 142-144.

[7] Andrianov I.V., The use of Padé approximation to eliminate nonuniformities of asymptotic expansion. *Fluid Dynamics*, vol. 19, No 3, 1984, 484-486.

[8] Andrianov I.V., Construction of simplified equation of nonlinear dynamics of plates and shallow shells by the averaging method. *PMM, J. Appl. Math Mech*, vol. 50, No 1, 1986, 126-129.

[9] Andrianov I.V., Constituting equations of structurally orthotropic cylindrical shells. *J. Appl. Mechs Technical Physics*, vol. 27, No 3, 1986, 468-472.

[10] Andrianov I.V., Application of Padé-approximants in perturbation methods. *Advances in Mechs*, vol. 14, No 2, 1991, 3-25.

[11] Andrianov I.V., Laplace transform inverse problem: application of two-point Padé approximant. *Appl. Math. Lett.*, vol. 5, No 4, 1992, 3-5.

[12] Andrianov I.V., Blawdziewicz J., Tokarzewski S., Effective conductivity for densely packed highly condensed cylinders. *Appl. Physics A*, vol. 59, 1994, 601-604.

[13] Andrianov I.V., Gristchak V.Z., Ivankov A.O., New asymptotic method for the natural, free and forced oscillations of rectangular plates with mixed boundary conditions. *Technische Mechanik*, vol. 14, No 3/4, 1994, 185-193.

[14] Andrianov I.V., Ivankov A.O., New asymptotic method for solving of mixed boundary value problem. *Intern. Series Numer. Maths*, vol. 106, 1992, 39-45.

[15] Andrianov I.V., Ivankov A.O., On the solution of plate bending mixed problems using modified technique of boundary conditions perturbation. *ZAMM*, vol. 73, No 2, 1993, 120-122.

[16] Andrianov I.V., Kholod E.G., Non-linear free vibration of shallow cylindrical shell by Bolotin's asymptotic method. *J. Sound Vibr.*, vol. 160, No 1, 1993, 594-603.

[17] Andrianov I.V., Krizhevsky G.A., Investigation of natural vibrations of circular and sector plates with consideration of geometric nonlinearity. *Mechs of Solid*, vol. 26, No 2, 1991, 143-148.

[18] Andrianov I.V., Krizhevsky G.A., Free vibration analysis of rectangular plates with structural inhomogenity. *J. Sound Vibr.*, vol. 162, No 2, 1993, 231-241.

[19] Andrianov I.V., Loboda V.V., Manevitch L.I., Formulation of boundary-value problems for simplified equations of the theory of eccentrically reinforced cylindrical shells. *Soviet Appl. Mechs*, vol. 11, No 7, 1976, 726-730.

[20] Andrianov I.V., Manevitch L.I., Calculation for the strain-stress state in an orthotropic strip stiffened by ribs. *Mechs of Solids*, vol. 10, No 4, 1975, 125-129.

[21] Andrianov I.V., Manevitch L.I., Approximate equations of axisymmetric vibrations of cylindrical shells. *Soviet Appl. Mechs*, vol. 17, No 8, 1982, 722-727.

[22] Andrianov I.V., Manevitch L.I., Asymptotology 1: Problems, ideas and results. *J. Natural Geometry*, vol. 2, No 2, 1992, 137-150.

[23] Andrianov I.V., Starushenko G.A., Solution of dynamic problems for perforated structures by the method of averaging. *J. Soviet Maths*, vol. 57, No 5, 1991, 3410-3412.

[24] Andrianov I.V., Starushenko G.A., Asymptotic methods in the theory of perforated membranes of nonhomogenous structures. *Engn. Trans.*, vol. 43, No 1-2, 1995, 5-18.

[25] Arimoto S., Learning control theory for robotic motion. *International Journal of Adaptive Control and Signal Processing*, vol. 4, 1990, 549-564.

[26] Atadan A.S., Huseyin K., Symmetric and flat bifurcations, an oscillatory phenomenon. *Acta Mechanica*, vol. 53, 1984, 213-232.

[27] Atadan A.S., Huseyin K., On the oscillatory instability of multiple-parameter systems. *Int. J. Engng. Sci.*, vol. 23, No 8, 1985, 857-873.

[28] Awrejcewicz J., Determination of the limits of the unstable zones of the unstationary non-linear mechanical systems. *Int. J. Non-Linear Mechanics*, vol. 23, No 1, 1988, 87-94.

[29] Awrejcewicz J., Determination of periodic oscillations in nonlinear autonomous discrete-continuous systems with delay. *Int. J. Solids Structures* vol. 27, No 7, 1991, 825-832.

[30] Awrejcewicz J., Vibration control of nonlinear discrete-continuous systems with delay. *Proc. 1st Intern. Conf. on Motion and Vibration Control*, Yokohama, September 7-12, 1992. Eds. Seto K., Yoshida K., Nonami K., 958-963.

[31] Awrejcewicz J., Stationary and nonstationary one-frequency periodic oscillations in nonlinear autonomous systems with time delay. *J. Tech. Phys.*, vol. 34, No 3, 1993, 275-297.

[32] Awrejcewicz J., Analytical condition for the existence of an implicit two-parameter family of periodic orbits in the resonance case. *J. Sound Vibr.*, vol. 170, No 3, 1994, 422-425.

[33] Awrejcewicz J., *Modified Poincaré method and implicit function theory*. In: Awrejcewicz J. (ed), *Nonlinear dynamics: new theoretical and applied results*. Akademie Verlag, Berlin, 1995, 215-229.

[34] Awrejcewicz J., Strange nonlinear behaviour governed by a set of four averaged amplitude equations. *Meccanica*, vol. 31, 1996, 347-361.

[35] Awrejcewicz J., Someya T., Analytical condition for the existence of two-parameter family of periodic orbits in the autonomous system. *J. Phys. Soc. Jap.*, vol. 60, No 3, 1991, 781-784.

[36] Awrejcewicz J., Someya T., Analytical condition for the existence of two-parameter family of quasiperiodic orbits in the autonomous system (Non-resonance case). *J. Phys. Soc. Jap.*, vol. 61, No 7, 1992, 2231-2234.

[37] Awrejcewicz J., Someya T., Periodic oscillations and two-parameter unfoldings in nonlinear discrete-continuous systems with delay. *J. Sound Vibr.*, vol. 158, No 1, 1992.

[38] Awrejcewicz J., Someya T., Analytical conditions for the existence of a two-parameter family of periodic orbits in nonautonomous dynamical systems. *Nonlinear Dynamics*, vol. 4, 1993, 39-50.

[39] Awrejcewicz J., Someya T., On introducing inertial forces into non-linear analysis of spatial structures. *J. Sound Vibr.*, vol. 103, No 3, 1993, 545-548.

[40] Aziz A., Lunardini V.J., Perturbation techniques in phase change heat transfer. *Appl. Mechs Rev.*, vol. 46, No 2, 1993, 29-68.

[41] Babcock C.D., Chen J.C., Nolinear vibration of cylindrical shell. *AIAA J.*, vol. 13, No 7, 1975, 868-876.

[42] Baker G.A., *Essential of Padé approximants.* Academic Press, New York, 1975.

[43] Baker G.A., Graves-Morris P., *Padé approximants.* Addison-Wesley Publ. Co., New York, 1981.

[44] Bakhalov N., Panasenko G., *Averaging processes in periodic media. Mathematical problems in mechanics of composite materials.* Kluwer Academic Publishers, Dordrecht, 1989.

[45] Barantsev R.G., Asymptotic versus classical mathematics. *Topics in Mathematical Analysis*, 1989, 49-64.

[46] Barenblatt G.I., *Similarity, self-similarity and intermediate asymptotics.* Plenum, New York, London, 1979.

[47] Barenblatt G.I., *Dimensional analysis.* Gordon and Breach, New York, London, 1987.

[48] Barenblatt G.I., Intermediate asymptotic, scaling laws and renormalization group in continuum mechanics. *Meccanica*, vol. 28, 1993, 177-183.

[49] Bender C.M., Milton K.A., Pinsky S.S., Simmon L.M.Jr., A new perturbative approach to nonlinear problems. *J. Math. Phys.*, vol. 30, No 7, 1989, 1447-1455.

[50] Bensoussan A., Lions J.-L., Papanicolaou G., *Asymptotic method in periodic structures.* North-Holland Publ. Co., New York, 1978.

[51] Berger H.M., A new approach to the analysis of large deflections of plates. *J. Appl. Mechs*, vol. 55, No 4, 1955, 465-472.

[52] Birkhoff G., Numerical fluid dynamics. *SIAM Review*, vol. 25, No 1, 1983, 1-24.

[53] Bodenschatz E., Pesch W., Kramer L., Structure and dynamics of dislocations in anisotropic patterns-forming systems. *Physica D*, vol. 32, 1988, 135-145.

[54] Boettcher S., Bender C.M., Nonperturbative square-well approximation to a quantum theory. *J. Math. Phys.*, vol. 31, No 11, 1990, 2579-2585.

[55] Bogoliubov N.N., Mitropolsky Yu.A., *Asymptotic method in the theory of nonlinear oscillations.* Gordon and Breach, London, 1985.

[56] Bolotin V.V., An asymptotic method for the study of the problem of eigenvalues of rectangular regions. *Problem of Continuum Mechs.* SIAM, Philadelphia, 1961, 56-58.

[57] Bruning J., Seeley R., Regular singular asymptotics. *Advances in Maths*, vol. 58, 1985, 133-148.

[58] Bucco D., Jones R., Masumdar J., The dynamic analysis of shallow spherical shells. *Trans. ASME, J. Appl. Mechs*, vol. 45 No 3, 1978, 690-691.

[59] Bush A.W., *Perturbation methods for engineers and scientists.* CRC Press, Boca Raton, 1992.

[60] Caillerie D., Thin elastic and periodic plates. *Math. Methods in App. Sc.*, vol. 6, 1984, 151-191.

[61] Chalbaud E., Martin P., Two-point quasifractional approximant in physics: Method improvement and application to $J_n(x)$. *J. Math. Phys.*, vol. 33, No 7, 1992, 2483-2486.

[62] Chen G., Goleman M.P., Zhon J., The eigenvalence between the wave propagation method and Bolotin's method in the asymptotic estimation of eigenfrequencies of a rectangular plate. *Wave Motion*, vol. 16, No 3, 1992, 285-297.

[63] Christensen R.M., *Mechanics of composite materials*. John Wiley & Sons, New York, 1979.

[64] Ciarlet P.G., *Plates and junctions in elastic multi-structures*. Masson, Paris, 1990.

[65] Ciarlet P.G., *Mathematical shell theory. A linearly elastic introduction*. Computional Mechs Advances, North-Holland, Amsterdam, 1994.

[66] Cole J.D., *Perturbation methods in applied mathematics*. Ginn (Blaisdell), Boston, 1968.

[67] Datta S., Large amplitude free vibration of irregular plates placed on elastic foundations. *Int. J. Nonlin. Mechs*, vol. 11, 1976, 337-345.

[68] Destuynder P., A classification of thin shell theories. *Acta Appl. Math.*, vol. 4, 1985, 15-63.

[69] Donnell L.H., *Beams, plates and shells*. McGraw-Hill, New York, 1976.

[70] Dowell E.H., On the nonlinear flectural vibration of rings. *AIAA J.*, No 5, 1967, 1508-1509.

[71] Dowell E.H., Ventres C.S., Modal equations for nonlinear flectural vibration of a cylindrical shell. *Int. J. Solid Str.*, No 4, 1968, 975-991.

[72] Draux A., On two-point Padé-type and two-point Padé approximants. *Ann. Mat. Pures et Appl.*, No 158, 1991, 99-124.

[73] Elishakoff I., Bolotin's dynamic edge effect method. *The Shock and Vibr. Digest*, vol. 8, No 1, 1976, 95-104.

[74] Elishakoff I., Sternberg A., Van Baten T.J., Vibrations of multispan all-round clamped stiffened plates by modified dynamic edge effect method. *Comp. Methods in Appl. Mechs and Engn.*, vol. 105, 1993, 211-223.

[75] Erofeyev V.I., Potapov A.I., Longitudinal strain waves in nonlinearity-elastic media with couple stresses. *Int. J. Nonlinear Mechs*, vol. 28, No 4, 1993, 483-488.

[76] Estrada R.P., Kanwal R.P., *Asymptotic analysis: a distributional approach*. Birkauser, Boston, Basel, Berlin, 1994.

[77] Evensen D.A., Some observations on the nonlinear vibration of thin cylindrical shell. *AIAA J.*, vol. 1, 1963, 2857-2858.

[78] Evensen D.A., A theoretical and experimental study of nonlinear flectural vibration of thin circular ring. *Trans. ASME, J. Appl. Mechs*, vol. 4, No 3, 1966, 553-560.

[79] Fomenko A.T., *Mathematical impressions*. AMS, Providence, 1990.

[80] Fredrichs K.O., Asymptotic phenomena in mathematical physics. *Bull. Amer. Math. Soc.*, vol. 61, 1955, 485-504.

[81] Frost P.A., Harper E.Y., Extended Padé procedure for constructing global approximations from asymptotic expansion: an explication with examples. *SIAM Rev.*, vol. 18, 1976, 62-91.

[82] Gol'denveizer A.L., *Theory of elastic thin shells*. Pergamon Press, New York, Oxford, London, Paris, 1961.

[83] Gol'denveizer A.L., Lidsky V.B., Tovstik P.E., *Free vibration of thin elastic shells*. Nauka, Moscow, 1979 (in Russian).

[84] Goodier J.N., McIvor I.K., The elastic cylindrical shell under nearly uniform radial impulse. *Trans. ASME, J. Appl. Mechs*, vol. 31, No 2, 1964, 259-266.

[85] Grundy R.E., Laplace transform inversion using two-point rational approximants. *J. Inst. Maths Applics.*, vol. 20, 1977, 299-306.

[86] Grundy R.E., The solution of Volterra integral equation of the convolution type using two-point rational approximants. *J. Inst. Maths Applics.*, vol. 22, 1978, 147-158.

[87] Grundy R.E., On the solution of nonlinear Volterra integral equations using two-point Padé approximants. *J. Inst. Maths Applics.*, vol. 22, 1978, 317-320.

[88] Guckenheimer J., Multiple bifurcation problems of codimension two. *SIAM J. of Math. Anal.*, 15, 1, 1981, 1-49.

[89] Guckenheimer J., Holmes P.J., *Nonlinear oscillations, dynamical systems and bifurcation of vector fields.* Springer, Berlin, Heidelberg, 1983.

[90] Han S., On the free vibration of a beam on a nonlinear elastic foundation. *Trans. ASME, J. Appl. Mechs*, vol. 32, No 2, 1965, 445-447.

[91] Hassard B.D., Kazarinoff N.D., Wan Y.H., *Theory and applications of the Hopf bifurcation.* Cambridge University Press, Cambridge, 1980.

[92] Heutemaker M.S., Frankel P.N., Gollub J.P., Convection patterns: time evolution of the wave-vector field. *Physical Review Letters*, vol. 54, No 13, 1985, 1369-1372.

[93] *Handbook of special functions with formulas graphs and mathematical tables.* Abramowitz M., Stegun I.A. (eds.). *Appl. Maths Ser.*, vol. 55, Dover, National Bureau of Standards, 1964.

[94] Hinch E.J., *Perturbation methods.* Cambridge University Press, Cambridge, 1991.

[95] Holms M.H., *Introduction to perturbation methods.* Springer, Berlin, Heidelberg, 1995.

[96] Humphreys D.S., Bodner S.R., Dynamic buckling of shallow shells under impulsive loading. *ACSE J. Engn. Mech. Div.*, vol. 88, No 2, 1962, 17-36.

[97] Humphreys D.S., Roth R.S., Zalters J., Experiments on dynamic buckling of shallow spherical shell under shock loading. *AIAA J.*, vol. 3, No 1, 1965, 33-39.

[98] Iooss G., Joseph D.D., *Elementary stability and bifurcation theory.* Springer, Berlin, Heidelberg, 1980.

[99] Jones W.B., Thron W.J., *Analytic theory and applications.* Addison-Wesley Publ. Co., London, 1980.

[100] Kaas-Petersen C., Continuation methods as the link between perturbation analysis and asymptotic analysis. *SIAM Rev.*, vol. 29, No 1, 1987, 115-120.

[101] Kalamkarov A.L., *Composite and reinforced elements of construction.* Wiley, Chichester, New York, 1992.

[102] Kamolphan D.-U., Mei C., Finite element method for forced nonlinear vibrations of circular plates. *Int. J. Math. Engn.*, vol. 29, No 9, 1986, 1715-1726.

[103] Kovrigin D.A., Potapov A.I., Nonlinear resonance interactions of longitudinal and flexural waves in a ring. *Soviet Physics Doklady*, vol. 34, No 4, 1989, 330-332.

[104] Kovrigin D.A., Potapov A.I., Nonlinear vibration of a thin ring. *Soviet Appl. Mechs*, vol. 25, No 3, 1989, 281-286.

[105] Kruskal M.D., *Asymptotology: Mathematical Models in Physical Science.* Prentice-Hall, Englewood Cliffs, N.J., 1963, 17-48.

[106] Ladygina Ye.V., Manevitch A.I., Free oscillations of a non-linear cubic system with two degrees of freedom and close natural frequencies. *PMM, J. Appl. Maths Mechs*, vol. 57, No 2, 257-266.

[107] Lau S.L., Cheung Y.K., Chen Chuhui, An alternative perturbation procedure of multiple scales for nonlinear dynamics systems. *Trans. ASME, J. Appl. Mechs*, vol. 56, No 3, 1989, 587-605.

[108] Lighthill M.J., A technique for rendering approximate solutions to physical problems uniformly valid. *Phil. Mag.*, vol. 40, 1179-1201.

[109] Lin S.S., Segel L.A., *Mathematical methods applied to deterministic problems in the natural sciences.* SIAM, Philadelphia, 1988.
[110] Lindberg H.E., Stress amplification in a ring caused by dynamic instability. *Trans. ASME, J. Appl. Mechs*, vol. 41, No 2, 1974, 392-400.
[111] Litvinov G.L., *Approximate construction of rational approximants and effect of self-correction of error.* Maths and Modeling, Puschino, 1990, 99-141 (in Russian).
[112] Lock M.H., Okubo S., Whittier J.S., Experiments on the snapping of a shallow dome under a step pressure load. *AIAA J.*, vol. 6, No 7, 1968, 1320-1326.
[113] Luke Y.L., Computations of coefficients in the polynominals of Padé approximants by solving systems of linear equations. *J. Comp. Appl. Maths*, vol. 6, No 3, 1980, 213-218.
[114] McCabe J.H., Murphy J.A., Continued fractions which correspond to power series expansions at two points. *J. Inst. Maths Applics.*, vol. 17, 1976, 233-247.
[115] Manevitch L.I., Pavlenko A.V., Koblik S.G., *Asymptotic methods in the theory of elasticity of orthotropic body.* Visha Shkola, Kiev-Donezk, 1979 (in Russian).
[116] Marsden J.E., McCraken M., The Hopf bifurcation and its applications. *Appl. Mathem. Sci.* 19, Springer, Berlin, Heidelberg, 1976.
[117] Martin P., Baker G.A. Jr., Two-point quasifractional approximant in physics. Truncation error. *J. Math. Phys.*, vol. 32, No 6, 1991, 1470-1477.
[118] Mikhlin Yu.V., Matching of local expansion in the theory of nonlinear vibrations, *J. Sound Vibr.*, vol. 182, No 4, 1995, 577-588.
[119] Nayfeh A.H., *Perturbation methods.* John Wiley & Sons, New York, 1973.
[120] Nayfeh A.H., *Introduction to perturbation techniques.* John Wiley & Sons, New York, 1981.
[121] Nayfeh A.H., *Problems in perturbations.* John Wiley & Sons, New York, 1985.
[122] Nayfeh A.H., Mook D.T., *Nonlinear oscillations.* John Wiley & Sons, New York, 1979.
[123] Nemeth G., Paris G., The Gibbs phenomenon in generalized Padé approximants. *J. Math. Phys.*, vol. 26, No 6, 1985, 1175-1178.
[124] Novozhilov V.V., *Theory of thin shells.* Wolters-Noordhoff, Groningen, 1970.
[125] Obraztsov I.F., Andrianov I.V., Nerubaylo B.V., Continuum approximation for high-frequency oscillations of a chain and composite equations. *Soviet Physics Doklady*, vol. 336, No 7, 1991, 522.
[126] O'Malley R.E., *Introduction to singular perturbation.* Academic Press, New York, 1974.
[127] Ortoleva P., Dynamical Padé approximants in the theory of perturbated and chaotic chemical center waves. *J. Chem. Phys.*, vol. 69, No 1, 1978, 300-307.
[128] Ortoleva P., Dynamical Padé approximants and behaviour singularities in non-linear physicochemical systems. *Lect. Notes Maths*, vol. 782, 1980, 255-264.
[129] Pesch W., Kramer L., Nonlinear analysis of spatial structures in two dimensional anisotropic pattern forming systems. *Zeitschrift für Physik B-Condensed Matter*, vol. 63, 1986, 121-130.
[130] Pierls R., Model-making in physics. *Contemporary Phys.*, vol. 21, No 1, 1980, 3-17.
[131] Reissner E., An asymptotic expansions for circular cylindrical shells. *Trans ASME, J. Appl. Mechs*, vol. 31, 1964, 245-252.
[132] Reissner E., Simmonds J.G., Asymptotic solutions of boundary value problems for elastic semi-infinite circular cylindrical shells. *J. Math. Phys.*, vol. 45, No 1, 1966.
[133] Rogers R.J., Pick R.J., On the dynamic spatial response of a heat-exchanger tube with intermittent baffle contact. *Nucl. Engn. Design*, No 36, 1976.

[134] Rogers R.J., Pick R.J., Factors associated with support plate forced due to heat-exchanger tube vibratory contact. *Nucl. Engn. Design*, No 44, 1977.

[135] Rosenberg R.M., On nonlinear vibrations of systems with many degrees of freedom. *Advances Appl. Mechs*, vol. 9, 1966, 156-243.

[136] Ross E.W.(Jr.), Asymptotic analysis of the axisymmetric vibrations of shells. *Trans ASME, J. Appl. Mechs*, vol. 33, No 1, 1966, 85-92.

[137] Rubanik W.D., *Vibration in complex quasi-linear systems with delay*. University Press, Minsk, 1985 (in Russian).

[138] Rutten H.S., *Theory and design of shells on the basis of asymptotic analysis*. Rutten & Kruisman, Consulting Engns, Rijswijk, 1973.

[139] Sanchez-Palencia E., *Non-homogeneous media and vibration theory*. Springer-Verlag, Berlin, Heidelberg, 1980.

[140] Sanders J.-L., Nonlinear theory for thin shell. *Quart. Appl. Mechs*, vol. 21, 1963, 21-36.

[141] Sanders J.A., Verhulst F., *Averaging methods in nonlinear dynamical systems*. Springer, Berlin, Heidelberg, 1985.

[142] Segel L.A., The importance of asymptotic analysis in applied mathematics. *Amer. Math. Monthly*, vol. 73, No 1, 1966, 7-14.

[143] Segel L.A., Distant side-walls cause slow amplitude modulation of cellular convection. *J. of Fluid Mechanics*, vol. 38, No 1, 1969, 203-224.

[144] Shanks D., Nonlinear transforms of divergent and slowly convergent sequences. *J. Maths & Phys.*, vol. 34, 1955, 1-42.

[145] Shinbrot T., Grebogi C., Ott E., Yorke J.A., Using small perturbations to control chaos. *Nature*, vol. 363 (June), 1993, 411-417.

[146] Slotine J.-J.E., Li W., *Applied nonlinear control*. Prentice-Hall, Englewood Cliffs, New Jersey, 1991.

[147] Smith D.A., Ford W.F., Acceleration of linear and logarithmic convergence. *SIAM J. Numer. Anal.*, vol. 16, No 2, 1979, 223-240.

[148] Srinivasan R.S., Thiruvenkatachar V., Free vibration of annual sector plates by an integral equation technique. *J. Sound Vibr.*, vol. 89, No 3, 1983, 425-432.

[149] Strickling J.A., Martinez J.E., Tillerson J.R., et al., Nonlinear dynamic analysis of shells of revolution by matrix displacement method. *AIAA J.*, vol. 9, No 4, 1971, 629-636.

[150] Surkin R.G., Zuev B.M., Stepanov S.G., Experimental investigation of dynamical stability of spherical segments. *Soviet Appl. Mechs*, vol. 3, No 8, 1967, 124-132.

[151] Szubshcik L.S., Stolyar A.M., Tsybulin B.G., Asymptotic integration of nonlinear equations of cylindrical panel vibrations. *PMM, J. Appl. Maths Mechs*, vol. 52, No 1, 1988, 511-518.

[152] *Tables of integral transformation*, vol. 1. McGraw-Hill, New York, 1954.

[153] Toda M., *Theory of nonlinear lattices*. Springer, Berlin, Heidelberg, 1981.

[154] Van Dyke M., *Perturbation methods in fluid mechanics*. The Parabolic Press, Stanford, California, 1975.

[155] Vasil'eva A.B., Butuzov V.F., Kalachev L.V., *The boundary function method for singular perturbation problems*. SIAM, Philadelphia, 1995.

[156] Vedenova E.G., Manevitch L.I., Nisichenko V.P., Lysenko S.A., Solitons with a substantially nonlinear one-dimensional chain. *J. Appl. Mechs Technical Physics*, vol. 25, No 6, 1984, 910-914.

[157] Vol'mir A.S., *Nonlinear plates and shells dynamics*. Nauka, Moscow, 1972 (in Russian).

[158] Volos N.P., Korol' I.You., Nonlinear vibrations of rectangular plates with various boundary constraints. *Soviet Appl. Mechs*, vol. 12, No 7, 1976, 101-106.

[159] Wah T., The normal modes of vibration of certain nonlinear continuous systems. *Trans ASME, J. Appl. Mechs*, vol. 31, No 1, 1964.
[160] Whitham G., *Linear and non-linear waves*. John Wiley & Sons, New York, 1974.
[161] Wong R., Distributional derivation of an asymptotic expansion. *Proc. Amer. Math. Soc.*, vol. 80, No 2, 1980, 266-270.
[162] Zarutsky V.A., Oscillations of ribbed shells. *Intern. Appl. J. Mechs.*, vol. 29, No 10, 1993, 837-841.
[163] Zeitounian R., *Asymptotic modeling of atmospheric flows*. Springer, Berlin, Heidelberg, 1990.
[164] Zemanian A.H., *Distribution theory and transform analysis*. McGraw-Hill, New York, 1965.
[165] Zhuravlev V.F., Klimov D.M., *Applied methods in the theory of oscillations*. Nauka, Moscow, 1988 (in Russian).

Detailed References (d)

[1] Amiro I.Ya., Zarutsky V.A., Studies of the dynamics of ribbed shells. *Soviet. Appl. Mechs*, vol. 17, No 11, 1981, 949-962.

[2] Andrianov I.V., Asymptotic solutions for nonlinear systems with high degrees of nonlinearity. *PMM, J. Appl. Maths Mechs*, vol. 57, No 5, 1993, 941-943.

[3] Andrianov I.V., A new asymptotic method of integrating the equations of quantum mechanics for strong coupling. *Physics Doklady*, vol. 38, No 2, 1993, 56-57.

[4] Andrianov I.V., Two-point Padé approximants in the mechanics of solids. *ZAMM*, vol. 74, No 4, 1994, 121-122.

[5] Andrianov I.V., Danishevskyi V.V., Asymptotic investigation of the nonlinear dynamic boundary value problem for rod. *Technische Mechanik*, vol. 15, No 1, 1995, 53-55.

[6] Andrianov I.V., Kholod E.G., Exact solution in the nonlinear theory of re-inforced with discrete stiffness plates. *Facta universitatis*, University of Niš, vol. 1, No 4, 1994, 389-400.

[7] Andrianov I.V., Kholod E.G., Chernetsky V.A., Asymptotic-based method for a plane elasticity mixed boundary eigenvalue problem. *J. Theor. Appl. Mechs*, vol. 3, No 32, 1994, 701-709.

[8] Andrianov I.V., Lesnichaya V.A., Manevitch L.I., *Homogenization methods in statics and dynamics of reinforced shells*. Nauka, Moscow, 1985 (in Russian).

[9] Andrianov I.V., Manevitch L.I., Homogenization method in the theory of shells. *Advances in Mechs*, vol. 6, No 3/4, 1983, 3-29 (in Russian).

[10] Andrianov I.V., Manevitch L.I., *Asymptotology: ideas, methods, results*. Aslan, Moscow, 1994 (in Russian).

[11] Andrianov I.V., Sedin V.L., Composition of simplified equations of nonlinear dynamics of plates and shells on the basis of homogenization method. *ZAMM*, vol. 67, No 7, 1988, 573-575.

[12] Andrianov I.V., Shevchenko V.V., Perforated plates and shells. *Asymptotic Methods in the System Theory*, Irkutsk, 1989, 217-243 (in Russian).

[13] Andrianov I.V., Starushenko G.A., Application of the averaging method for the calculation of perforated plates. *Soviet Appl. Mechs*, vol. 24, No 4, 1988, 410-415.

[14] Awrejcewicz J., Vibration system: rotor with self-excited support. *Proceedings of Int. Conf. on Rotor Dynamics*, Tokyo, Sept. 14-17, 1986, 517-522.

[15] Awrejcewicz J., Hopf bifurcations in Duffing oscillator. *PAN American Congress of Appl. Mech.*, Rio de Janeiro, 1989, 640-643.

[16] Awrejcewicz J., The analytical method to detect Hopf bifurcation solutions in the unstationary nonlinear systems. *J. Sound Vibr.*, vol. 129, No 1, 1989, 175-178.

[17] Awrejcewicz J., Parametric and self-excited vibrations included by friction in a system with three-degree-of-freedom. *KSME Journal*, vol. 4, No. 2, 1990, 156-166.

[18] Awrejcewicz J., Analysis of double Hopf bifurcations. *Nonlinear Vibration Problems*, vol. 24, 1991, 123-140.

[19] Awrejcewicz J., Analysis of the biparameter Hopf bifurcation. *Nonlinear Vibration Problems*, vol. 24, 1991, 63-76.

[20] Awrejcewicz J., On the Hopf bifurcation. *Nonlinear Vibration Problems*, vol. 24, 1991, 15-31.

[21] Awrejcewicz J., Nonlinear oscillations of the string caused by the electromagnetic field. In: *Proceedings of the Third Polish-Japanese Seminar on Modelling and Control of Electromagnetic Phenomena*, Kazimierz, Poland, April 19-21, 1993, 125-128.

[22] Awrejcewicz J., Analytical and numerical study of nonlinear behaviour of the electromechanical systems. *J. Tech. Phys.*, vol. 34, No 1, 1993, 57-68.

[23] Awrejcewicz J., *Nonlinear dynamics of machines*. Technical University Press, Łódź, 1994 (in Polish).

[24] Awrejcewicz J., Strange nonlinear behaviour of the electromechanical system. *Machine Dynamics Problems*, vol. 7, 1994, 7-10.

[25] Awrejcewicz J., Strange nonlinear dynamical behaviour of a string type generator with a time delay amplifier. *Proc. Ninth World Congress on the TMM*, Milan, Italy, August 30-31/Sept 1-2, 1995, 913-916.

[26] Awrejcewicz J., *Vibration of discrete deterministic systems*. WNT Publishers in Science and Technology, Warsaw, 1996 (in Polish).

[27] Awrejcewicz J., Tomczak K., Lamarque C.-H., Controlling system with impact. *Proceed. of the Int. Conf. on Nonlinearity, Bifurcation and Chaos*, Łódź-Dobieszków, Sept. 16-18, 1996, 73-76.

[28] Babich V.M., Buldirev V.S., The art of asymptotic. *Vestnik Leningrad Univ. Maths*, vol. 10, 1982, 227-235.

[29] Bauer S.M., Fillipov S.B., Smirnov A.L., Tovstik P.E., Asymptotic methods in mechanics with applications to thin plates and shells. Asymptotic methods in Mechanics. *CRM Proc. and Lect. Notes*, vol. 3, 1994, 3-140.

[30] Belov M.A., Tyrulis T.T., *Asymptotic methods of integral transforms inverse*. Zinatne, Riga, 1985 (in Russian).

[31] Bender C.M., Milton K.A., Boettcher S., A new perturbative approach to nonlinear partial differential equations. *J. Math. Phys.*, vol. 32, No 11, 1991, 3031-3038.

[32] Bogdanovich A.E., Zarutsky V.A., Oscillations of ribbed shells. *Soviet Appl. Mechs*, vol. 27, No 10, 1991, 1001-1006.

[33] Elsgolc L.E., *Introduction to the theory of differential equations with a delay*. Nauka, Moscow, 1964 (in Russian).

[34] Estrada R.P., Kanwal R.P., A distributional theory for asymptotic expansions. *Proc. R. Soc. Lond. A*, vol. 428, 1990, 399-430.

[35] Estrada R.P., Kanwal R.P., Taylor expansions for distribution. *Math. Methods in Appl. Sc.*, vol. 16, 1993, 297-304.

[36] Evkin A.Yu., A new approach to the asymptotic integration of the equations of shallow convex shells in the post-critical stage. *PMM, J. of Appl. Maths Mechs*, vol. 53, No 1, 1986, 92-96.

[37] Filatov A.N., Sharova L.V., *Integral inequalities and theory of nonlinear oscillations*. Nauka, Moscow, 1976 (in Russian).

[38] Fok V.A., *Problems of difraction and spreading of electromagnetic waves*. Nauka, Moscow, 1970 (in Russian).

[39] Gonchar A.A., About uniform convergence of the diagonal Padé approximants. *Matemat. Sbornik*, vol. 118, No 4, 1982, 535-556.

[40] Grigoluk E.I., Phylshtinsky L.A., *Perforated plates and shells.* Nauka, Moscow, 1970 (in Russian).

[41] Gryboś R., *Theory of impact in discrete mechanical systems.* PWN, Warsaw, 1969 (in Polish).

[42] Hamada M., Ota T., Fundamental frequencies of simply supported but partially clamped square plates. *Bull. Japan Soc. Mech. Eng.*, vol. 6, No 23, 1963, 397-403.

[43] Hara S., Yamamoto Y., Omata T., Nokano M., Repetitive control system: a new type servo system for periodic exogenous signals. *IEEE Transactions on Automatic Control*, vol. 33, No. 7, 1988, 659-668.

[44] Huang N.C., Axisymmetric dynamic snap-through of elastic clamped shallow spherical shells. *Trans. ASME, J. Appl. Mechs*, vol. 7, No 2, 1969, 215-220.

[45] Huseyin K., Atadan A.S., On the analysis of Hopf bifurcation. *Int J. Engng. Sci.*, vol. 21, No 3, 1983, 247-262.

[46] Huseyin K., Atadan A.S., On generalized Hopf bifurcations. *J. Dyn. Sys., Measur. and Contr.*, vol. 106, 1984, 327-334.

[47] Infeld E., Rowlands G., Lenkowska-Czerwinska T., *On introducing dynamics into the theory of nucleation of coherent structures* (to be published).

[48] Kantorovitch L.V., Krylov V.I., *Approximate methods of higher analysis.* Nauka, Moscow, 1949 (in Russian).

[49] Karmishin A.V., Zhukov A.I., Kolosov V.G., et al., *Methods of dynamic calculation and testing for thin-walled structures.* Mashinostroyenie, Moscow, 1990 (in Russian).

[50] Kayuk Ya.F., *Some problems of the methods splitting by parameters.* Naukova Dumka, Kiev, 1991 (in Russian).

[51] Koiter W.T., On the nonlinear theory of thin elastic shells. *Proc. Kon. Ned. Ak. Wet.*, ser B, vol. 69, No 1, 1966, 1-54.

[52] Krodkiewski J.M., Faragher J.-S., Stability of the periodic motion of nonlinear systems. *Proceedings of the International Conference on Vibration and Noise*, Venice, 1995.

[53] Kryloff V.I., Skoblya N.S., *Methods of approximate Fourier and Laplace transform inverse.* Nauka, Moscow, 1974 (in Russian).

[54] Lyapunov A.M., *The general problem of the stability of motion.* Gostehizdat, Moscow, 1950.

[55] Manevitch A.I., *Stability and optimal design of reinforced shells.* Visha Shkola, Kiev-Donetzk, 1972 (in Russian).

[56] Manevitch L.I., Mikhlin Yu.V., Pilipchuk V.N., *The method of normal oscillations for essentially nonlinear systems.* Nauka, Moscow, 1989 (in Russian).

[57] Marchuk G.I., Agoshkov V.I., Conjugate operators and algorithms of perturbation in non-linear problems: 1. Principles of construction of conjugate operators. *Soviet J. Num. Anal. and Math. Modeling*, No 1, 1988, 21-46.

[58] Marchuk G.I., Agoshkov V.I., Conjugate operators and algorithms of perturbation in non-linear problems: 2. Perturbation algorithms. *Soviet J. Num. Anal. and Math. Modeling*, No 2, 1988, 115-136.

[59] Mitropol'sky Yu.A., Homa G.P., Gromak M.I., *Asymptotic methods of investigation of quasiwave equations of hyperbolic type.* Naukova Dumka, Kiev, 1991 (in Russian).

[60] Nayfeh A.H., Numerical-perturbation methods in mechanics. *Comput. and Struct.*, vol. 30, No 1-2, 1988, 185-204.

[61] Newell A.C., Whitehead J.A., Finite bandwidth, finite amplitude convection. *J. of Fluid Mechanics*, vol. 38, No 2, 1969, 279-303.

[62] Obraztsov I.F., Nerubaylo B.V., Andrianov I.V., *Asymptotic methods in the theory of thin-walled structures.* Mashinostroyenie, Moscow, 1991 (in Russian).

[63] Peterka F., Laws of impact motion of mechanical systems with one degree of freedom. Part I and II. *Acta Technica CSAV*, No. 5, 1974, 462-473, 569-580.

[64] Potier-Ferry M., Bensaadi M.H., Computation of periodic solutions by using Padé approximants. *1st European Nonl. Osc. Conf.* Abstracts, 1993, 14.

[65] Proskuriakov A.P., *Poincaré method in the theory of nonlinear oscillations.* Nauka, Moscow, 1977 (in Russian).

[66] Pyragas K., Continuous control of chaos by self-controlling feedback. *Physics Letters, A* vol. 170, Iss. 6, 1992, 421-428.

[67] Semerdjiev Kh., Trigonometric Padé approximants and Gibbs phenomenon. *Rep. United & Inst. Nucler Res.*, NP5 - 12484, Dubna, 1919, 10 (in Russian).

[68] Slepyan A.I., Yakovlev U.S., *Integral transforms in the nonstationary mechanical problems.* Shipbuilding Publ., Leningrad, 1980 (in Russian).

[69] Therapos C.P., Diamessis J.E., Approximate Padé approximants with applications to rational approximation and linear-order reduction. *Proc. of the IEEE*, vol. 72, No 12, 1984, 1811-1813.

[70] Tzeitlin A.I., Glikman B.G., Elastic rectangular plate bending under mixed boundary conditions. *Proceed. Central Res. Inst. Civil Engn. Str. Res. Str. Dynamics*, vol. 17, Moscow, 1971, 123-142 (in Russian).

[71] Youncef-Toumi K., Wu S.-T., Input/output linearization using time delay control. *Journal of Dynamics Systems*, Measurement and Control, vol. 114 (Mar), 1992, 10-19.

Index

Springer
and the
environment

At Springer we firmly believe that an international science publisher has a special obligation to the environment, and our corporate policies consistently reflect this conviction.
We also expect our business partners – paper mills, printers, packaging manufacturers, etc. – to commit themselves to using materials and production processes that do not harm the environment. The paper in this book is made from low- or no-chlorine pulp and is acid free, in conformance with international standards for paper permanency.

Springer